MODERN SPECTROSCOPY

Second Edition

J. Michael Hollas
University of Reading

JOHN WILEY & SONS
Chichester · New York · Brisbane · Toronto · Singapore

Other Wiley Editorial Offices

John Wiley & Sons, Inc., 605 Third Avenue,
New York, NY 10158–0012, USA

Jacaranda Wiley Ltd, 33 Park Road, Milton,
Queensland 4064, Australia

John Wiley & Sons (Canada) Ltd, 22 Worcester Road,
Rexdale, Ontario M9W 1L1, Canada

John Wiley & Sons (SEA) Pte Ltd, 37 Jalan Pemimpin #05-04,
Block B, Union Industrial Building, Singapore 2057

Library of Congress Cataloging-in-Publication Data

Hollas, J. Michael (John Michael)
  Modern spectroscopy / J. Michael Hollas.—2nd ed.
    p.   cm.
  Includes bibliographical references and index.
  ISBN 0-471-93076-8 : —ISBN 0-471-93077-6 (pbk.)
  1. Spectrum analysis.   I. Title.
  QC451.H65   1992
  535.8′4—dc20
                                                                91-3400
                                                                CIP

A catalogue record for this book is available from the British Library

Typeset by Techset Composition Ltd, Salisbury, Wiltshire
Printed and bound in Great Britain by
Biddles Ltd, Guildford and King's Lynn

Contents

3 GENERAL FEATURES OF EXPERIMENTAL METHODS

4 MOLECULAR SYMMETRY

5 ROTATIONAL SPECTROSCOPY

Preface to First Edition

Modern Spectroscopy has been written to fulfil a need for an up-to-date text on spectroscopy. It is aimed primarily at a typical undergraduate audience in chemistry, chemical physics, or physics in the United Kingdom and at both undergraduate and graduate student audiences elsewhere.

Spectroscopy covers a very wide area which has been widened further since the mid-1960s by the development of lasers and such techniques as photoelectron spectroscopy and other closely related spectroscopies. The importance of spectroscopy in the physical and chemical processes going on in planets, stars, comets, and the interstellar medium has continued to grow as a result of the use of satellites and the building of radiotelescopes for the microwave and millimetre wave regions.

In planning a book of this type I encountered three major problems. The first is that of covering the analytical as well as the more fundamental aspects of the subject. The importance of the applications of spectroscopy to analytical chemistry cannot be overstated but the use of many of the available techniques does not necessarily require a detailed understanding of the processes involved. I have tried to refer to experimental methods and analytical applications where relevant.

The second problem relates to the inclusion, or otherwise, of molecular symmetry arguments. There is no avoiding the fact that an understanding of molecular symmetry presents a hurdle (although I think it is a low one) which must be surmounted if selection rules in vibrational and electronic spectroscopy of polyatomic molecules are to be understood. This book surmounts the hurdle in Chapter 4 which is devoted to molecular symmetry but which treats the subject in a non-mathematical way. For those lecturers and students who wish to leave out this chapter much of the subsequent material can be understood but, in some areas, in a less satisfying way.

The third problem also concerns the choice of whether to leave out certain material. In a book of this size it is not possible to cover all branches of spectroscopy. Such decisions are difficult ones but I have chosen not to include spin resonance spectroscopy (n.m.r. and e.s.r.), nuclear quadrupole resonance spectroscopy (n.q.r.), and Mössbauer spectroscopy. The exclusion of these areas, which have been well covered in other texts, has been caused, I suppose, by the inclusion, in Chapter 8, of photoelectron spectroscopy (ultraviolet and X-ray), Auger electron spectroscopy, and extended X-ray absorption fine structure, including applications to studies of solid surfaces, and, in Chapter 9, the theory and some examples of lasers and some of their uses in spectroscopy. Most of the material in these two chapters will not be found in comparable texts but is of very great importance in spectroscopy today.

My understanding of spectroscopy owes much to having been fortunate in working in and discussing the subject with Professor I. M. Mills, Dr A. G. Robiette, Professor J. A. Pople, Professor D. H. Whiffen, Dr J. K. G. Watson, Dr G. Herzberg, Dr A. E. Douglas, Dr D. A. Ramsay, Professor D. P. Craig, Professor J. H. Callomon, and Professor G. W. King (in more or less reverse historical order), and I am grateful to all of them.

When my previous book *High Resolution Spectroscopy* was published by Butterworths in 1982 I had it in mind to make some of the subject matter contained in it more accessible to students at a later date. This is what I have tried to do in *Modern Spectroscopy* and I would like to express my appreciation to Butterworths for allowing me to use some textual material and, particularly, many of the figures from *High Resolution Spectroscopy*. New figures were very competently drawn by Mr M. R. Barton.

Although I have not included *High Resolution Spectroscopy* in the bibliography of any of the chapters it is recommended as further reading on all topics.

Mr A. R. Bacon helped greatly with the page proof reading and I would like to thank him very much for his careful work. Finally, I would like to express my sincere thanks to Mrs A. Gillett for making such a very good job of typing the manuscript.

J. Michael Hollas

Preface to Second Edition

A new edition of any book presents an opportunity which an author welcomes for several reasons. It is a chance to respond to constructive criticisms of the previous edition which he thinks are valid. New material can be introduced which may be useful to teachers and students in the light of the way the subject, and the teaching of the subject, has developed in the intervening years. Last, and certainly not least, there is an opportunity to correct any errors which had escaped the author's notice.

Fourier transformation techniques in spectroscopy are now quite common—the latest to arrive on the scene is Fourier transform Raman spectroscopy. In Chapter 3 I have expanded considerably the discussion of these techniques and included Fourier transform Raman spectroscopy for the first time.

In teaching students about Fourier transform techniques I find it easier to introduce the subject by using radiofrequency radiation, for which the variations of the signal with time can be readily detected—as happens in an ordinary radio. Fourier transformation of the radiofrequency signal, which the radio itself carries out, is quite easy to visualize without going deeply into the mathematics. The use of a Michelson interferometer in the infrared, visible or ultraviolet regions is necessary because of the inability of a detector to respond to these higher frequencies, but I think the way in which it gets over this problem is rather subtle. In this second edition I have discussed Fourier transformation relating, first, to radiofrequency and then to higher frequency radiation.

In the first edition of *Modern Spectroscopy* I tried to go some way towards bridging the gulf which often seems to exist between high resolution spectroscopy and low resolution, often analytical, spectroscopy. In this edition I have gone further by including X-ray fluorescence spectroscopy and inductively coupled plasma atomic emission spectroscopy, both of which are used almost entirely for analytical purposes. I think it is important that the user understands the processes going on in any analytical spectroscopic technique that he or she might be using.

In Chapter 4, on molecular symmetry, I have added two new sections. One of these concerns the relationship between symmetry and chirality, which is of great importance in synthetic organic chemistry. The other relates to the connection between the symmetry of a molecule and whether it has a permanent dipole moment.

In the chapter on vibrational spectroscopy (Chapter 6) I have expanded the discussions of inversion, ring-puckering and torsional vibrations, including some model potential functions. These types of vibration are very important in the determination of molecular structure.

The development of lasers has continued in the last few years and I have included discussions of two more in this edition. These are the alexandrite and titanium–sapphire lasers. Both are solid state and, unusually, tunable over quite wide wavelength ranges. The titanium–sapphire laser is probably the most promising for general use because of its wider range of tunability and the fact that it can be operated in a CW or pulsed mode.

Laser spectroscopy is such a wide subject, with many ingenious experiments using one or two, CW or pulsed lasers to study atomic or molecular structure or dynamics, that it is difficult to do justice to it at the level at which *Modern Spectroscopy* is aimed. In this edition I have expanded the section on supersonic jet spectroscopy which is an extremely important and wide-ranging field.

I would like to thank Professor I. M. Mills for the material he provided for Figure 3.14(b) and Figure 3.16 and Dr P. Hollins for help in the production of Figures 3.7(a), 3.8(a), 3.9(a) and 3.10(a). The spectrum in Figure 9.36 will be published in a paper by Dr J. M. Hollas and Dr P. F. Taday.

J. Michael Hollas

Units, Dimensions, and Conventions

Throughout the book I have adhered to the SI system of units, with a few exceptions. The angstrom (Å) unit, where $1\,\text{Å} = 10^{-10}$ m, seems to be persisting generally when quoting bond lengths, which are of the order of 1 Å. I have continued this usage but, when quoting wavelengths in the visible and near-ultraviolet regions, I have used the nanometre, when $1\,\text{nm} = 10\,\text{Å}$. The angstrom is still used sometimes in this context but it seems just as convenient to write, say, 352.3 nm as 3523 Å.

In photoelectron and related spectroscopies, ionization energies are measured. For many years such energies have been quoted in electron volts, where $1\,\text{eV} = 1.602\,177\,38 \times 10^{-19}$ J, and I have continued to use this unit.

Pressure measurements are not often quoted in the text but the unit of Torr, where $1\,\text{Torr} = 1\,\text{mmHg} = 133.322\,387$ Pa, is a convenient practical unit and appears occasionally.

Dimensions are physical quantities such as mass, length, and time and examples of units corresponding to these dimensions are the gram (g), metre (m), and second (s). If, for example, something has a mass of 3.5 g then we write

$$m = 3.5\,\text{g}$$

Units, here the gram, can be treated algebraically so that, if we divide both sides by 'g', we get

$$m/\text{g} = 3.5$$

The right-hand side is now a pure number and, if we wish to plot mass, in grams, against, say, volume on a graph we label the mass axis 'm/g' so that the values marked along the axis are pure numbers. Similarly, if we wish to tabulate a series of masses, we put 'm/g' at the head of a column of what are now pure numbers. The old style of using '$m(\text{g})$' is now seen to be incorrect as, algebraically, it could be interpreted only as $m \times \text{g}$ rather than $m \div \text{g}$ which we require.

An issue which is still only just being resolved concerns the use of the word 'wavenumber'. Whereas the frequency ν of electromagnetic radiation is related to the wavelength λ by

$$v = \frac{c}{\lambda}$$

where c is the speed of light, the wavenumber $\tilde{v}$ is simply its reciprocal:

$$\tilde{v} = \frac{1}{\lambda}$$

Since c has dimensions of LT^{-1} and λ those of L, frequency has dimensions of T^{-1} and often has units of s^{-1} (or hertz). On the other hand, wavenumber has dimensions of L^{-1} and often has units of cm^{-1}. Therefore

$$v = 15.3 \text{ s}^{-1} \text{ (or hertz)}$$

is, in words, 'the frequency is 15.3 reciprocal seconds (or second-minus-one or hertz)', and

$$\tilde{v} = 20.6 \text{ cm}^{-1}$$

is, in words, 'the wavenumber is 20.6 reciprocal centimetres (or centimetre-minus-one)'. All of this seems simple and straightforward but the fact is that many of us would put the second equation, in words, as 'the frequency is 20.6 wavenumbers'. This is quite illogical but very common—although not, I hope, in this book.

Another illogicality is the very common use of the symbols A, B, and C for rotational constants irrespective of whether they have dimensions of frequency or wavenumber. It is bad practice to do this, but although a few have used $\tilde{A}$, $\tilde{B}$, and $\tilde{C}$ to imply dimensions of wavenumber, this excellent idea has only rarely been put into practice and, regretfully, I go along with a very large majority and use A, B and C whatever their dimensions.

The starting points for many conventions in spectroscopy are the paper by R. S. Mulliken in the *Journal of Chemical Physics* (**23**, 1997, 1955) and the books of G. Herzberg. Apart from straightforward recommendations of symbols for physical quantities which are generally adhered to, there are rather more contentious recommendations. These include the labelling of cartesian axes in discussions of molecular symmetry and the numbering of vibrations in a polyatomic molecule which are often, but not always, used. In such cases it is important that any author makes it clear what convention *is* being used.

The case of vibrational numbering in, say, fluorobenzene illustrates the point that we must be flexible when it may be helpful. Many of the vibrations of fluorobenzene strongly resemble those of benzene. In 1934, before the Mulliken recommendations of 1955, E. B. Wilson had devised a numbering scheme for the thirty vibrations of benzene. This was so well established by 1955 that its use has tended to continue ever since. In fluorobenzene there is the further complication that, although Mulliken's system provides it with its own numbering scheme, it is useful very often to use the same number for a benzene-like vibration as it has in benzene itself—for which there is a choice of Mulliken's or Wilson's numbering! Clearly not all problems of conventions have been solved, and some are not really soluble, but we should all try to make it clear to any reader just what choice we have made.

One very useful convention which was proposed by J. C. D. Brand, J. H. Callomon, and J. K. G. Watson in 1963 is applicable to electronic spectra of polyatomic molecules, and I have used it throughout this book. In this system 32_1^2, for example, refers to a vibrational transition, in an electronic band system, from $v = 1$ in the lower to $v = 2$ in the upper electronic state where the vibration concerned is the one whose conventional number is 32. It is a very neat system compared to, for example, $(001) - (100)$ which is still frequently used for triatomics to indicate a transition from the $v = 1$ level in v_1 in the lower electronic state to the $v = 1$ level in v_3 in the upper electronic state. The general symbolism in this system is $(v_1' v_2' v_3') - (v_1'' v_2'' v_3'')$. The alternative $3_0^1 1_1^0$ label is much more compact but is little used for such small molecules. For consistency I have used this compact symbolism throughout.

Although it is less often done, I have used an analogous symbolism for pure vibrational transitions for the sake of consistency. Here $N_{v''}^{v'}$ refers to a vibrational (infrared or Raman) transition from a lower state with vibrational quantum number v'' to an upper state v' in the vibration numbered N.

FUNDAMENTAL CONSTANTS

Quantity	Symbol	Value and units[†]
Speed of light (*in vacuo*)	c	$2.997\,924\,58 \times 10^8$ m s^{-1} (exactly)
Vacuum permeability	μ_0	$4\pi \times 10^{-7}$ H m^{-1} (exactly)
Vacuum permittivity	$\varepsilon_0 (= \mu_0^{-1} c^{-2})$	$8.854\,187\,816 \times 10^{-12}$ F m^{-1}
Charge on proton	e	$1.602\,177\,33(49) \times 10^{-19}$ C
Planck constant	h	$6.626\,075\,5(40) \times 10^{-34}$ J s
Molar gas constant	R	$8.314\,510(70)$ J mol^{-1} K^{-1}
Avogadro constant	N_A, L	$6.022\,136\,7(36) \times 10^{23}$ mol^{-1}
Boltzmann constant	$k (= R N_A^{-1})$	$1.380\,658(12) \times 10^{-23}$ J K^{-1}
Atomic mass unit	$u (= 10^{-3}$ kg mol$^{-1} N_A^{-1})$	$1.660\,540\,2(10) \times 10^{-27}$ kg
Rest mass of electron	m_e	$9.109\,389\,7(54) \times 10^{-31}$ kg
Rest mass of proton	m_p	$1.672\,623\,1(10) \times 10^{-27}$ kg
Rydberg constant	R_∞	$1.097\,373\,153\,4(13) \times 10^7$ m^{-1}
Bohr radius	a_0	$5.291\,772\,49(24) \times 10^{-11}$ m
Bohr magneton	$\mu_B [= e\hbar (2m_e)^{-1}]$	$9.274\,015\,4(31) \times 10^{-24}$ J T^{-1}
Nuclear magneton	μ_N	$5.050\,786\,6(17) \times 10^{-27}$ J T^{-1}
Electron magnetic moment	μ_e	$9.284\,770\,1(31) \times 10^{-24}$ J T^{-1}
g-Factor for free electron	$\frac{1}{2}g_e (= \mu_e \mu_B^{-1})$	$1.001\,159\,652\,193(10)$

[†] Values taken from *Quantities, Units and Symbols in Physical Chemistry*, International Union of Pure and Applied Chemistry, Blackwells Scientific Publications (1988). The uncertainties in the final digits are given in parentheses.

USEFUL CONVERSION FACTORS

Unit	cm^{-1}	MHz	kJ	eV	kJ mol^{-1}
1 cm^{-1}	1	29 979.25	$1.986\,45 \times 10^{-26}$	$1.239\,84 \times 10^{-4}$	$1.196\,27 \times 10^{-2}$
1 MHz	$3.335\,64 \times 10^{-5}$	1	$6.626\,08 \times 10^{-31}$	$4.135\,67 \times 10^{-9}$	$3.990\,31 \times 10^{-7}$
1 kJ	$5.034\,11 \times 10^{25}$	$1.509\,19 \times 10^{30}$	1	$6.241\,51 \times 10^{21}$	$6.022\,14 \times 10^{23}$
1 eV	8065.54	$2.417\,99 \times 10^{8}$	$1.602\,18 \times 10^{-22}$	1	96.485
1 kJ mol^{-1}	83.593 5	$2.506\,07 \times 10^{6}$	$1.660\,54 \times 10^{-24}$	$1.036\,43 \times 10^{-2}$	1

CHAPTER 1

Some Important Results in Quantum Mechanics

1.1 Spectroscopy and Quantum Mechanics

Spectroscopy is basically an experimental subject and is concerned with the absorption, emission, or scattering of electromagnetic radiation by atoms or molecules. As we shall see in Chapter 3, electromagnetic radiation covers a wide wavelength range from radio waves to γ-rays and the atoms or molecules may be in the gas, liquid, or solid phase or, of great importance in surface chemistry, adsorbed on a solid surface.

Quantum mechanics, on the other hand, is a theoretical subject relating to many aspects of chemistry and physics, but particularly to spectroscopy.

Experimental methods of spectroscopy began in the more accessible visible region of the electromagnetic spectrum where the eye could be used as the detector. In 1665 Newton had started his famous experiments on the dispersion of white light into a range of colours using a triangular glass prism. However it was not until about 1860 that Bunsen and Kirchhoff began to develop the prism spectroscope as an integrated unit for use as an analytical instrument. Early applications were the observation of the emission spectra of various samples in a flame, the origin of flame tests for various elements, and of the sun.

The visible spectrum of atomic hydrogen had been observed both in the solar spectrum and in an electrical discharge in molecular hydrogen many years earlier, but it was not until 1885 that Balmer fitted the resulting series of lines to a mathematical formula. In this way began the close relationship between experiment and theory in spectroscopy, the experiments providing the results and the relevant theory attempting to explain them and to predict results in related experiments. However theory ran increasingly into trouble so long as it was based on classical newtonian mechanics until, from 1926 onwards, the development by Schrödinger of quantum mechanics. Even after this breakthrough, the importance of which cannot be overstressed, it is not, I think, unfair to say that theory tended to limp along behind experiment. Data from spectroscopic experiments, except for those on the simplest atoms and mole-

1

cules, were easily able to outstrip the predictions of theory which was almost always limited by the approximations which had to be made in order that the calculations would be manageable. It was only from about 1960 onwards that the situation changed as a result of the availability of large, fast computers requiring many fewer approximations to be made. Nowadays it is not uncommon for predictions to be made of spectroscopic and structural properties of fairly small molecules which are comparable in accuracy with those obtainable from experiment.

Although spectroscopy and quantum mechanics are closely interrelated it is nevertheless the case that there is still a tendency to teach the subjects separately while drawing attention to the obvious overlap areas. This is the attitude I shall adopt in this book, which is concerned primarily with the techniques of spectroscopy and the interpretation of the data which accrue. References to texts on quantum mechanics are given in the bibliography at the end of this chapter.

1.2 The Evolution of Quantum Theory

During the late nineteenth century evidence began to accumulate that classical newtonian mechanics, which was completely successful on a macroscopic scale, was unsuccessful when applied to problems on an atomic scale.

In 1885 Balmer was able to fit the discrete wavelengths λ of part of the emission spectrum of the hydrogen atom, now called the Balmer series and illustrated in Figure 1.1, to the empirical formula

$$\lambda = \frac{n'^2 G}{n'^2 - 4} \tag{1.1}$$

where G is a constant and $n' = 3, 4, 5, \ldots$. In this figure the wavenumber[†] $\tilde{\nu}$ and the wavelength λ are used: the two are related by

$$\tilde{\nu} = \frac{1}{\lambda} \tag{1.2}$$

Using the relationship

$$\nu = \frac{c}{\lambda} \tag{1.3}$$

where ν is the frequency and c the speed of light in a vacuum, equation (1.1) becomes

$$\nu = R_H \left(\frac{1}{2^2} - \frac{1}{n'^2} \right) \tag{1.4}$$

[†] See Units, Dimensions and Conventions on p. xvii.

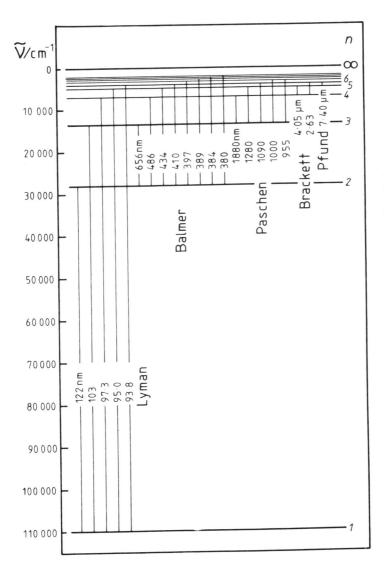

Figure 1.1 Energy levels and observed transitions of the hydrogen atom.

in which R_H is the Rydberg constant for hydrogen. This equation, and even the fact that the spectrum is discrete rather than continuous, is completely at variance with classical mechanics.

Another phenomenon which was inexplicable in classical terms was the photoelectric effect discovered by Hertz in 1887. When ultraviolet light falls on an alkali metal surface, electrons are ejected from the surface only when the frequency of the radiation reaches the threshold frequency v_t for the metal. As the frequency is increased, the kinetic energy of the ejected electrons, known as photoelectrons, increases linearly with v, as Figure 1.2 shows. As we shall see in Chapter 8 the photoelectric effect forms the basis of relatively new branches of spectroscopy such as photoelectron and Auger spectroscopy.

The explanation of the hydrogen atom spectrum and the photoelectric effect, together with other anomalous observations such as the behaviour of the molar heat capacity C_v of a solid at temperatures close to 0 K and the frequency distribution of black body radiation, originated with Planck. In 1900 he proposed that the microscopic oscillators, of which a black body is made up, have an oscillation frequency v related to the energy E of the emitted radiation by

$$E = nhv \tag{1.5}$$

where n is an integer; h is known as the Planck constant and its presently accepted value is

$$h = (6.626\,0755 \pm 0.000\,004\,0) \times 10^{-34}\,\text{J s} \tag{1.6}$$

The energy E is said to be quantized in discrete packets, or quanta, each of energy hv. It is because of the extremely small value of h that quantization of energy in macroscopic systems had escaped notice, but, of course, it applies to all systems.

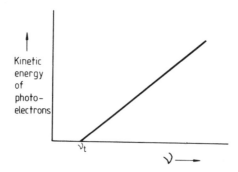

Figure 1.2   Variation of kinetic energy of photoelectrons with the frequency of incident radiation.

Einstein, in 1906, applied this theory to the photoelectric effect and showed that

$$hv = \tfrac{1}{2}m_e v^2 + I \tag{1.7}$$

where hv is the energy associated with a quantum, which Lewis in 1924 called a photon, of the incident radiation, $\tfrac{1}{2}m_e v^2$ is the kinetic energy of the photo-electron ejected with velocity v, and I is the ionization energy[†] of the metal surface.

In 1913 Bohr amalgamated classical and quantum mechanics in explaining the observation of not only the Balmer series but the Lyman, Paschen, Brackett, Pfund, etc., series in the hydrogen atom emission spectrum illustrated in Figure 1.1. Bohr assumed empirically that the electron can move only in specific circular orbits around the nucleus and that the angular momentum p_θ, for an angle of rotation θ, is given by

$$p_\theta = \frac{nh}{2\pi} \tag{1.8}$$

where $n = 1,2,3,\ldots$ and defines the particular orbit. Energy is emitted or absorbed when the electron moves from an orbit with higher n to one of lower n, or vice versa. The energy E_n of the electron can be shown by classical mechanics to be

$$E_n = -\frac{\mu e^4}{8h^2\varepsilon_0^2}\left(\frac{1}{n^2}\right) \tag{1.9}$$

where $\mu = m_e m_p/(m_e + m_p)$ is the reduced mass of the system of electron e plus proton p, e is the electronic charge, and ε_0 the permittivity of a vacuum. These are the energy levels in Figure 1.1 except that $\bar{v} = E_n/hc$, rather than E_n, is plotted. The zero of energy is taken to correspond to $n = \infty$ at which level the atom is ionized[‡]. The energy levels are discrete below $n = \infty$ but continuous above it since the electron can be ejected with any amount of kinetic energy.

When the electron transfers from, say, a lower n'' to an upper n' orbit[§] the energy ΔE required is, from equation (1.9),

$$\Delta E = \frac{\mu e^4}{8h^2\varepsilon_0^2}\left(\frac{1}{n''^2} - \frac{1}{n'^2}\right) \tag{1.10}$$

[†] I is often, but strictly incorrectly, referred to as the ionization potential although, for a solid, it is usually called the work function.
[‡] Note that, when $A \rightarrow A^+ + e$, it is the atom A, not the electron, that is ionized.
[§] The use of single (') and double (") primes to indicate the upper and lower states respectively of a transition is general in spectroscopy and will be used throughout the book.

or, since $\Delta E = h\nu$, we have, in terms of frequency,

$$\nu = \frac{\mu e^4}{8h^3\varepsilon_0^2}\left(\frac{1}{n''^2} - \frac{1}{n'^2}\right) \tag{1.11}$$

Comparison with the empirical equation (1.4) shows that $R_H = \mu e^4/8h^3\varepsilon_0^2$ and that $n'' = 2$ for the Balmer series. Similarly $n'' = 1,3,4$, and 5 for the Lyman, Paschen, Brackett, and Pfund series although it is important to realize that there is an infinite number of series. Many series with high n'' have been observed, by techniques of radioastronomy, in the interstellar medium where there is a large amount of atomic hydrogen. For example, the $(n' = 167) - (n'' = 166)$ transition[†] has been observed with $\nu = 1.425$ GHz ($\lambda = 21.04$ cm).

The Rydberg constant from equation (1.11) has dimensions of frequency but is more often quoted with dimensions of wavenumber when

$$\tilde{R}_H = \frac{\mu e^4}{8h^3\varepsilon_0^2 c} = (1.096\ 775\ 830\ 6 \pm 0.000\ 000\ 001\ 3) \times 10^7\ \text{m}^{-1} \tag{1.12}$$

a very accurately determined constant.

Planck's quantum theory was very successful in explaining the hydrogen atom spectrum, the wavelength distribution of black body radiation, the photoelectric effect, and the low temperature heat capacities of solids, but it also gave rise to apparent anomalies. One of these concerned the photoelectric effect in which the ultraviolet light falling on an alkali metal surface behaves as if it consists of particles whereas the phenomena of interference and diffraction of light are explained by its wave nature. This dual wave–particle nature, which applies not only to light but also to any particle or radiation, was resolved by de Broglie in 1924. He proposed that

$$p = \frac{h}{\lambda} \tag{1.13}$$

relating the momentum p in the particle picture to the wavelength λ in the wave picture. This equation led, for example, to the important prediction that a beam of electrons, travelling with uniform velocity and, therefore, momentum, should show wave-like properties. In 1925 Davisson and Germer confirmed this by showing that the surface of crystalline nickel reflected and diffracted a mono-chromatic electron beam. Their experiment formed the basis of the LEED (low energy electron diffraction) technique for investigating structure near the surface of crystalline materials. Further experiments showed that transmission of an electron beam through a thin metal foil also resulted in diffraction. Using a

---

[†] The use of U–L to indicate a transition between an upper state U and a lower state L is general in spectroscopy and will be used throughout the book.

gaseous, rather than a solid, sample the technique of electron diffraction is an important method of determining molecular geometry in a way which is complementary to spectroscopic methods.

Now that the wave and particle pictures were reconciled it became clear why the electron in the hydrogen atom may be only in particular orbits with angular momentum given by equation (1.8). In the wave picture the circumference $2\pi r$ of an orbit of radius r must contain an integral number of wavelengths

$$n\lambda = 2\pi r \tag{1.14}$$

where $n = 1, 2, 3,\ldots, \infty$, for a standing wave to be set up. Such a wave is illustrated in Figure 1.3(a) for $n = 6$ whereas Figure 1.3(b) shows how a travelling wave results when n is not an integer: the wave interferes with itself and is destroyed.

The picture of the electron in an orbit as a standing wave does, however, pose the important question of where the electron, regarded as a particle, is. We shall consider the answer to this for the case of an electron travelling with constant velocity in a direction x. The de Broglie picture of this is of a wave with a specific wavelength travelling in the x direction as in Figure 1.4(a), and it is clear that we cannot specify where the electron is.

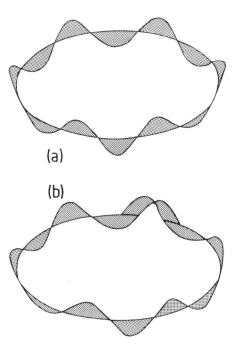

(a)

(b)

Figure 1.3 (a) A standing wave for an electron in an orbit with $n = 6$. (b) A travelling wave results when n is not an integer.

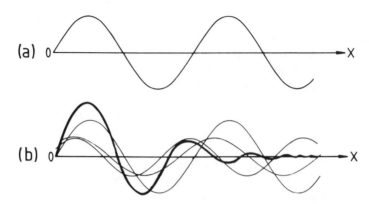

Figure 1.4 (a) The wave due to an electron travelling with specific velocity in the x direction. (b) Superposition of waves of different wavelengths reinforcing each other near to $x = 0$.

At the other extreme we can consider the electron as a particle which can be observed as a scintillation on a phosphorescent screen. Figure 1.4(b) shows how, if there is a large number of waves of different wavelengths and amplitudes travelling in the x direction, they may reinforce each other at a particular value of x, x_s say, and cancel each other elsewhere. This superposition at x_s is called a wave packet and we can say the electron is behaving as if it were a particle at x_s.

For the situation illustrated in Figure 1.4(a) the momentum p_x of the electron is certain but the position x of the electron is completely uncertain whereas, for that in Figure 1.4(b), x is certain but the wavelength, and therefore p_x, is uncertain. In 1927 Heisenberg proposed that, in general, the uncertainties Δp_x and Δx, in p_x and x, are related by

$$\Delta p_x \Delta x \geqslant \hbar \qquad (1.15)$$

which is known as the Heinsenberg uncertainty principle. In this equation $\hbar(= h/2\pi)$ is used: this quantity occurs often in quantum mechanics and spectroscopy and is a convenient abbreviation. We can see from equation (1.15) that, in the extreme wave picture, $\Delta p_x = 0$ and $\Delta x = \infty$ and, in the extreme particle picture, $\Delta x = 0$ and $\Delta p_x = \infty$.

Another important form of the uncertainty principle is

$$\Delta t \Delta E \geqslant \hbar \qquad (1.16)$$

relating the uncertainties in time t and energy E. This shows that, if we know the energy of a state exactly, $\Delta E = 0$ and $\Delta t = \infty$. Such a state does not change with time and is known as a stationary state.

These arguments regarding the reconciliation of the wave and particle pictures of the electron apply similarly to any other small particle such as a positron, neutron, or proton. They also parallel similar arguments applied to the nature of light following Young's experiment in 1807, in which he observed interference fringes when the same source of light illuminated two close pinholes. The wave picture is required to explain such phenomena as interference and diffraction whereas the particle (photon) picture is satisfactory in solving problems in geometrical optics.

1.3 The Schrödinger Equation and Some of its Solutions

It is not the intention that this book should be a primary reference on quantum mechanics: such references are given in the bibliography at the end of this chapter. Nevertheless it is necessary at this stage to take a brief tour through the development of the Schrödinger equation and some of its solutions which are vital to the interpretation of atomic and molecular spectra.

1.3.1 The Schrödinger Equation

The Schrödinger equation cannot be subjected to firm proof but was put forward as a postulate, based on the analogy between the wave nature of light and of the electron. The equation was justified by the remarkable successes of its applications.

Just as a travelling light wave can be represented by a function of its amplitude at a particular position and time, so it was proposed that a wave function $\Psi(x,y,z,t)$, a function of position and time, describes the amplitude of an electron wave. In 1926 Born related the wave and particle views by saying that we should not speak of a particle being at a particular point at a particular time but of the probability of finding the particle there. This probability is given by $\Psi^*\Psi$ where Ψ^* is the complex conjugate of Ψ obtained by replacing all $i(= \sqrt{-1})$ in Ψ by $-i$. It follows that

$$\int \Psi^*\Psi \, d\tau = 1 \tag{1.17}$$

where $d\tau$ is the volume element $dxdydz$, because the probability of finding the electron anywhere in space is unity. It seemed reasonable also that this probability is independent of time:

$$\frac{\partial\left(\int \Psi^*\Psi \, d\tau\right)}{\partial t} = 0 \tag{1.18}$$

which was assumed in the non-relativistic quantum mechanics of Schrödinger developed in 1926. Dirac, in 1928, showed that, when relativity is taken into account, this is not quite true, but we shall not be concerned with the effects of relativity.

The form postulated for the wave function is

$$\Psi = b \exp\left(\frac{iA}{h}\right) \tag{1.19}$$

where b is a constant and A is the action which is related to the kinetic energy T and the potential energy V by

$$-\frac{\partial A}{\partial t} = T + V = H \tag{1.20}$$

where H, the sum of the kinetic and potential energies, is known as the hamiltonian in classical mechanics. From equations (1.19) and (1.20) it follows that

$$H\Psi = i\hbar \frac{\partial \Psi}{\partial t} \tag{1.21}$$

Schrödinger postulated that the form of the hamiltonian in quantum mechanics is obtained by replacing the kinetic energy in equation (1.20) giving

$$H = -\frac{\hbar^2}{2m} \nabla^2 + V \tag{1.22}$$

The symbol ∇ is called 'del' and in cartesian coordinates ∇^2, known as the laplacian, is given by

$$\nabla^2 = \frac{\partial^2}{\partial x^2} + \frac{\partial^2}{\partial y^2} + \frac{\partial^2}{\partial z^2} \tag{1.23}$$

From equations (1.21) and (1.22) we obtain the time-dependent Schrödinger equation

$$-\frac{\hbar^2}{2m} \nabla^2 \Psi + V\Psi = i\hbar \frac{\partial \Psi}{\partial t} \tag{1.24}$$

but, since we shall be dealing mostly with standing waves, it is the time-independent part which will concern us most.

We consider, for ease of manipulation, the wave travelling in the x direction and assume that $\Psi(x,t)$ can be factorized into a time-dependent part $\theta(t)$ and a time-independent part $\psi(x)$, giving

$$\Psi(x,t) = \psi(x)\,\theta(t) \tag{1.25}$$

Combination of equation (1.24), for a one-dimensional system, and equation

(1.25) gives

$$-\frac{\hbar^2}{2m\psi(x)}\frac{\partial^2\psi(x)}{\partial x^2} + V(x) = \frac{i\hbar}{\theta(t)}\frac{\partial\theta(t)}{\partial t} \tag{1.26}$$

Since the left-hand side is a function of x only and the right-hand side a function of t only, they must both be constant. Since they have the same dimensions as $V(x)$, i.e. energy, we put them equal to E. For the left-hand side this gives

$$-\frac{\hbar^2}{2m}\frac{\partial^2\psi(x)}{\partial x^2} + V(x)\psi(x) = E\psi(x) \tag{1.27}$$

which is the one-dimensional, time-independent Schrödinger equation, often simply called the wave equation. It can be rewritten in the general form

$$H\psi = E\psi \tag{1.28}$$

where H is the hamiltonian of equation (1.22) but for one dimension only. Since H contains $\partial/\partial x$ it is, in mathematical terms, an operator as, for example, d/dx is an operator which operates on x^2 to give $2x$. As a result, equation (1.28) appears deceptively simple. It means that operating on ψ by H gives the result of ψ multiplied by an energy E. A simple example is provided by

$$\frac{d}{dx}\exp(2x) = 2\exp(2x) \tag{1.29}$$

an equation of the same form as equation (1.28).

For the quantum mechanical results that we require we shall be concerned only with stationary states, known sometimes as eigenstates. The wave functions for these states may be referred to as eigenfunctions and the associated energies E as the eigenvalues.

The details of the methods of solving the Schrödinger equation for ψ and E for various systems do not concern us here but may be found in books listed in the bibliography. We require only the results, some of which will now be discussed.

1.3.2 The Hydrogen Atom

The hydrogen atom, consisting of a proton and only one electron, occupies a very important position in the development of quantum mechanics because the Schrödinger equation may be solved exactly for this system. This is true also for the hydrogen-like atomic ions He^+, Li^{2+}, Be^{3+}, etc., and simple one-electron molecular ions such as H_2^+.

In the quantum mechanical picture of the hydrogen atom the total energy E_n is quantized having exactly the same values as in equation (1.9) derived from classical mechanics. The angular momentum of the electron in a particular

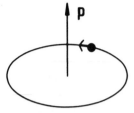

Figure 1.5 Direction of the angular momentum vector p for an electron in an orbit.

orbit, or orbital as it is called in quantum mechanics, may also take only discrete values. This orbital angular momentum p is a vector[†] and is defined, therefore, by its magnitude *and* direction. In the classical picture of an electron circulating in an orbit in the direction shown in Figure 1.5 the direction of the corresponding vector, which is also shown, is given by the right-hand screw rule. If we now introduce some kind of directionality into space, by a magnetic field for example, p can take only certain orientations with respect to that direction so that the component of p in that direction can take only certain, discrete values. This phenomenon is referred to as space quantization which arises in a semi-classical treatment also, but in a way which is quantitatively incorrect. The effect is known as the Zeeman effect.

For the hydrogen atom, the hamiltonian of equation (1.22) becomes

$$H = -\frac{\hbar^2}{2\mu}\nabla^2 - \frac{e^2}{4\pi\varepsilon_0 r} \tag{1.30}$$

The second term on the right-hand side is the coulombic potential energy for the attraction between charges $-e$ and $+e$ a distance r apart. The first term contains the reduced mass $\mu\ (= m_e m_p/(m_e + m_p))$ for the system of an electron of mass m_e and a proton of mass m_p. It also contains the laplacian ∇^2 which is here defined as

$$\nabla^2 = \frac{1}{r^2 \sin\theta}\left[\sin\theta\frac{\partial}{\partial r}\left(r^2\frac{\partial}{\partial r}\right) + \frac{\partial}{\partial\theta}\left(\sin\theta\frac{\partial}{\partial\theta}\right) + \frac{1}{\sin\phi}\frac{\partial^2}{\partial\phi^2}\right] \tag{1.31}$$

where r, the distance of a point P from the origin, θ, the co-latitude, and ϕ, the azimuth, are spherical polar coordinates illustrated in Figure 1.6. It may be surprising, judging from the complexity of equation (1.31), but the solutions of the Schrödinger equation are much simpler than if we use cartesian coordinates. In particular, the wave functions corresponding to the hamiltonian of equation (1.30) may be factorized:

$$\psi(r, \theta, \phi) = R_{n\ell}(r)Y_{\ell m_\ell}(\theta,\phi) \tag{1.32}$$

[†] A vector quantity is indicated by bold italic type and its magnitude by italic type.

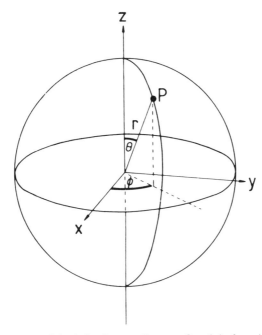

Figure 1.6 Spherical polar coordinates r, θ, and ϕ of a point P.

The $Y_{\ell m_\ell}$ functions are known as the angular wave functions or, because they describe the distribution of ψ over the surface of a sphere of radius r, spherical harmonics. The quantum number $n = 1,2,3,\ldots,\infty$ and is the same as in the Bohr theory, ℓ is the azimuthal quantum number associated with the discrete orbital angular momentum values, and m_ℓ is known as the magnetic quantum number which results from the space quantization of the orbital angular momentum. These quantum numbers can take the values

$$\ell = 0, 1, 2, \ldots, (n-1) \tag{1.33}$$

$$m_\ell = 0, \pm 1, \pm 2, \ldots, \pm \ell \tag{1.34}$$

The function $Y_{\ell m_\ell}(\theta,\phi)$ in equation (1.32) can be factorized further to give

$$Y_{\ell m_\ell}(\theta,\phi) = (2\pi)^{-1/2}\Theta_{\ell m_\ell}(\theta)\exp(im_\ell\phi) \tag{1.35}$$

The $\Theta_{\ell m_\ell}$ functions are the associated Legendre polynomials of which a few are given in Table 1.1. They are independent of Z, the nuclear charge number, and therefore the same for all one-electron atoms.

Solution of the Schrödinger equation for $R_{n\ell}(r)$, known as the radial wave functions since they are functions only of r, follows a well-known mathematical procedure to produce the solutions known as the associated Laguerre functions,

Table 1.1 Some $\Theta_{\ell m_\ell}$ wave functions for hydrogen and hydrogen-like atoms.

ℓ	m_ℓ	$\Theta_{\ell m_\ell}(\theta)$	ℓ	m_ℓ	$\Theta_{\ell m_\ell}$
0	0	$\dfrac{1}{2^{1/2}}$	2	0	$\dfrac{10^{1/2}}{4}(3\cos^2\theta - 1)$
1	0	$\dfrac{6^{1/2}}{2}\cos\theta$	2	± 1	$\dfrac{15^{1/2}}{2}\sin\theta\cos\theta$
1	± 1	$\dfrac{3^{1/2}}{2}\sin\theta$	2	± 2	$\dfrac{15^{1/2}}{4}\sin^2\theta$

of which a few are given in Table 1.2. The radius of the Bohr orbit for $n = 1$ is given by

$$a_0 = \frac{\hbar^2 4\pi\varepsilon_0}{\mu e^2 Z} \tag{1.36}$$

For hydrogen, $a_0 = 0.529$ Å and a useful quantity ρ is related to r by

$$\rho = \frac{Zr}{a_0} \tag{1.37}$$

Orbitals are labelled according to the values of n and ℓ. The rather curious symbols $s, p, d, f, g, \ldots$ indicate values of $\ell = 0, 1, 2, 3, 4, \ldots$, and the reason for this will be given in Section 7.1.3. Thus we speak of $1s$, $2s$, $2p$, $3s$, $3p$, $3d$, etc., orbitals where the 1, 2, 3, etc., refer to the value of n.

There are three useful ways of representing $R_{n\ell}$ graphically:

1. Plot $R_{n\ell}$ against ρ (or r). This is done in Figure 1.7(a) where it can be seen that R_{10} and R_{21} are always positive but R_{20} changes from positive to negative and, at one value of ρ, is zero.

Table 1.2 Some $R_{n\ell}$ wave functions for hydrogen and hydrogen-like atoms.

n	ℓ	$R_{n\ell}(r)$
1	0	$\left(\dfrac{Z}{a_0}\right)^{3/2} 2\exp(-\rho)$
2	0	$\left(\dfrac{Z}{a_0}\right)^{3/2} \dfrac{1}{2^{1/2}}\left(1 - \dfrac{\rho}{2}\right)\exp\left(-\dfrac{\rho}{2}\right)$
2	1	$\left(\dfrac{Z}{a_0}\right)^{3/2} \left(\dfrac{1}{2}\right)\dfrac{1}{6^{1/2}}\rho\exp\left(-\dfrac{\rho}{2}\right)$

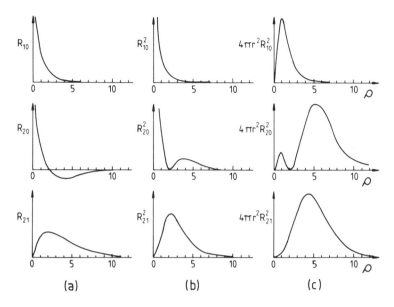

Figure 1.7 Plots of (a) the radial wave function $R_{n\ell}$, (b) the radial probability distribution function $R_{n\ell}^2$, and (c) the radial charge density function $4\pi r^2 R_{n\ell}^2$ against ρ.

2. Plot $R_{n\ell}^2$ against ρ (or r), as shown in Figure 1.7(b). Since $R_{n\ell}^2 \, dr$ is the probability of finding the electron between r and $r + dr$ this plot represents the radial probability distribution of the electron.

3. Plot $4\pi r^2 R_{n\ell}^2$ against ρ (or r), as shown in Figure 1.7(c). The quantity $4\pi r^2 R_{n\ell}^2$ is called the radial charge density and is the probability of finding the electron in a volume element consisting of a thin spherical shell of thickness dr, radius r, and volume $4\pi r^2 dr$.

Diagrammatic representations of the $Y_{\ell m_\ell}$ functions in equation (1.35) cannot be made until we convert them from imaginary into real functions. Exceptions are the functions with $m_\ell = 0$ which are already real.

In the absence of an electric or magnetic field all the $Y_{\ell m_\ell}$ functions with $\ell \neq 0$ are $(2\ell + 1)$-fold degenerate, which means that there are $(2\ell + 1)$ functions, each having one of the $(2\ell + 1)$ possible values of m_ℓ, with the same energy. It is a property of degenerate functions that linear combinations of them are also solutions of the Schrödinger equation. For example, just as $\psi_{2p,1}$ and $\psi_{2p,-1}$ are solutions, so are

$$\psi_{2p_x} = 2^{-1/2}(\psi_{2p,1} + \psi_{2p,-1}) \tag{1.38}$$

$$\psi_{2p_y} = -2^{-1/2}i(\psi_{2p,1} - \psi_{2p,-1})$$

From equations (1.32), (1.35), and (1.38), together with the $\Theta_{\ell m_\ell}$ functions in Table 1.1, it follows that

$$\psi_{2p_x} = \frac{1}{2(4\pi)^{1/2}} R_{21}(r)\, 3^{1/2} \sin\theta[\exp(i\phi) + \exp(-i\phi)] \tag{1.39}$$

$$\psi_{2p_y} = \frac{1}{2(4\pi)^{1/2}} iR_{21}(r)\, 3^{1/2} \sin\theta[\exp(i\phi) - \exp(-i\phi)]$$

However, since

$$\exp(i\phi) + \exp(-i\phi) = 2\cos\phi \tag{1.40}$$

$$\exp(i\phi) - \exp(-i\phi) = 2i\sin\phi$$

equations (1.39) become

$$\psi_{2p_x} = \frac{1}{(4\pi)^{1/2}} R_{21}(r)3^{1/2} \sin\theta\cos\phi \tag{1.41}$$

$$\psi_{2p_y} = \frac{1}{(4\pi)^{1/2}} R_{21}(r)3^{1/2} \sin\theta\sin\phi \tag{1.42}$$

In addition, the third degenerate $\psi_{2p,0}$ wave function is always real and we label it ψ_{2p_z} where

$$\psi_{2p_z} = \frac{1}{(4\pi)^{1/2}} R_{21}(r)3^{1/2} \cos\theta \tag{1.43}$$

All the ψ_{ns} wave functions are always real and so are the ψ_{nd}, ψ_{nf}, etc., wave functions for $m_\ell = 0$. However, for ψ_{nd} with $m_\ell = \pm 1$ or ± 2, it is necessary to form linear combinations of the imaginary wave functions $\psi_{nd,1}$ and $\psi_{nd,-1}$, or $\psi_{nd,2}$ and $\psi_{nd,-2}$ to obtain real functions. The ψ_{nd} orbital wave functions for any $n > 2$ are distinguished by subscripts nd_{z^2} ($m_\ell = 0$), nd_{xz} and nd_{yz} ($m_\ell = \pm 1$), and nd_{xy} and $nd_{x^2-y^2}$ ($m_\ell = \pm 2$). There are seven nf orbitals for any $n > 3$ but we shall not consider them here.

In Figure 1.8 the real $Y_{\ell m_\ell}$ wave functions for the $1s$, $2p$, and $3d$ orbitals are plotted in the form of polar diagrams, the construction of which may be illustrated by the simple case of the $2p_z$ orbital. The wave function in equation (1.43) is independent of ϕ and is simply proportional to $\cos\theta$. The polar diagram consists of points on a surface obtained by marking off, on lines drawn outwards from the nucleus in all directions, distances proportional to $|\cos\theta|$ at a constant value of $R_{21}(r)$. The resulting surface consists of two touching spheres, as shown in Figure 1.8 which shows polar diagrams for all $1s$, $2p$, and $3d$ orbitals.

For all orbitals except $1s$ there are regions in space where $\psi(r,\theta,\phi) = 0$ because either $Y_{\ell m_\ell} = 0$ or $R_{n\ell} = 0$. In these regions the electron density is zero and we call them nodal surfaces or, simply, nodes. For example, the $2p_z$ orbital has

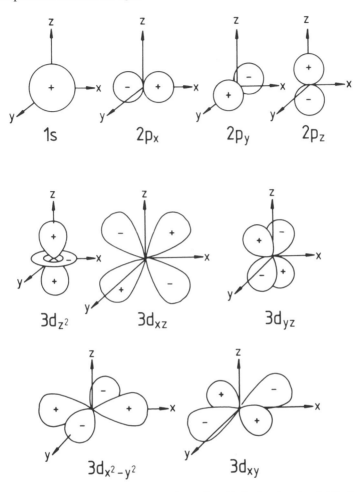

Figure 1.8 Polar diagrams for 1s, 2p, and 3d atomic orbitals showing the distributions of the angular wave functions.

a nodal plane, the xy plane, while each of the 3d orbitals has two nodal planes. In general there are ℓ such angular nodes where $Y_{\ell m_\ell} = 0$. The 2s orbital has one spherical nodal plane, or radial node, as Figure 1.7 shows. In general, there are $(n - 1)$ radial nodes for an ns orbital (or n if we count the one at infinity).

Quantum mechanical solution results in the same expression for the energy levels, given in equation (1.9), as in the Bohr theory—indeed, it must do since the Bohr theory agrees exactly with experiment, except for the fine structure of the spectrum which our present quantum mechanical treatment has not explained either. There are far-reaching differences in the quantum mechanical treatment. Some of these are embodied in Figures 1.7 and 1.8 which portray the electron as smeared out in various patterns of probability which always approach zero as r tends to infinity. The probability distribution also contains nodal surfaces where there is zero probability of finding the electron. All of this is a far cry from the classical picture of the electron orbiting round the nucleus like the moon orbiting round the earth.

Unlike the total energy, the quantum mechanical value P_ℓ of the orbital angular momentum is significantly different from that in the Bohr theory given in equation (1.8). It is now given by

$$P_\ell = [\ell(\ell + 1)]^{1/2}\hbar \qquad (1.44)$$

where $\ell = 0, 1, 2, \ldots (n - 1)$, as in equation (1.33).

An effect of space quantization of orbital angular momentum may be observed if a magnetic field is introduced along what we now identify as the z axis. The orbital angular momentum vector P, of magnitude P_ℓ, may take up only certain orientations such that the component $(P_\ell)_z$ along the z axis is given by

$$(P_\ell)_z = m_\ell \hbar \qquad (1.45)$$

where $m_\ell = 0, \pm 1, \pm 2, \ldots, \pm \ell$, as in equation (1.34). This is illustrated in Figure 1.9 for an electron in a d orbital ($\ell = 3$).

1.3.3 Electron Spin and Nuclear Spin Angular Momentum

In the classical picture of an electron orbiting round the nucleus it would not surprise us to discover that the electron and the nucleus could each spin on its own axis, just like the earth and the moon, and that each has an angular momentum associated with spinning. Unfortunately, although quantum mechanical treatment gives rise to two new angular momenta, one associated with the electron and one with the nucleus, this simple physical picture breaks down. When we think, for example, of the wave rather than the particle picture of the electron this breakdown is not surprising. However, in spite of this it is still usual to speak of electron spin and nuclear spin.

From a quantum mechanical treatment the magnitude of the angular momentum P_s due to the 'spin' of one electron, whether it is in the hydrogen atom or any other atom, is given by

$$P_s = [s(s + 1)]^{1/2}\hbar \qquad (1.46)$$

where the electron spin quantum number s can take the value $\frac{1}{2}$ *only*. In fact this result cannot be derived from the Schrödinger equation but only from Dirac's relativistic quantum mechanics. Space quantization of this angular momentum results in the component

$$(P_s)_z = m_s \hbar \qquad (1.47)$$

where $m_s = \pm \frac{1}{2}$ only. This is illustrated in Figure 1.10.

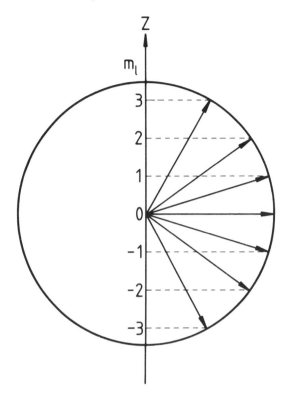

Figure 1.9 Space quantization of orbital angular momentum for $\ell = 3$.

Similarly, the magnitude of the angular momentum P_I due to nuclear spin is given by

$$P_I = [I(I + 1)]^{1/2}\hbar \qquad (1.48)$$

The nuclear spin quantum number I may be zero, half-integer, or integer depending on the nucleus concerned. Nuclei contain protons and neutrons, for each of which $I = \frac{1}{2}$. The way in which the spin angular momenta of these

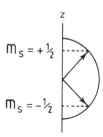

Figure 1.10 Space quantization of electron spin angular momentum.

Table 1.3 Some values of the nuclear spin quantum number I.

Nucleus	I	Nucleus	I	Nucleus	I	Nucleus	I
^{1}H	1/2	^{12}C	0	^{16}O	0	^{30}Si	0
^{2}H	1	^{13}C	1/2	^{19}F	1/2	^{31}P	1/2
^{10}B	3	^{14}N	1	^{28}Si	0	^{35}Cl	3/2
^{11}B	3/2	^{15}N	1/2	^{29}Si	1/2	^{37}Cl	3/2

particles couple together determines the value of I, a few being given in Table 1.3. It is essential for nuclear magnetic resonance spectroscopy that $I \neq 0$ for the nuclei being studied so that, for example, ^{12}C nuclei are of no use in this respect but, increasingly in pulsed Fourier transform n.m.r. spectroscopy, ^{13}C with $I = \frac{1}{2}$ is studied in natural abundance (1.1 per cent).

1.3.4 The Born–Oppenheimer Approximation

Just as for an atom, the hamiltonian H for a diatomic or polyatomic molecule is the sum of the kinetic energy T, or its quantum mechanical equivalent, and the potential energy V, as in equation (1.20). In a molecule the kinetic energy T consists of contributions T_e and T_n from the motions of the electrons and nuclei respectively. The potential energy comprises two terms, V_{ee} and V_{nn}, due to coulombic repulsions between the electrons and between the nuclei and a third term V_{en} due to attractive forces between the electrons and nuclei, giving

$$H = T_e + T_n + V_{en} + V_{ee} + V_{nn} \qquad (1.49)$$

For fixed nuclei $T_n = 0$ and V_{nn} is constant, and there is a set of electronic wave functions ψ_e which satisfy

$$H_e \psi_e = E_e \psi_e \qquad (1.50)$$

where

$$H_e = T_e + V_{en} + V_{ee} \qquad (1.51)$$

Since H_e depends on nuclear coordinates, because of the V_{en} term, so do ψ_e and E_e but, in the Born–Oppenheimer approximation proposed in 1927, it is assumed that vibrating nuclei move so slowly compared to electrons that ψ_e and E_e involve the nuclear coordinates as parameters only. The result for a diatomic molecule is that a curve (like that in Figure 1.13, for example) of potential energy against internuclear distance r (or the displacement from equilibrium) can be drawn for a particular electronic state in which T_e and V_{ee} are constant.

The Born–Oppenheimer approximation is valid because the electrons adjust instantaneously to any nuclear motion: they are said to follow the nuclei. For this reason E_e can be treated as part of the potential field in which the nuclei move so that

$$H_n = T_n + V_{nn} + E_e \tag{1.52}$$

and the Schrödinger equation for nuclear motion is

$$H_n\psi_n = E_n\psi_n \tag{1.53}$$

It follows from the Born–Oppenheimer approximation that the total wave function ψ can be factorized:

$$\psi = \psi_e(q,Q)\psi_n(Q) \tag{1.54}$$

where the q are electron coordinates and ψ_e is a function of nuclear coordinates Q as well as q. It follows from equation (1.54) that

$$E = E_e + E_n \tag{1.55}$$

The wave function ψ_n can be factorized further into a vibrational part ψ_v and a rotational part ψ_r:

$$\psi_n = \psi_v\psi_r \tag{1.56}$$

for the same reasons, perhaps unexpectedly, that the hydrogen atom wave function $\psi(r,\theta,\phi)$ can be factorized into $R_{n\ell}(r)$ and $Y_{\ell m_\ell}(\theta,\phi)$ as in equation (1.32). From equation (1.56) it follows that

$$E_n = E_v + E_r \tag{1.57}$$

so that

$$\psi = \psi_e\psi_v\psi_r \tag{1.58}$$

and

$$E = E_e + E_v + E_r \tag{1.59}$$

If any atoms have nuclear spin then this part of the total wave function can be factorized and the energy treated additively. It is for these reasons that we can treat electronic, vibrational, rotational, and n.m.r. spectroscopy separately.

1.3.5 The Rigid Rotor

A useful approximate model for the end-over-end rotation of a diatomic molecule is that of the rigid rotor in which the bond joining the nuclei is regarded as a rigid, weightless rod, as shown in Figure 1.11(a). The angular momentum is given by

$$P_J = [J(J + 1)]^{1/2}\hbar \tag{1.60}$$

where the rotational quantum number $J = 0, 1, 2, \ldots$. In general, J is asso-

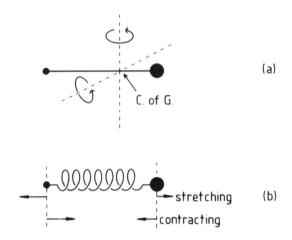

Figure 1.11 (a) Rotation of a heteronuclear diatomic molecule about axes perpendicular to the bond and through the centre of gravity. (b) Vibration of the same molecule.

ciated with total angular momentum excluding nuclear spin, i.e. rotational + orbital + electron spin, but, when there is no orbital or electron spin angular momentum, it refers simply to rotation.

Just as with other angular momenta there is space quantization of rotational angular momentum so that the z component is given by

$$(P_J)_z = M_J \hbar \tag{1.61}$$

where $M_J = J, J - 1, \ldots, -J$. Therefore, in the absence of an electric or magnetic field, each rotational energy level is $(2J + 1)$-fold degenerate.

Solution of the Schrödinger equation for a rigid rotor shows that the rotational energy E_r is quantized with values

$$E_r = \frac{h^2}{8\pi^2 I} J(J + 1) \tag{1.62}$$

where the moment of inertia $I = \mu r^2$ in which r is the internuclear distance and $\mu = m_1 m_2/(m_1 + m_2)$ is the reduced mass for nuclei of masses m_1 and m_2. Figure 1.12 shows how the rotational energy levels given by equation (1.62) diverge with increasing J.

1.3.6 The Harmonic Oscillator

Figure 1.11(b) illustrates the ball-and-spring model which is adequate for an approximate treatment of vibration of a diatomic molecule. For small displacements the stretching and compression of the bond, represented by the spring,

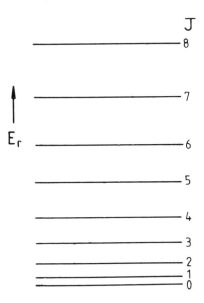

Figure 1.12 A set of rotational energy levels E_r.

obeys Hooke's law:

$$\text{Restoring force} = -\frac{\mathrm{d}V(x)}{\mathrm{d}x} = -kx \tag{1.63}$$

where V is the potential energy, k is the force constant whose magnitude reflects the strength of the bond, and $x(=r-r_e)$ is the displacement from the equilibrium bond length r_e. Integrating this equation gives

$$V(x) = \tfrac{1}{2}kx^2 \tag{1.64}$$

In Figure 1.13, $V(r)$ is plotted against r and illustrates the parabolic relationship.

The quantum mechanical hamiltonian for a one-dimensional harmonic oscillator is given by

$$H = -\frac{\hbar^2}{2\mu}\frac{\mathrm{d}^2}{\mathrm{d}x^2} + \tfrac{1}{2}kx^2 \tag{1.65}$$

where μ is the reduced mass of the nuclei. The Schrödinger equation (see equation 1.27) becomes

$$\frac{\mathrm{d}^2\psi_v}{\mathrm{d}x^2} + \left(\frac{2\mu E_v}{\hbar^2} - \frac{\mu kx^2}{\hbar^2}\right)\psi_v = 0 \tag{1.66}$$

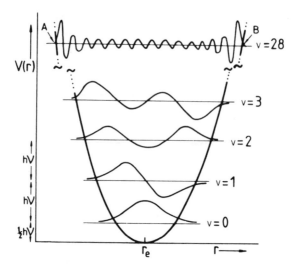

Figure 1.13 Plot of $V(r)$ against r for the harmonic oscillator model for vibration. A few energy levels and wave functions are shown.

from which it can be shown that

$$E_v = h\nu(v + \tfrac{1}{2}) \tag{1.67}$$

where ν is the classical vibration frequency given by

$$\nu = \frac{1}{2\pi}\left(\frac{k}{\mu}\right)^{1/2} \tag{1.68}$$

As expected the frequency increases with k (the stiffness of the bond) and decreases with μ. More commonly, though, we use vibration wavenumber ω, rather than frequency, where[†]

$$E_v = hc\omega(v + \tfrac{1}{2}) \tag{1.69}$$

The vibrational quantum number v can take the values 0, 1, 2, ...

  Equation (1.69) shows the vibrational levels to be equally spaced, by $hc\omega$, and that the $v = 0$ level has an energy $\tfrac{1}{2}hc\omega$, known as the zero-point energy. This is the minimum energy the molecule may have even at the absolute zero of temperature and is a consequence of the uncertainty principle.

  Each point of intersection of an energy level with the curve corresponds to a classical turning point of a vibration where the velocity of the nuclei is zero and all the energy is in the form of potential energy. This contrasts with the mid-point of each energy level where all the energy is kinetic energy.

[†] See page 129.

Table 1.4 Hermite polynomials for $v = 0$ to 5.

v	$H_v(y)$	v	$H_v(y)$
0	1	3	$8y^3 - 12y$
1	$2y$	4	$16y^4 - 48y^2 + 12$
2	$4y^2 - 2$	5	$32y^5 - 160y^3 + 120y$

The wave functions ψ_v resulting from solution of equation (1.66) are

$$\psi_v = \left(\frac{1}{2^v v! \pi^{1/2}}\right)^{1/2} H_v(y) \exp\left(-\frac{y^2}{2}\right) \tag{1.70}$$

where the $H_v(y)$ are known as Hermite polynomials and

$$y = \left(\frac{4\pi^2 v\mu}{h}\right)^{1/2} (r - r_e) \tag{1.71}$$

A few are given in Table 1.4 and some ψ_v are plotted in Figure 1.13 where each energy level corresponds to $\psi_v = 0$. There are several important features to note regarding these wave functions:

1. They penetrate regions outside the parabola which would be forbidden in a classical system.
2. As v increases, the two points where ψ_v^2, the vibrational probability, has a maximum value occur nearer to the classical turning points. This is illustrated for $v = 28$ for which A and B are the classical turning points contrasting with $v = 0$ for which the maximum probability is at the mid-point of the level.
3. The wavelength of the ripples in ψ_v increases away from the classical turning points. This is more apparent as v increases and is pronounced for $v = 28$.

Questions

1. Using equations (1.11) and (1.12) calculate the wavenumbers of the first two (lowest n'') members of the Balmer series of the hydrogen atom. Convert these to wavelengths.
2. Using equation (1.7) calculate the velocity of photoelectrons ejected from a sodium metal surface, with a work function of 2.46 eV, by ultraviolet light of wavelength 250 nm.
3. Show that the wavelength λ associated with a beam of electrons accelerated by a potential difference V is given by $\lambda = h(2eVm_e)^{-1/2}$. Calculate the wavelength of such a beam for an accelerating potential difference of 36.20 kV.

4. From the Heisenberg uncertainty principle as stated in equation (1.16) estimate, in cm^{-1} and Hz, the wavenumber and frequency spread of pulsed radiation with a pulse length of 30 fs, typical of a very short pulse from a visible laser, and of 6 µs, typical of pulsed radiofrequency radiation used in a pulsed Fourier transform n.m.r. experiment.

5. Using equation (1.36) calculate, to six significant figures, the radius of the $n = 1$ Bohr orbit for Be^{3+}.

6. Using equation (1.62) calculate the rotational energy levels, in joules, for $J = 0$ to 4 for ^{12}C ^{16}O and ^{13}C ^{16}O. Convert these to units of cm^{-1}.

7. Given that the vibration wavenumbers of the molecules HCl, SO and PN are 2991, 1149 and 1337 cm^{-1} respectively calculate, from equation (1.68), their force constants and hence comment on the comparative bond strengths.

Bibliography

Atkins, P. W. (1970). *Molecular Quantum Mechanics*, Oxford University Press, Oxford.

Atkins, P. W. (1974). *Quanta*, Oxford University Press, Oxford.

Bockhoff, F. J. (1969). *Elements of Quantum Theory*, Addison-Wesley, Reading, Massachusetts.

Feynman, R. P., Leighton, R. B., and Sands, M. (1965). *The Feynman Lectures on Physics*, Addison-Wesley, Reading, Massachusetts.

Jørgensen, P., and Oddershede, J. (1983). *Problems in Quantum Chemistry*, Addison-Wesley, Reading, Massachusetts.

Kauzmann, W. (1957). *Quantum Chemistry*, Academic Press, New York.

Landau, L. D., and Lifshitz, E. M. (1959). *Quantum Mechanics*, Pergamon Press, Oxford.

Pauling, L., and Wilson, E. B. (1935). *Introduction to Quantum Mechanics*, McGraw-Hill, New York.

Schutte, C. J. H. (1968). *The Wave Mechanics of Atoms, Molecules and Ions*, Arnold, London.

CHAPTER 2 ——————————————————————

Electromagnetic Radiation and Its Interaction with Atoms and Molecules

2.1 Electromagnetic Radiation

Electromagnetic radiation includes, in addition to what we commonly refer to as 'light', radiation of longer and shorter wavelengths (see Section 3.1). As the name implies it contains both an electric and a magnetic component which are best illustrated by considering plane-polarized (also known as linearly polarized) radiation. Figure 2.1 illustrates such radiation travelling along the x axis. The electric component of the radiation is in the form of an oscillating electric field of strength E and the magnetic component is in the form of an oscillating magnetic field of strength H. These oscillating fields are at right angles to each other as shown and, if the directions of the vectors E and H are y and z respectively, then

$$E_y = A \sin(2\pi\nu t - kx)$$
$$H_z = A \sin(2\pi\nu t - kx)$$

$$(2.1)$$

where A is the amplitude. Therefore the fields oscillate sinusoidally with a frequency of $2\pi\nu$ and, because k is the same for each component, they are in-phase.

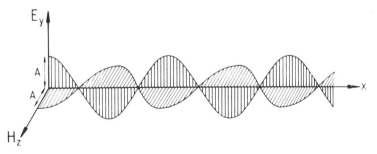

Figure 2.1   Plane-polarized electromagnetic radiation travelling along the x axis.

The plane of polarization is conventionally taken to be the plane containing the direction of E and that of propagation; in Figure 2.1 this is the xy plane. The reason for this choice is that interaction of electromagnetic radiation with matter is more commonly through the electric component.

2.2 Absorption and Emission of Radiation

In Figure 2.2(a) states m and n of an atom or molecule are stationary states, so-called because they are time-independent. This pair of states may be, for example, electronic, vibrational, or rotational. We consider the three processes which may occur when such a two-state system is subjected to radiation of frequency v, or wavenumber $\tilde{v}$, corresponding to the energy separation ΔE where

$$\Delta E = E_n - E_m = hv = hc\tilde{v} \qquad (2.2)$$

These processes are:

1. *Induced absorption*, in which the molecule (or atom) M absorbs a quantum of radiation and is excited from m to n:

$$M + hc\tilde{v} \rightarrow M^* \qquad (2.3)$$

   This is the familiar absorption process illustrated by the appearance of an aqueous solution of copper sulphate as blue due to the absorption of the complementary colour, red, by the solution.
2. *Spontaneous emission*, in which M^* (in state n) spontaneously emits a quantum of radiation:

$$M^* \rightarrow M + hc\tilde{v} \qquad (2.4)$$

   Almost all emission that we usually encounter, such as that from a sodium vapour or tungsten filament lamp, is of the spontaneous type.

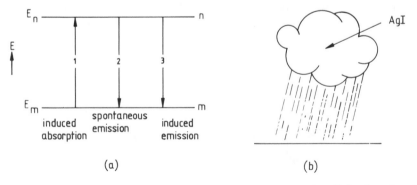

(a)   (b)

Figure 2.2 (a) Absorption and emission processes between states m and n. (b) Seeding of a rain cloud to induce a shower of rain.

3. *Induced, or stimulated, emission.* This is a different type of emission process from that of type 2 in that a quantum of radiation of wavenumber $\tilde{v}$ given by equation (2.2) is required to induce, or stimulate, M* to go from n to m. The process is represented by

$$M^* + hc\tilde{v} \rightarrow M + 2hc\tilde{v} \qquad (2.5)$$

and may seem rather unusual to anyone used only to the spontaneous emission process. There is a useful analogy between induced emission, requiring the presence of radiation of the correct wavenumber for it to occur, and the seeding of a cloud with silver iodide crystals in order to induce it to shed a shower of rain which it would not otherwise do, as shown in Figure 2.2(b). The silver iodide plays the part of the quantum of radiation necessary to induce the process. The reason why the absorption process is strictly referred to as induced absorption may now be appreciated since, of course, it requires the presence of radiation of wavenumber $\tilde{v}$ in order to occur.

The rate of change of population N_n of state n due to induced absorption is given by

$$\frac{dN_n}{dt} = N_m B_{mn} \rho(\tilde{v}) \qquad (2.6)$$

where B_{mn} is a so-called Einstein coefficient and $\rho(\tilde{v})$, the spectral radiation density, is given by

$$\rho(\tilde{v}) = \frac{8\pi hc\tilde{v}^3}{\exp(hc\tilde{v}/kT) - 1} \qquad (2.7)$$

Similarly, induced emission changes the population N_n by

$$\frac{dN_n}{dt} = -N_n B_{nm} \rho(\tilde{v}) \qquad (2.8)$$

where B_{nm} is the Einstein coefficient for this process and is equal to B_{mn}. For spontaneous emission

$$\frac{dN_n}{dt} = -N_n A_{nm} \qquad (2.9)$$

where A_{nm} is another Einstein coefficient and the absence of $\rho(\tilde{v})$ indicates a spontaneous process. In the presence of radiation of wavenumber $\tilde{v}$ all three processes are going on at once and, when the populations have reached their equilibrium values,

$$\frac{dN_n}{dt} = (N_m - N_n)B_{nm}\rho(\tilde{v}) - N_n A_{nm} = 0 \qquad (2.10)$$

At equilibrium N_n and N_m are related, through the Boltzmann distribution law, by

$$\frac{N_n}{N_m} = \frac{g_n}{g_m}\exp\left(-\frac{\Delta E}{kT}\right) = \exp\left(-\frac{\Delta E}{kT}\right) \tag{2.11}$$

if the degrees of degeneracy g_n and g_m of states n and m are the same. Putting this relationship and the expression $\rho(\tilde{\nu})$, given in equation (2.7), into equation (2.10) gives the result

$$A_{nm} = 8\pi hc\tilde{\nu}^3 B_{nm} \tag{2.12}$$

This equation illustrates the important point that spontaneous emission increases rapidly relative to induced emission as $\tilde{\nu}$ increases. Since lasers (Section 9.1) operate entirely by induced emission the equation is particularly relevant to laser design.

The Einstein coefficients are related to the wave functions ψ_m and ψ_n of the combining states through the transition moment $\boldsymbol{R}^{nm}$, a vector quantity given by

$$\boldsymbol{R}^{nm} = \int \psi_n^* \boldsymbol{\mu} \psi_m \, d\tau \tag{2.13}$$

for interaction with the electric component of the radiation. The quantity $\boldsymbol{\mu}$ is the electric dipole moment operator and

$$\boldsymbol{\mu} = \sum_i q_i \boldsymbol{r}_i \tag{2.14}$$

where q_i and $\boldsymbol{r}_i$ are the charge and position vector of the ith particle (electron or nucleus). The transition moment can be thought of as the oscillating electric dipole moment due to the transition. Figure 2.3 shows the π and π^* molecular orbitals of ethylene and, if an electron is promoted from π to π^* in an electronic transition, there is a corresponding non-zero transition moment. This example illustrates the important point that a transition moment may be non-zero even though the permanent electric dipole moment is zero in both the states m and n.

The square of the magnitude of $\boldsymbol{R}^{nm}$ is the transition probability and is related

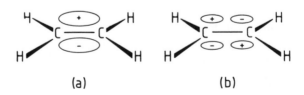

(a)   (b)

Figure 2.3   (a) A π and (b) a π^* molecular orbital of ethylene.

to B_{nm} by

$$B_{nm} = \frac{8\pi^3}{(4\pi\varepsilon_0)3h^2} |R^{nm}|^2 \qquad (2.15)$$

and the following illustrates how B_{nm} may be found experimentally in the case of an absorption experiment.

The experiment is illustrated in Figure 2.4(a) where radiation of intensity I_0 falls on the absorption cell of length ℓ containing absorbing material of concentration c in the liquid phase. The radiation emerges with intensity I and scanning the radiation through an appropriate wavenumber range of the absorption, say $\tilde{v}_1$ to $\tilde{v}_2$, and measuring I_0/I produces the absorption spectrum typically measured as absorbance, defined as $\log_{10}(I_0/I)$. According to the Beer–Lambert law the absorbance A is proportional to c and ℓ so that

$$A = \log_{10}\left(\frac{I_0}{I}\right) = \varepsilon(\tilde{v})c\ell \qquad (2.16)$$

where ε is a function of $\tilde{v}$ and is the molar absorption coefficient or molar absorptivity (although it used to be known as the molar extinction coefficient which will be encountered in older texts). Since A is dimensionless, ε has dimensions of (concentration × length)$^{-1}$ and the units are very often mol^{-1} dm^3 cm^{-1}. The spectrum is shown in Figure 2.4(b). The quantity ε_{max} corresponds to the maximum value of A and is sometimes used as a measure of the total absorption intensity, but the spectrum in Figure 2.4(c), which has

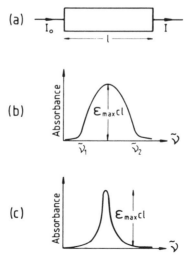

Figure 2.4 (a) An absorption experiment. (b) A broad and (c) a narrow absorption band with the same ε_{max}.

the same ε_{max} but a much lower integrated intensity, illustrates the dangers of using ε_{max}. What we should do is to integrate the area under the curve; then, provided $N_n \ll N_m$ so that decay of state n by induced emission is negligible,

$$\int_{\tilde{\nu}_1}^{\tilde{\nu}_2} \varepsilon(\tilde{\nu})d\tilde{\nu} = \frac{N_A h\tilde{\nu}_{nm}B_{nm}}{\ln 10} \qquad (2.17)$$

where $\tilde{\nu}_{nm}$ is the average wavenumber of the absorption and N_A is the Avogadro constant.

If the absorption is due to an electronic transition f_{nm}, the oscillator strength, is often used to quantify the intensity and is related to the area under the curve by

$$f_{nm} = \frac{4\varepsilon_0 m_e c^2 \ln 10}{N_A e^2} \int_{\tilde{\nu}_1}^{\tilde{\nu}_2} \varepsilon(\tilde{\nu})d\tilde{\nu} \qquad (2.18)$$

The quantity f_{nm} is dimensionless and is the ratio of the strength of the transition to that of an electric dipole transition between two states of an electron oscillating in three dimensions in a simple harmonic way, and its maximum value is usually 1.

The transition probability $|R^{nm}|^2$ is related to selection rules in spectroscopy: it is zero for a forbidden transition and non-zero for an allowed transition. By 'forbidden' or 'allowed' we shall mostly be referring to electric dipole selection rules, i.e. to transitions occurring through interaction with the electric vector of the radiation.

The electric dipole moment operator μ has components along the cartesian axes:

$$\mu_x = \sum_i q_i x_i; \qquad \mu_y = \sum_i q_i y_i; \qquad \mu_z = \sum_i q_i z_i \qquad (2.19)$$

where q_i and x_i are the charge and x coordinate of the ith particle, etc. Similarly the transition moment can be resolved into three components such that

$$R_x^{nm} = \int \psi_n^* \mu_x \psi_m \, dx; \qquad R_y^{nm} = \int \psi_n^* \mu_y \psi_m \, dy; \qquad R_z^{nm} = \int \psi_n^* \mu_z \psi_m \, dz \qquad (2.20)$$

and the transition probability is related to these by

$$|R^{nm}|^2 = (R_x^{nm})^2 + (R_y^{nm})^2 + (R_z^{nm})^2 \qquad (2.21)$$

2.3 Line Width

The ubiquitous use of the word 'line' to describe an experimentally observed transition goes back to the early days of observations of visible spectra with spectroscopes in which the lines observed in, say, the spectrum of a sodium flame are images, formed at various wavelengths, of the entrance slit. Although,

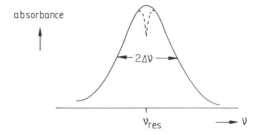

Figure 2.5 Typical (gaussian) absorption line showing the HWHM of Δv and (dashed curve) a Lamb dip.

nowadays, observations tend to be in the form of a plot of some measure of the intensity of the transition against wavelength, frequency, or wavenumber, we still refer to peaks in such a spectrum as lines.

Figure 2.5 shows, for a sample in the gas phase, a typical absorption line with an HWHM (half-width at half-maximum) of Δv and a characteristic line shape. The line is not infinitely narrow even if we assume that the instrument used for observation has not imposed any broadening of its own. We shall consider three important factors which may contribute to the line width and shape.

2.3.1 Natural Line Broadening

If state n in Figure 2.2(a) is populated in excess of its Boltzmann population by absorption, the species M* in this state will decay to the lower state until the Boltzmann population is regained. The decay is a first-order process so that

$$-\frac{\mathrm{d}N_n}{\mathrm{d}t} = kN_n \tag{2.22}$$

where k is the first-order rate constant and

$$\frac{1}{k} = \tau \tag{2.23}$$

Here τ is the time taken for N_n to fall to $1/e$ of its initial value (e is the base of natural logarithms) and is referred to as the lifetime of state n. If spontaneous emission is the *only* process by which M* decays, comparison with equation (2.9) shows that

$$k = A_{nm} \tag{2.24}$$

The Heisenberg uncertainty principle in the form

$$\tau \Delta E \geqslant \hbar \tag{2.25}$$

relates the lifetime to the smearing out, in terms of energy, of the state n. This equation illustrates the point that state n has an exactly defined energy only if τ is infinite, but, since this is never the case, all energy levels are smeared out to some extent with resulting line broadening.

Combining equations (2.12) and (2.15) relates A_{nm} to the transition probability

$$A_{nm} = \frac{64\pi^4 v^3}{(4\pi\varepsilon_0)3hc^3} |R^{nm}|^2 \tag{2.26}$$

so that, from equation (2.25),

$$\Delta v \geqslant \frac{32\pi^3 v^3}{(4\pi\varepsilon_0)3hc^3} |R^{nm}|^2 \tag{2.27}$$

The dependence of Δv, the frequency spread, on v^3 results in a much larger value for an excited electronic state, typically 30 MHz, than for an excited rotational state, typically 10^{-4} to 10^{-5} Hz, because of the much greater v for an excited electronic state.

Equation (2.27) illustrates what is called the natural line broadening. Since each atom or molecule behaves identically in this respect it is an example of homogeneous line broadening which results in a characteristic lorentzian line shape.

Natural line broadening is usually very small compared to other causes of broadening. However, not only is it of considerable theoretical importance but, in the ingenious technique of Lamb dip spectroscopy (see Section 2.3.4.2), observations may be made of spectra in which all other sources of broadening are removed.

2.3.2 Doppler Broadening

Whether radiation is being absorbed or emitted the frequency at which it takes place depends on the velocity of the atom or molecule relative to the detector. This is for the same reason that an observer hears the whistle of a train travelling towards him as having a frequency apparently higher than it really is, and lower when it is travelling away from him. The effect is known as the Doppler effect.

If an atom or molecule is travelling towards the detector with a velocity v_a, then the frequency v_a at which a transition is observed to occur is related to the actual transition frequency v in a stationary atom or molecule by

$$v_a = v\left(1 - \frac{v_a}{c}\right)^{-1} \tag{2.28}$$

where c is the speed of light. Because of the usual Maxwell velocity distribution there is a spread of values of v_a and a characteristic line broadening given by

$$\Delta v = \frac{v}{c}\left(\frac{2kT\ln 2}{m}\right)^{1/2} \qquad (2.29)$$

where m is the mass of the atom or molecule, this Δv is normally far greater than the natural line width. The broadening is inhomogeneous, since not all atoms or molecules in a particular sample behave in the same way, and results in a line shape known as gaussian.

2.3.3 Pressure Broadening

When collisions occur between gas phase atoms or molecules there is an exchange of energy which leads effectively to a broadening of energy levels. If τ is the mean time between collisions and each collision results in a transition between two states there is a line broadening Δv of the transition where

$$\Delta v = (2\pi\tau)^{-1} \qquad (2.30)$$

derived from the uncertainty principle of equation (1.16). This broadening is, like natural line broadening, homogeneous and usually produces a lorentzian line shape except for transitions at low frequencies when an unsymmetrical line shape results.

2.3.4 Removal of Line Broadening

Of the three types of broadening which have been discussed that due to the natural line width is, under normal conditions, much the smallest and it is the removal, or the decrease, of the effects of only Doppler and pressure broadening which can be achieved.

Except at very low frequencies pressure broadening may be removed simply by working at a sufficiently low pressure. Doppler broadening may be reduced or removed by two general methods which will be discussed briefly.

2.3.4.1 *Effusive atomic or molecular beams*

An effusive beam of atoms or molecules (see Ramsey in the bibliography) is produced by pumping them through a narrow slit, typically 20 μm wide and 1 cm long, with a pressure of a few torr on the source side of the slit. The beam may be further collimated by suitable apertures along it.

Such beams have many uses including some important ones in spectroscopy. In particular, pressure broadening of spectral lines is removed in an effusive beam and, if observations are made perpendicular to the direction of the beam, Doppler broadening is considerably reduced because the velocity component in the direction of observation is very small.

2.3.4.2 Lamb dip spectroscopy

In 1969 Costain devised a very elegant method of eliminating Doppler broadening without using an effusive beam.

Figure 2.5 shows a Doppler-broadened line and Figure 2.6 illustrates an absorption experiment in which the source radiation is reflected back through the cell, by a reflector R, to the detector. We assume that the source radiation is narrow compared to the line width, a condition which may be satisfied with, for example, microwave or laser radiation. Tuning the source to a frequency v_a, higher than the resonance frequency v_{res} at the line centre, results in only molecules like 1 and 2 in Figure 2.6, which have a velocity component v_a away from the source (see equation 2.28), absorbing radiation. On the return journey of the radiation back to the detector a new group of molecules like 6 and 7, which have a velocity component $-v_a$ away from the source, absorb at, say v_a^-. Because of this, when the radiation is tuned to v_a, for example, the number of molecules in the lower state m of the transition and having a velocity component v_a away from the source is depleted. This is referred to as hole burning in the usual Maxwell velocity distribution of molecules in state m.

Molecules like 3, 4, and 5 in Figure 2.6, which have a zero velocity component away from the source, behave uniquely in that they absorb radiation of the same frequency v_{res} whether the radiation is travelling towards or away from R, and this may result in what is called saturation.

The equilibrium Boltzmann population of level m in Figure 2.2(a) is higher than that of n, as equation (2.11) shows. If the system is exposed to radiation of energy $E_n - E_m$ of such an intensity that the populations become equal, no further absorption can take place: this state of affairs is referred to as saturation. We can now see that, if saturation occurs for the set of molecules 3, 4, and 5 while the radiation is travelling towards R, no further absorption takes place as it travels back from R. The result is that a dip in the absorbance curve is observed at v_{res}, as indicated in Figure 2.5. This is known as a Lamb dip, an effect which was predicted by Lamb in 1964. The width of the dip is the natural line width and observation of the dip results in much greater accuracy of measurement of v_{res}.

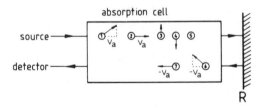

Figure 2.6 Three typical groups of molecules, with velocities v_a, 0, and $-v_a$, towards the source in a Lamb dip experiment.

Saturation is clearly achieved more readily if states m and n are close together, as is the case for microwave or n.m.r. transitions, but, even if they are far apart, a laser source may be sufficiently powerful to cause saturation.

Questions

1. The number of collisions z which a molecule in the gas phase makes per unit time, when only one species is present, is given by

$$z = \pi d^2 \left(\frac{8kT}{\pi m}\right)^{1/2} \frac{p}{kT}$$

where d is the collision diameter, m the molecular mass, T the temperature, and p the pressure. For benzene at 1 Torr and 293 K, and assuming $d = 5$ Å, calculate z and hence the pressure broadening Δv, in hertz, of observed transitions.

2. Calculate in hertz the broadening Δv of transitions in HCN at 25°C due to the Doppler effect in regions of the spectrum typical of rotational transitions (10 cm^{-1}), vibrational transitions (1500 cm^{-1}), and electronic transitions $(60\,000 \text{ cm}^{-1})$.

3. As a function of frequency the spectral radiation density is given by

$$\rho(v) = \frac{8\pi h v^3}{c^3} \frac{1}{\exp(hv/kT) - 1}$$

Calculate typical values in the microwave $(v = 50 \text{ GHz})$ and near-ultraviolet $(\tilde{v} = 30\,000 \text{ cm}^{-1})$ regions.

4. Calculate the ratio of molecules in a typical excited rotational, vibrational, and electronic energy level to that in the lowest energy state at 25 and 1000°C taking the levels to be 30, 1000 and 40 000 cm^{-1} respectively above the lowest energy state.

Bibliography

Ramsey, N. F. (1956). *Molecular Beams*, Oxford University Press, Oxford.
Townes, C. H., and Schawlow, A. L. (1955). *Microwave Spectroscopy*, McGraw-Hill, New York.

CHAPTER 3

General Features of Experimental Methods

3.1 The Electromagnetic Spectrum

In a vacuum all electromagnetic radiation travels at the same speed, the speed of light c, and may be characterized by its wavelength λ, in air or vacuum, or by its wavenumber $\tilde{v}$ or frequency v, both conventionally in a vacuum, where

$$\lambda_{\text{vac}} = \frac{c}{v} = \frac{1}{\tilde{v}} \tag{3.1}$$

Figure 3.1 illustrates the extent of the electromagnetic spectrum from low energy radiowave to high energy γ-ray radiation. The division into the various named regions does not imply any fundamental differences but is useful to indicate

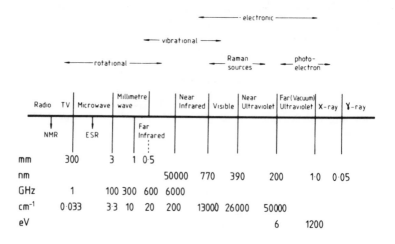

Figure 3.1 Regions of the electromagnetic spectrum.

that different experimental techniques are used. Indications of region boundaries, which should not be regarded as clear cut, are given in wavelength (mm or nm), frequency (GHz) and wavenumber (cm^{-1}). In addition, in the high energy regions the energy is indicated in electron volts (eV) where

$$1 \text{ eV} = hc(8065.54 \text{ cm}^{-1}) = h(2.417\,99 \times 10^{14}\,\text{s}^{-1}) \tag{3.2}$$

Also indicated in Figure 3.1 are the processes which may occur in an atom or molecule exposed to the radiation. A molecule may undergo rotational, vibrational, electronic, or ionization processes, in order of increasing energy: typical ranges are indicated. A molecule may also scatter light in a Raman process and the light source for such an experiment is usually in the visible or near-ultraviolet region (see, however, Section 5.3.1). An atom may undergo only an electronic transition or ionization since it has no rotational or vibrational degrees of freedom. Nuclear magnetic resonance (n.m.r.) and electron spin resonance (e.s.r.) processes involve transitions between nuclear spin and electron spin levels, respectively, but these spectroscopies necessitate the sample being between the poles of a magnet and will not be covered in this book.

3.2   General Components of an Absorption Experiment

Emission spectroscopy is largely confined to the visible and ultraviolet regions where spectra may be produced in an arc or discharge or by laser excitation. Absorption spectroscopy is, generally speaking, a more frequently used technique in all regions of the spectrum and it is for this reason that we shall concentrate rather more on absorption.

Figure 3.2 shows schematically the four main components, source, cell, dispersing element, and detector, of an absorption experiment. The source is ideally a continuum in which radiation is emitted over a wide wavelength range with uniform intensity. The absorption cell containing the sample must have windows made from a material which transmits the radiation and it must also be long enough for the absorbance (equation 2.16) to be sufficiently high.

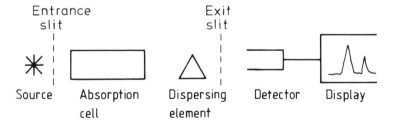

Figure 3.2   The components of a typical absorption experiment.

The choice of phase of the sample is important. Generally, in high resolution spectroscopy (see Section 3.3.1 for a discussion of resolution), the sample is in the gas phase at a pressure which is sufficiently low to avoid pressure broadening (see Section 2.3.3). In the liquid phase all rotational structure is lost and vibrational structure is considerably broadened but is important in, for example, the use of infrared spectroscopy as an analytical tool or obtaining oscillator strengths (see equation 2.18) by measuring the area under an electronic absorption curve. In the solid phase, in which the sample may be a pure crystal, mixed crystal, or solid solution in, say, a frozen noble gas, rotational motion is quenched as the molecules are held rigidly. Vibrational and electronic transitions are generally broad at normal temperatures but may be dramatically sharpened at liquid helium temperature (*ca* 4 K).

The dispersing element to be described in Section 3.3 splits up the radiation into its component wavelengths and is likely to be a prism, diffraction grating, or interferometer, but microwave and millimetre wave spectroscopy do not require such an element.

The detector must be sensitive to the radiation falling on it and the spectrum is very often displayed on a chart recorder. The spectrum may be a plot of absorbance or percent transmittance ($100I/I_0$—see equation 2.16) as a function of frequency or wavenumber displayed linearly along the chart paper. Wavelength is not normally used because, unlike frequency and wavenumber, it is not proportional to energy. Wavelength relates to the optics rather than the spectroscopy of the experiment.

3.3 Dispersing Elements

3.3.1 Prisms

Although prisms, as dispersing elements, have been largely superseded by diffraction gratings and interferometers they still have uses in spectroscopy and they also illustrate some important general points regarding dispersion and resolution.

Figure 3.3 shows a prism of base length b with one face filled with radiation from a white light source made parallel by lens L_1. The prism disperses and

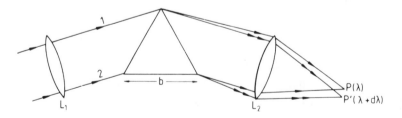

Figure 3.3 Dispersion by a prism.

resolves the radiation which is focused by L_2 onto a detector. If wavelengths λ and $\lambda + d\lambda$ are just observably separated then $d\lambda$, or the corresponding frequency interval dv or wavenumber interval $d\tilde{v}$, is the resolution[†] which is obtained. The resolving power R of a dispersing element is defined as

$$R = \frac{\lambda}{d\lambda} = \frac{v}{dv} = \frac{\tilde{v}}{d\tilde{v}} \qquad (3.3)$$

and, for a prism,

$$R = b\frac{dn}{d\lambda} \qquad (3.4)$$

provided that the incident beam is sufficiently wide that the extreme rays 1 and 2 fall on the edges of the face of the prism. In equation (3.4) n is the refractive index of the prism material and, for high resolving power, $dn/d\lambda$ should be large. This happens as we approach a wavelength at which the material absorbs radiation. For glass, absorption occurs at $\lambda < ca$ 360 nm and, therefore, the resolving power is greatest in the blue and violet regions, whereas quartz absorbs at $\lambda < ca$ 185 nm and its resolving power is greatest in the 300 to 200 nm region and rather low in the visible region.

If, in Figure 3.3, P and P′ are a distance $d\ell$ apart then the linear dispersion is defined as $d\ell/d\lambda$ whereas the angular dispersion is defined as $d\theta/d\lambda$, where $d\theta$ is the difference between the angles at which the rays going to P and P′ emerge from L_2.

We have said that wavelengths λ and $\lambda + d\lambda$ are just observably separated, or resolved, but we have not established a criterion for judging that this is so. It was Rayleigh who proposed such a criterion and it is illustrated in Figure 3.4. If we imagine a narrow entrance slit before lens L_1 in Figure 3.3 then the detector sees not just an image of the slit at wavelength λ but a diffraction pattern as shown in Figure 3.4(a) with intensity minima at $\pm\lambda, \pm2\lambda, \ldots$ from the central maximum. Rayleigh suggested that if the two diffraction patterns corresponding to points P and P′ in Figure 3.3 are such that the central maximum of P′ is no closer to that of P than the first minimum of P, as in Figure 3.4(b), then P and P′ are said to be just resolved.

It is important to realize that a line in a spectrum is an image of the entrance slit formed at a particular wavelength, the image being a diffraction pattern like that in Figure 3.4(a). Therefore, as the slit is widened the main peak in the diffraction pattern is widened and the resolution of which the dispersing element is capable is effectively reduced. On the other hand, when the slit width is reduced, there comes a point when the observed line width is not reduced any

---

[†] We refer to an observation as 'high resolution' or 'low resolution' when $d\lambda$, dv or $d\tilde{v}$ is small or large respectively.

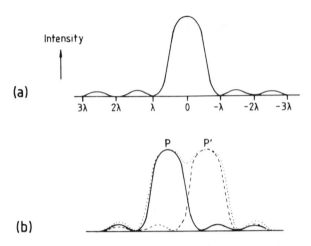

Figure 3.4 (a) Diffraction pattern produced by a narrow slit. (b) The Rayleigh criterion for resolution.

further even though the dispersing element may be capable of extremely high resolution. This point is likely to occur when the line width is limited by pressure or Doppler broadening (see Section 2.3.2 and 2.3.3).

3.3.2 Diffraction Gratings

A diffraction grating consists of a series of parallel grooves ruled on a hard glassy or metallic material. The grooves are extremely closely spaced, a spacing of the order of 1 μm being not unusual. Gratings are usually coated on the ruled surface with a reflecting material such as aluminium so that the grating acts as a mirror as well. The surface may be plane or concave, the latter type serving to focus as well as disperse and reflect the light falling on it.

Figure 3.5 shows how white light falling on the surface of a reflection grating G is dispersed. The general equation for diffraction by a grating is

$$m\lambda = d(\sin i + \sin \theta) \tag{3.5}$$

where i and θ are the angles of incidence and reflection, both measured from the normal to the surface, d is the groove spacing, λ the wavelength, and m ($= 0, 1, 2, \ldots$) the order of diffraction. For normal incidence

$$m\lambda = d \sin \theta \tag{3.6}$$

The angular dispersion produced by the grating is given by

$$\frac{d\theta}{d\lambda} = \frac{m}{d \cos \theta} \tag{3.7}$$

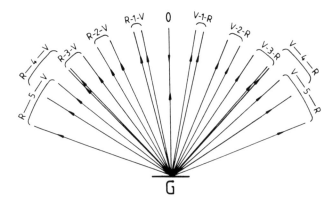

Figure 3.5 Various orders of diffraction from a plane reflection grating G.

and the figure shows how it increases with the order, the dispersion shown being from violet (V) to red (R) in each order. The resolving power R (equation 3.3) of a grating is given by

$$R = mN \qquad (3.8)$$

where N is the total number of grooves but, of course, all the grooves must receive the incident light if the resolving power is to be fully realized. When we require high dispersion and high resolution from a grating it is clear, from equations (3.7) and (3.8), that we should use as high an order as possible. Figure 3.5 shows that in higher orders there is an increasing problem of overlapping with adjacent orders. This may be avoided by filtering or pre-dispersion, with a small prism or grating, of the incident light.

If we are using only one order of diffraction it is very wasteful to reject the radiation diffracted in other orders and also that in the same order but on the other side of the incident beam. The radiation can be diffracted preferentially close to a particular angle by using a blazed grating. Grating grooves are ruled by a diamond and would normally be symmetrically V-shaped but, if they are ruled so that each has a long and a short side, as in Figure 3.6, then reflection will be most efficient when the incident and diffracted beams make an angle ϕ to the normal N to the grating surface as shown: the beams are normal to the long side of the groove. The angle ϕ is called the blaze angle and when the grating is used with equal angles of incidence and reflection, as in the figure, equation (3.5) becomes

$$m\lambda = 2d \sin \theta \qquad (3.9)$$

Diffraction gratings may be made by a holographic process, but blaze characteristics cannot be controlled and their efficiency is low in the infrared. They are mostly used for low order work in the visible and near-ultraviolet.

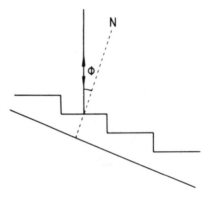

Figure 3.6 Use of a blazed grating at the blaze angle ϕ.

3.3.3 Fourier Transformation and Interferometers

A thin layer of oil on water often shows regions of different colours and this illustrates a third method of dispersing light. White light falling on the oil is reflected backwards and forwards within the layer, a part of the beam emerging from the surface each time. The various emerging beams may interfere constructively or destructively with each other, depending on the wavelength, and give rise to the different colours observed. This principle is made use of in an interferometer which is used to disperse infrared, visible or ultraviolet radiation, but is especially important for infrared radiation.

We shall discuss infrared interferometers in Section 3.3.3.2 but, before this, we need to understand the principles of Fourier transformation since this is involved in the very important step in treating the signal from an interferometer. This signal does not resemble the kind of spectrum that we usually obtain from a spectrometer and it is the modification of the signal to the intensity-versus-wavelength type of spectrum which we require that involves Fourier transformation. The principles of Fourier transformation are probably easier to understand when applied to longer wavelength radiofrequency radiation than to shorter wavelength infrared, visible or ultraviolet radiation. We shall consider radiofrequency radiation first.

3.3.3.1 *Radiofrequency radiation*

Consider radiofrequency radiation with a frequency of 100 MHz, or a wavelength of 3 m. Such radiation is used, for example, in a nuclear magnetic resonance (n.m.r.) spectrometer and as a carrier of FM (or VHF) radio signals. Frequencies typical of radiofrequency radiation are relatively low compared to, say, near infrared radiation of frequency 10 000 GHz (see Figure 3.1) and there is no problem in producing detectors which can respond sufficiently rapidly to determine the frequency of a radiofrequency signal directly. If we consider a source of monochromatic radiofrequency radiation, the detector responds to the electric field E which is oscillating at a frequency v, as in Figure 2.1. The variation of the signal detected, $f(t)$, with time t is shown in Figure 3.7(a) and is said to be the spectrum in the time domain. In this case it is the spectrum of a monochromatic source.

We are more used to spectra recorded, not in the time domain, but in the frequency (or wavenumber or wavelength) domain in which the detector signal is plotted against, say, frequency v rather than time t. We can easily see that the time domain spectrum in Figure 3.7(a) corresponds to the frequency domain

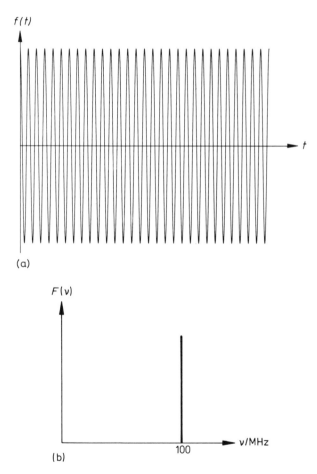

Figure 3.7 (a) The time domain spectrum and (b) the corresponding frequency domain spectrum for radiation of a single frequency.

spectrum in Figure 3.7(b) in which the signal $F(v)$ is plotted against frequency. The monochromatic source produces a single line spectrum.

The process of going from the time domain spectrum $f(t)$ to the frequency domain spectrum $F(v)$ is known as Fourier transformation. In this case the frequency of the line, say 100 MHz, in Figure 3.7(b) is simply the value of v which appears in the equation

$$f(t) = A \cos 2\pi v t \qquad (3.10)$$

for the time domain spectrum, where A is the amplitude of $f(t)$.

If the source now emits radiofrequency radiation consisting of two frequencies, v and $\frac{1}{4}v$, of the same amplitude then

$$f(t) = A[\cos 2\pi v t + \cos 2\pi (\tfrac{1}{4}v)t] \qquad (3.11)$$

and this is shown in Figure 3.8(a). The corresponding frequency domain spectrum is illustrated in Figure 3.8(b) which shows two equally intense lines, one at 100 MHz and the other at 25 MHz.

The sum of the two cosine waves in Figure 3.8(a) shows a beating between the waves. In general, if the two frequencies are v_1 and v_2, the beat frequency, v_B, is given by

$$v_B = |v_1 - v_2| \qquad (3.12)$$

In the example in Figure 3.8, v_B is 75 MHz.

Figure 3.9(a) shows a time domain spectrum corresponding to the frequency domain spectrum in Figure 3.9(b) in which there are two lines, at 25 and 100 MHz, with the latter having half the intensity of the former, so that

$$f(t) = \tfrac{1}{2}A \cos 2\pi v t + A \cos 2\pi (\tfrac{1}{4}v)t \qquad (3.13)$$

Conceptually, the problem of going from the time domain spectra in Figures 3.7(a)–3.9(a) to the frequency domain spectra in Figures 3.7(b)–3.9(b) is straightforward, at least in these cases because we knew the result before we started. Nevertheless, we can still visualize the breaking down of any time domain spectrum, however complex and irregular in appearance, into its component waves, each with its characteristic frequency and amplitude. Although we can visualize it, the process of Fourier transformation which actually carries it out is a mathematically complex operation which requires a computer. The mathematical principles will be discussed only briefly here.

In general, the time domain spectrum can be expressed as

$$f(t) = \int_{-\infty}^{+\infty} F(v) \exp(i2\pi v t)dv \qquad (3.14)$$

where $i = \sqrt{-1}$ and $F(v)$ is the frequency domain spectrum we require but, because

$$\exp(i\phi t) = \cos \phi t + i \sin \phi t \qquad (3.15)$$

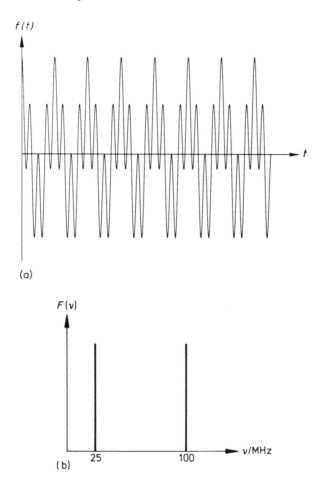

Figure 3.8 (a) The time domain spectrum and (b) the corresponding frequency domain spectrum for radiation of two different frequencies with a 1:1 intensity ratio.

equation (3.14) becomes

$$f(t) = \int_{-\infty}^{+\infty} F(v)(\cos 2\pi vt + i \sin 2\pi vt)dv \qquad (3.16)$$

For our purposes we can neglect the imaginary part of equation (3.16), i.e. $i \sin 2\pi vt$, and then it is apparent that $f(t)$ is a sum of cosine waves as we had originally supposed. Fourier transformation allows us to go from $f(t)$ to $F(v)$ by the relationship

$$F(v) = \int_{-\infty}^{+\infty} f(t) \exp(-i2\pi vt)dt \qquad (3.17)$$

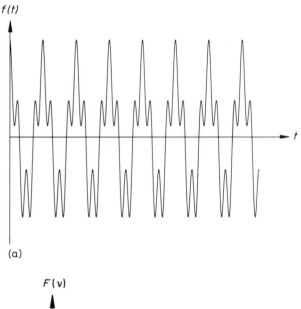

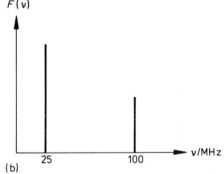

Figure 3.9 (a) The time domain spectrum and (b) the corresponding frequency domain spectrum for radiation of two different frequencies with a 2:1 intensity ratio.

or, using the fact that

$$\exp(-i\phi t) = \cos \phi t - i \sin \phi t \qquad (3.18)$$

we get

$$F(v) = \int_{-\infty}^{+\infty} f(t)(\cos 2\pi vt - i \sin 2\pi vt)dt \qquad (3.19)$$

where, again, we can neglect the imaginary part, $i \sin 2\pi vt$.

The computer digitizes the time domain spectrum $f(t)$ and carries out the Fourier transformation to give a digitized $F(v)$. Then digital-to-analogue conversion gives the frequency domain spectrum $F(v)$ in the analogue form in which we require it.

In the case of a radio operating in the FM wavelength band, or indeed any wavelength band, the aerial receives a signal which contains all the transmitted frequencies. What the radio does is, effectively, to Fourier transform the signal so that we can tune in to any of the frequencies without interference from any others.

There is one important point, however, that we have neglected so far. Real spectra in the frequency domain do not look like those in Figures 3.7(b)–3.9(b): the lines in the spectra are not stick-like and infinitely sharp but have width and shape.

If the radiofrequency spectrum is due to emission of radiation between pairs of states, for example nuclear spin states in n.m.r. spectroscopy, the width of a line is a consequence of the lifetime, τ, of the upper, emitting state. The lifetime and the energy spread, ΔE, of the upper state are related through the uncertainty principle (see equation (1.16)) by

$$\tau \Delta E \geqslant \hbar \qquad (3.20)$$

or, since $\Delta E = h\Delta v$,

$$\Delta v \geqslant \frac{1}{2\pi\tau} \qquad (3.21)$$

The effect of the lifetime of the upper state can be seen in the frequency domain spectrum by irradiating the sample with a short pulse of radiofrequency radiation and observing the decay of the signal. Figure 3.10(a) shows a similar time domain spectrum to that in Figure 3.8(a) but with a time decay super-imposed on it. The Fourier transformed spectrum, in Figure 3.10(b), shows two lines having widths given by equation (3.21).

The result in equation (3.21) also shows that if $\tau = \infty$, as in the examples in Figures 3.7(a)–3.9(a), then $\Delta v = 0$ and the lines are infinitely sharp.

Fourier transform spectroscopy in the radiofrequency region has been applied most importantly in pulsed Fourier transform n.m.r. spectroscopy, which is not a subject which will be treated in detail here.[†] More recently it has become a very important technique when applied to microwave spectroscopy.

There are now alternative ways for us to record an emission spectrum such as that in Figure 3.8. We can either record the time domain spectrum and Fourier transform to the frequency domain spectrum or obtain the frequency domain spectrum directly, in the more usual way, by scanning through the frequency range and recording the signal at the detector. However, there is an

---

[†] See, for example, Abraham, R. J., Fisher, J. and Loftus, P. (1988). *Introduction to NMR Spectroscopy*, Wiley, Chichester.

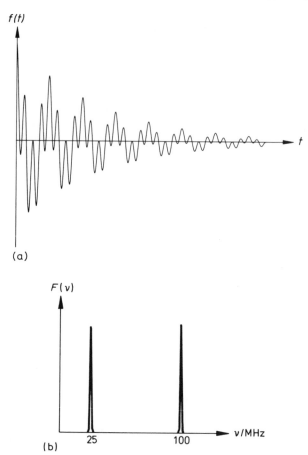

Figure 3.10 (a) Time domain and (b) frequency domain spectra corresponding to those in Figure 3.8 but in which the two lines are broadened.

important advantage in recording the time domain spectrum: all the frequencies in the spectrum are recorded all the time. This is known as the multiplex or Fellgett advantage and results in a comparable spectrum being obtained in a much shorter time. Consequently the Fourier transform (FT) technique can be used, for example, to obtain spectra of transient species formed during a chemical reaction.

3.3.3.2 Infrared, visible and ultraviolet radiation

For radiofrequency and microwave radiation there are detectors which can respond sufficiently quickly to the low frequencies (< 100 GHz) involved and

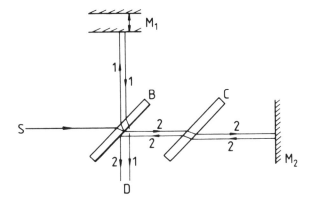

Figure 3.11   Michelson interferometer.

record the time domain spectrum directly. For infrared, visible and ultraviolet radiation the frequencies involved are so high (>600 GHz) that this is no longer possible. Instead an interferometer is used and the spectrum is recorded in the length domain rather than the frequency domain. Because the technique has been used mostly in the far-, mid- and near-infrared regions of the spectrum the instrument used is usually called a Fourier transform infrared (FTIR) spectrometer although it can easily be modified to operate in the visible and ultraviolet regions.

The most important component of an FTIR spectrometer is an interferometer based on the original design by Michelson in 1891 and shown in Figure 3.11.

For simplicity we consider a source S of monochromatic radiation entering the interferometer. A ray from this source strikes the beamsplitter B which is coated on its second surface with a material which makes it half transmitting (ray 2) and half reflecting (ray 1). Ray 1 is then reflected by the movable mirror M_1 back through B to a detector D while ray 2 is reflected by the fixed mirror M_2 and part is then reflected by B to the detector. C is a compensating plate introduced so that each of rays 1 and 2 has passed twice through the same length of the material from which both B and C are made. When they reach D rays 1 and 2 have traversed different paths with a path difference δ. This is called the retardation and its magnitude depends on the position of M_1. Figure 3.12(a) shows that, if $\delta = 0, \lambda, 2\lambda, \ldots$, the two rays interfere constructively at the detector while Figure 3.12(b) shows that, if $\delta = \lambda/2, 3\lambda/2, 5\lambda/2, \ldots$, they interfere destructively and no signal is detected. Therefore, if δ is smoothly changed from zero the detected signal intensity $I(\delta)$ changes like a cosine function as in Figure 3.13.

In a more usual emission experiment the source contains many wavelengths, the detector sees intensity due to many cosine waves of different wavelengths,

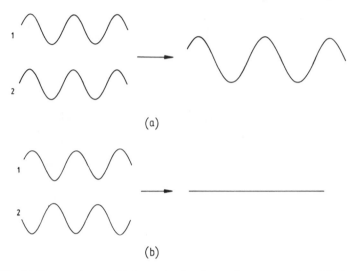

(a)

(b)

Figure 3.12 (a) Constructive and (b) destructive interference between rays 1 and 2 of monochromatic radiation.

and the detected intensity is of the form

$$I(\delta) = \int_0^\infty B(\tilde{v}) \cos 2\pi\tilde{v}\delta \, d\tilde{v} \qquad (3.22)$$

where $\tilde{v}$ is the wavenumber of the radiation and $B(\tilde{v})$ is the source intensity at that wavenumber (neglecting small corrections due to variable beamsplitter efficiency and detector response). A plot of $B(\tilde{v})$ against $\tilde{v}$ is the dispersed spectrum of the source and is obtained by the procedure of Fourier transformation discussed in Section 3.3.3.1, giving

$$B(\tilde{v}) = 2 \int_0^\infty I(\delta) \cos 2\pi\tilde{v}\delta \, d\delta \qquad (3.23)$$

The majority of infrared spectra are obtained by an absorption rather than an emission process and, as a result, the change of signal intensity $I(\delta)$ with retardation δ appears very different from that in Figure 3.13.

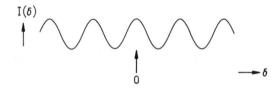

Figure 3.13 Change of signal intensity $I(\delta)$ with retardation δ.

Consider an infrared source emitting a wide range of wavenumbers for use in an absorption experiment. Figure 3.14(a) shows how the idealized wave-number domain spectrum of this source, emitting continuously between $\tilde{\nu}_1$ and $\tilde{\nu}_2$, might appear. We can regard this spectrum as comprising a very large number of wavenumbers between $\tilde{\nu}_1$ and $\tilde{\nu}_2$ so that the corresponding detector signal, as a function of retardation, will be the result of adding together very many cosine waves of different wavelengths. The signal is large at $\delta = 0$ since all the waves are in-phase but elsewhere they are out-of-phase, interfere with each other and produce total cancellation of the signal. The intense signal at $\delta = 0$ is known as the centre burst and is shown in Figure 3.14(b). (Because of slight dispersion by the beamsplitter B the waves at $\delta = 0$ are not quite in-phase resulting in an asymmetry of the centre burst about $\delta = 0$.)

If a single sharp absorption occurs at a wavenumber $\tilde{\nu}_a$, as shown in the wavenumber domain spectrum in Figure 3.15, the cosine wave corresponding to $\tilde{\nu}_a$ is not cancelled out and remains in the $I(\delta)$ versus δ plot, or interfero-

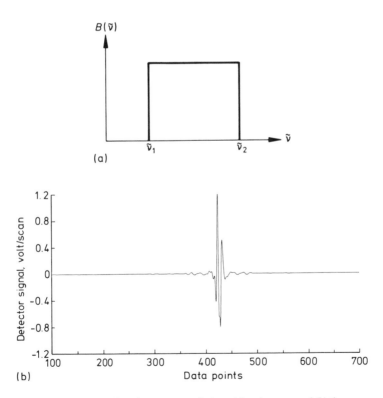

(a)

(b)

Figure 3.14 (a) Wavenumber domain spectrum of a broad band source and (b) the corresponding interferogram.

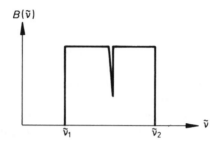

Figure 3.15 Wavenumber domain spectrum of a broad band source with a narrow absorption.

gram as it is often called. For a more complex set of absorptions the pattern of uncancelled cosine waves becomes more intense and irregular.

Figure 3.16(a) shows an interferogram resulting from the infrared absorption spectrum of air in the 400–3400 cm^{-1} region. The Fourier transformed spectrum in Figure 3.16(b) shows strong absorption bands due to CO_2 and H_2O, that of H_2O showing much fine structure because it is a lighter molecule.

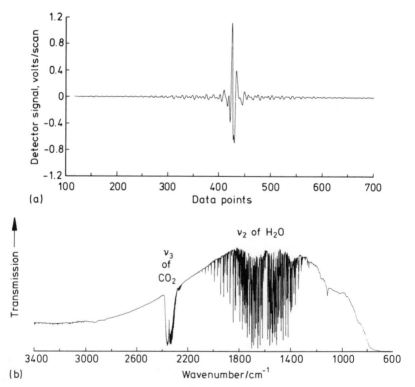

Figure 3.16 (a) Infrared interferogram of the absorption spectrum of air in the 400–3400 cm^{-1} region and (b) the Fourier transformed spectrum.

The interferogram is digitized before Fourier transformation by a dedicated computer. Figure 3.16(a) shows that, in this example, there are 600 data points in the part of the interferogram shown. The computer is limited in the number of data points it can handle and the user has a choice of having the data points closer together, and neglecting the outer regions of the interferogram, which gives a wider wavenumber range but lower resolution, or having the data points further apart, which gives higher resolution but a narrower wavenumber range. The resolution $\Delta\tilde{\nu}$ is always determined by how much of the interferogram can be observed which, in turn, depends on the maximum displacement δ of the mirror M_1. The resolution is given by

$$\Delta\tilde{\nu} = \frac{1}{\delta_{max}} \tag{3.24}$$

One of the main design problems in an FTIR spectrometer is to obtain accurate, uniform translation of M_1 over distances δ_{max} which may be as large as 1 m in a high resolution interferometer.

As in all Fourier transform methods in spectroscopy, the FTIR spectrometer benefits greatly from the multiplex, or Fellgett, advantage of detecting a broad band of radiation (a wide wavenumber range) all the time. By comparison, a spectrometer which disperses the radiation with a prism or diffraction grating detects, at any instant, only that narrow band of radiation which the orientation of the prism or grating allows to fall on the detector, as in the type of infrared spectrometer described in Section 3.6.

In addition to the multiplex advantage an FTIR spectrometer also has the advantage of a greater proportion of the source radiation passing through the instrument. The reason for this is that the narrow entrance slit (see Figure 3.2), which severely restricts the radiation throughput in a prism or grating spectrometer, is replaced by a circular aperture of larger area. This throughput advantage is known as the Jacquinot advantage.

3.4 Components of Absorption Experiments in Various Regions of the Spectrum

Table 3.1 summarizes the details of typical sources, absorption cells, dispersing elements, and detectors used in different regions of the electromagnetic spectrum.

3.4.1 Microwave and Millimetre Wave

In the microwave region tunable monochromatic radiation is produced by klystrons, each one being tunable over a relatively small frequency range, or a backward wave oscillator, tunable over a much larger range. Both are electronic devices. Absorption experiments are usually carried out in the gas phase and

Table 3.1 Elements of an absorption experiment in various regions of the spectrum.

Region	Source	Absorption cell windows	Dispersing element	Detector
Microwave	Klystron Backward wave oscillator	Mica	None	Crystal diode
Millimetre wave	Klystron (frequency multiplied) Backward wave oscillator	Mica Polymer	None	As for microwave or far-infrared
Far-infrared	Mercury arc	Polymer	Grating Interferometer	Golay cell Thermocouple Bolometer Pyroelectric
Mid- and near-infrared	Nernst filament Globar	NaCl or KBr	Grating Interferometer	As for far-infrared Photoconductive Photomultiplier Photodiode
Visible	Tungsten filament Xenon arc	Glass	Prism Grating Interferometer	Photographic plate As for visible
Near-ultraviolet	Deuterium discharge Xenon arc	Quartz	Prism Grating Interferometer	As for visible
Far-ultraviolet	Microwave discharge in noble gases Lyman discharge	LiF (or no windows)	Grating	As for visible

mica windows, which transmit in this region, are placed on either end of the absorption cell which may be several metres in length. Stark modulation, in which an electric field is applied between a metal plate or septum, down the centre of the cell, and the cell walls, is often used to increase sensitivity. This also allows the measurement of the dipole moment of the absorbing molecule. Unusually, no dispersing element is needed as the source radiation is monochromatic. The detector is a crystal diode rectifier.

Millimetre wave radiation may also be generated by a klystron or backward wave oscillator but, since klystrons produce only microwave radiation, the frequency must be multiplied with consequent loss of power. Windows on the absorption cell are of the same material as used for the microwave or far-infrared regions, depending on the frequency range: this also applies to the detector. No dispersing element is required as the radiation is monochromatic.

Both microwave and millimetre wave radiation can be channelled in any direction by a waveguide made from metal tubing of rectangular cross-section, the dimensions depending on the frequency range. The absorption cell is also made from waveguide tubing.

3.4.2 Far-infrared

In the far-infrared region strong absorption by the water vapour normally present in air necessitates either continuously flushing the whole optical line with dry nitrogen or, preferably, evacuation.

Sources of radiation are all of lower than ideal intensity. One of the most commonly used is a mercury discharge in a quartz envelope, most of the higher wavenumber radiation coming from the quartz rather than from the discharge plasma.

The windows of the absorption cell are made from polymer material such as polyethylene, poly(ethylene terephthalate) (Terylene), or polystyrene.

The dispersing element is commonly a Michelson type of interferometer (Section 3.3.3.2), with a beamsplitter made from a transmitting material, but a plane diffraction grating may also be employed. Various methods of employing a plane grating are used but the Czerny–Turner system, shown in Figure 3.17, is one of the most common. The radiation from the source S passes through the absorption cell and enters at the slit S_1, which is separated by its focal length from the concave mirror M_1. This mirror is turned so as to reflect the radiation, now a parallel beam, onto the grating G which disperses it and reflects it to M_2. This second mirror focuses the dispersed radiation onto the exit slit S_2 and thence to the detector D. The spectrum is scanned by smoothly rotating the grating.

A commonly used detector is a Golay cell in which there is a far-infrared absorbing material, such as aluminium deposited on collodion, inside the entrance window of the cell. The aluminium absorbs the radiation, heats up,

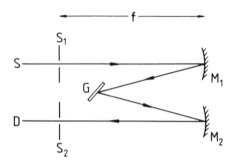

Figure 3.17 The Czerny–Turner grating mounting.

and transfers the heat to xenon gas contained in the cell. As the temperature of the gas varies the curvature of a flexible mirror of antimony-coated collodion, forming a part of the cell, changes. Reflection of a light beam from this mirror, which is on the outside of the Golay cell, indicates its curvature and therefore the intensity of radiation absorbed by the cell.

Thermocouples, bolometers, pyroelectric and semiconductor detectors are also used. The first three are basically resistance thermometers. A semi-conductor detector counts photons falling on it by measuring the change in conductivity due to electrons being excited from the valence band into the conduction band.

3.4.3 Near- and Mid-infrared

Evacuation is not necessary in this region and sources are much less of a problem than in the far-infrared. A heated black body emits strongly in the near-infrared and a Nernst filament, consisting of a mixture of rare earth oxides, or a silicon carbide Globar emulate a black body quite well.

Cell windows are usually made from sodium chloride, which transmits down to $700 \, \text{cm}^{-1}$, or potassium bromide, down to $400 \, \text{cm}^{-1}$. Concave mirrors, front-surface coated with aluminium, silver, or gold, may be used to multiply-reflect the radiation many times across the cell for higher absorbance by a gas phase sample. If the sample is in solution the solvent must be carefully chosen. No single solvent is transparent throughout the region but, for example, carbon disulphide is transparent except for the region 1400 to $1700 \, \text{cm}^{-1}$ where tetrachloroethylene is transparent. A solid sample may be ground into a fine power and, for wavenumbers less than $1300 \, \text{cm}^{-1}$, made into a mull with Nujol (a liquid paraffin), or, for observations over the whole range, ground together with potassium bromide and compressed under vacuum to give a KBr disk.

The dispersing element is usually a diffraction grating or an interferometer with a beamsplitter made from quartz or calcium fluoride coated with silicon or germanium.

Detectors are similar in type to those for the far-infrared, namely thermo-couples, bolometers, Golay cells, or photoconductive semiconductors.

3.4.4 Visible and Near-ultraviolet

Conventional, but not very intense, sources for these regions are a tungsten or tungsten–iodine filament lamp for the visible and a deuterium discharge lamp (more intense than a hydrogen discharge lamp), in a quartz envelope, for the near-ultraviolet. For both regions a much more intense source is a high pressure xenon arc lamp in which an arc (usually direct current) is struck between two tungsten poles about 1 mm to 1 cm apart in xenon gas at about 20 atm pressure and contained in a quartz envelope. Such a lamp produces radiation down to about 200 nm.

Useful transparent materials for cell windows, lenses, etc., are Pyrex glass for the visible and fused quartz for the visible and near-ultraviolet.

Dispersing elements may be either prisms (glass for the visible, quartz for the near-ultraviolet) or, more often, diffraction gratings for which a Czerny–Turner mounting, shown in Figure 3.17, may be used.

Detectors used are mostly photomultipliers in which photons fall on a metal surface, such as caesium, which then emits electrons (photoelectrons—see Section 1.2). These electrons are subjected to an accelerating voltage and fall on a second surface releasing secondary electrons, the process being repeated several times to give a large current amplification. Also used are photographic plates and arrays of photodiodes, both of which have the multiplex advantage of detecting a large range of wavelengths all the time.

3.4.5 Far-ultraviolet

As for the far-infrared, absorption by air in the far-ultraviolet necessitates evacuation of the optical path from source to detector. In this region it is oxygen which absorbs, being opaque below 185 nm.

Sources of far-ultraviolet radiation cause something of a problem. A deuterium discharge lamp emits down to 160 nm, a high voltage spark discharge in helium produces radiation from 100 to 60 nm, and microwave-induced discharges in argon, krypton or xenon cover the range 200 to 105 nm. A much larger continuum range, from the visible down to about 30 nm, is provided by a Lyman source in which a large condenser is repetitively discharged through a low pressure gas contained in a glass capillary. The most ideal source of far-ultraviolet radiation is the synchrotron radiation source which will be discussed in Section 8.1.1.

Lithium fluoride is transparent down to about 105 nm, below which window-less systems must be used with differential pumping.

The dispersing element is a diffraction grating preferably used under conditions of grazing incidence (ϕ in equation 3.9 about 89°) to improve the reflectance. The grating may also be concave to avoid the use of a focusing mirror.

Photomultipliers or photographic plates may be employed as detectors.

3.5 Other Experimental Techniques

3.5.1 Attenuated Total Reflectance (ATR) Spectroscopy and Reflection-Absorption Infrared Spectroscopy (RAIRS)

The attenuated total reflectance (ATR) technique is used commonly in the near-infrared for obtaining absorption spectra of thin films and opaque materials. The sample, of refractive index n_1, is placed in direct contact with a material which is transparent in the region of interest, such as thallium bromide/thallium iodide (known as KRS-5), silver chloride or germanium, of relatively high refractive index n_2 so that $n_2 \gg n_1$. Then, as Figure 3.18 shows, radiation falling on the interface may be totally reflected, provided the angle i is greater than a certain critical value. However, the radiation penetrates the sample to a depth of about 20 μm and may be absorbed by it. As a wavelength at which absorption takes place is approached the refractive index n_1, which is wavelength-dependent, changes rapidly, and the greater is this change the greater is the degree of attenuation of the radiation. Therefore the intensity of the reflected light varies with wavelength in a way resembling an absorption spectrum. In practice multiple internal reflections are used to enhance the attenuation.

Whereas ATR spectroscopy is most commonly applied in obtaining infrared absorption spectra of opaque materials, reflection–absorption infrared spectroscopy (RAIRS) is usually used to obtain the absorption spectrum of a thin layer of material adsorbed on an opaque metal surface. An example would be carbon monoxide adsorbed on copper. The metal surface may be either in the form of a film or, of great importance in the study of catalysts, one of the particular crystal faces of the metal.

The radiation from an infrared source is directed at a very small (grazing) angle of incidence onto the surface. The reflected and then dispersed light gives

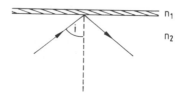

Figure 3.18 Total reflection of radiation in a medium of refractive index n_2 by a thin film of refractive index n_1, where $n_2 \gg n_1$.

the absorption spectrum of the adsorbed material. Interpretation of the spectrum provides information on the way in which the material is adsorbed. For example, it is possible to distinguish carbon monoxide molecules lying perpendicular to the metal surface from those lying parallel to it. This orientation of the adsorbate may change with the degree of surface coverage.

3.5.2 Atomic Absorption Spectroscopy

Atomic absorption spectroscopy (AAS) is complementary to atomic emission spectroscopy (see Section 3.5.3) and became possible for a wide range of atoms in the mid-1950s.

The main problem in this technique is getting the atoms into the vapour phase, bearing in mind the typically low volatility of many materials to be analysed. The method used is to spray, in a very fine mist, a liquid molecular sample containing the atom concerned into a high temperature flame. Air mixed with coal gas, propane or acetylene, or nitrous oxide mixed with acetylene produce flames in the temperature range 2100 to 3200 K, the higher temperature being necessary for such refractory elements as Al, Si, V, Ti, and Be.

The source radiation which passes through the flame is not a continuum, as would normally be used in absorption spectroscopy, but a hollow cathode lamp. The lamp, shown in Figure 3.19, contains a tungsten anode, a cup-shaped cathode made from the material to be investigated, and a carrier gas, such as neon, at about 5 Torr. When a voltage is applied a coloured discharge appears and the positive column, in which mainly neutral atom emission occurs, is confined to the inside of the cathode by choice of voltage and carrier gas pressure. The radiation emitted, apart from that from the carrier gas, is from the atom to be observed in absorption. Figure 3.20 shows just two atomic emission lines λ_1 and λ_2 from the hollow cathode. If, for example, λ_1 represents a transition from an excited state to the *ground* state of the atom then, as shown, this can be absorbed by the flame. However, if λ_2 represents a transition between two excited states, it cannot be absorbed since, even at a flame temperature of 3000 K, there is almost no population of any excited states because the separations ΔE from the ground state are so large (see equation 2.11). The

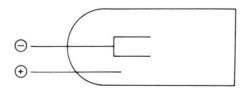

Figure 3.19 A hollow cathode lamp.

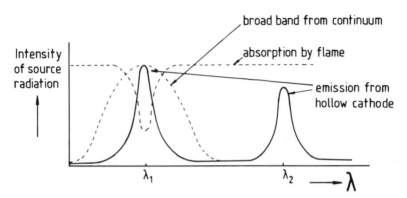

Figure 3.20 The principle of atomic absorption spectroscopy.

dispersing element of the spectrometer is set so that the photoelectric detector receives radiation of wavelength λ_1 only. Calibration is achieved by spraying a mist of a solution, containing a known concentration of the atom concerned, into the flame.

The advantage of using a hollow cathode rather than a broad-band continuum is illustrated in Figure 3.20. By using a continuum, sensitivity would be lost because only a relatively small amount of radiation would be absorbed.

3.5.3 Inductively Coupled Plasma Atomic Emission Spectroscopy

Emission spectroscopy is a very useful analytical technique in determining the elemental composition of a sample. The emission may be produced in an electrical arc or spark but, since the mid 1960s, an inductively coupled plasma has been used increasingly.

For inductively coupled plasma atomic emission spectroscopy (ICP–AES) the sample is normally in solution but may be a fine particulate solid or even a gas. If it is a solution, this is nebulized, resulting in a fine spray or aerosol, in flowing argon gas. The aerosol is introduced into a plasma torch which is illustrated in Figure 3.21.

The torch consists of three concentric quartz tubes. The argon aerosol passes up the central tube. There is an auxiliary supply of argon in the outer tube to provide cooling and, between these, there is a further supply of flowing argon. Radiofrequency radiation, with a frequency of 25–60 MHz and a power of 0.5–2.0 kW, is supplied through a copper coil around the outer tube. A high voltage spark is applied to initiate the plasma flame which is then maintained through the inductive heating of the gas by the radiofrequency radiation. The flow of argon in the central tube can be varied to adjust the height of the flame.

The temperature of the plasma in the region of observation is typically 7000–8000 K and all molecules contained in the aerosol sample are atomized.

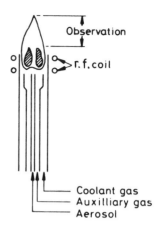

Figure 3.21 A plasma torch for inductively coupled plasma atomic emission spectroscopy.

The majority of the atoms are also singly ionized and many of the ions are produced in various excited electronic states. The radiation emitted from these excited ions is then analysed by a spectrometer which usually operates in the 800–190 nm region, evacuation being necessary if wavelengths lower than 190 nm are important. The wavelengths are dispersed and resolved by a diffraction grating.

There are two general types of spectrometer. In a scanning spectrometer there is a fixed photomultiplier detector and the grating is rotated smoothly so that a single detector covers the complete range of wavelengths in which the emission spectrum occurs. In the polychromator, on the other hand, the grating is fixed and a number of detectors are arranged at positions corresponding to the emission wavelengths of the atoms or ions being determined. In this way simultaneous, multielement analysis can be achieved.

A wider range of elements is covered by inductively coupled plasma atomic emission spectroscopy than by atomic absorption spectroscopy. All elements, except argon, can be determined with an inductively coupled plasma but there are some difficulties associated with He, Ne, Kr, Xe, F, Cl, Br, O, and N.

The detection limits of $1-100\ \mu g\ dm^{-3}$ are similar for both techniques.

3.5.4 Flash Photolysis

Transient species, existing for periods of time of the order of a microsecond or a nanosecond, may be produced by photolysis using far-ultraviolet radiation. Electronic spectroscopy is one of the most sensitive methods for detecting such species, whether they are produced in the solid, liquid, or the gas phase, but a special technique, that of flash photolysis devised by Norrish and Porter in 1949, is necessary.

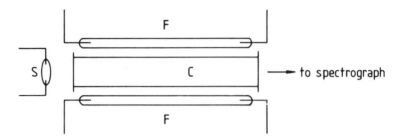

Figure 3.22 Principal components of a flash photolysis absorption experiment.

Figure 3.22 illustrates the general principles applied to a gas phase absorption experiment. The flashlamps F contain electrodes inside quartz envelopes containing a noble gas. High capacity condensers are discharged between the electrodes and visible and ultraviolet radiation passes from F through the walls of the quartz cell C. The parent molecule in C is photolysed, for example $NH_3 \rightarrow NH_2 + H$, by the far-ultraviolet radiation produced in the flash. In order to observe the absorption spectrum of, say, NH_2 at its optimum concentration a pulse of a continuum source S is initiated which passes through the cell to the spectrograph (photographic recording) or spectrometer (photoelectric recording) which records the spectrum. The process of photolysis flash followed by source flash may be repeated many times in order to accumulate the spectrum and a liquid or gaseous sample may be flowed through the cell. An important variable parameter is the delay time between the photolysis and source flashes and is typically in the region 1 to 1000 μs depending on the species and the conditions in the cell.

Pulsed lasers (Chapter 9) may be used both for photolysis and as a source. Since the pulses can be extremely short, of the order of a few picoseconds or less, species with comparably short lifetimes, such as an atom or molecule in a short-lived excited electronic state, may be investigated.

3.6 Typical Recording Spectrophotometers for the Near- and Mid-infrared, Visible, and Near-ultraviolet Regions

Perhaps the most common type of spectrometer that we encounter in the laboratory is the double-beam spectrophotometer for use in the mid- or near-infrared, visible, or near-ultraviolet region. 'Double beam' refers to the facility of having two beams of continuum radiation, one for a sample cell containing, say, a solution of A in a solvent B and the other for a reference cell containing only B. The absorption spectrum recorded with cells of identical pathlengths in each beam is that of solution-minus-solvent, in this case that of A alone. It is important to realize, however, that if the solvent is absorbing

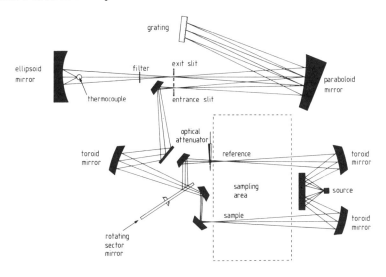

Figure 3.23 A typical double-beam recording mid- and near-infrared spectrophotometer.

nearly all of the radiation at a particular wavelength the value of the absorbance for solution-minus-solvent cannot be meaningful.

Figure 3.23 illustrates the layout of a typical mid- and near-infrared spectrophotometer. The source radiation is split into the reference and sample beams by two toroidal mirrors (concave but with different curvature in two perpendicular directions—like a section of the side of a barrel). Eventually they strike the rotating sector mirror which has alternate regions which either transmit the sample beam or reflect the reference beam towards the next toroidal mirror. This receives, alternately, light from the reference and sample beams which are reflected to the diffraction grating by a parabolic mirror. Such a mirror makes parallel all radiation emanating from the focus of the parabola whichever part of the mirror it strikes. The width of the entrance and exit slits may determine the resolution when they are relatively wide but when they are narrowed it is, eventually, the number of grooves on the grating which limits the resolution (equations 3.3 and 3.8). The detector receives radiation alternately from the sample and reference beams and phase-sensitive treatment of the signal allows these two to be separated.

In the method of recording in Figure 3.23 the absorbance in the sample beam relative to that in the reference beam is obtained by inserting an optical attenuator into the reference beam which is attenuated until the two detected signals are equal. This optical null method has the disadvantage of reducing drastically the intensity in the reference beam when the sample is absorbing strongly. In the infrared this method has been largely replaced by ratio-recording in which the intensity in the reference beam, I_0, and that in the

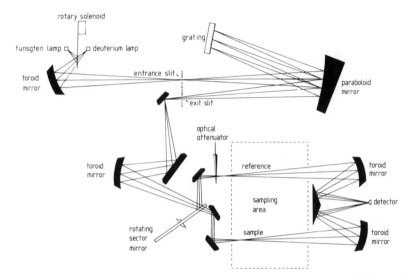

Figure 3.24 A typical double-beam recording visible–near-ultraviolet spectrophotometer.

sample beam, I, are ratioed by a microprocessor to give the absorbance A of the sample relative to that of the reference by equation (2.16). This method is particularly advantageous when absorbance is high, i.e. when the transmittance is low.

Figure 3.24 illustrates a typical optical layout for a visible–near-ultraviolet double-beam recording spectrophotometer. The arrangement is very similar to that in Figure 3.23 for an infrared instrument except that the optical line is reversed in that the positions of source and detector are interchanged. A mirror is rotated in order to change from the tungsten to the deuterium lamp for near-ultraviolet operation.

Typical regions covered by these instruments are 200 to 5000 cm^{-1} (50 to 2 μm) in the infrared and 11 100 to 51 300 cm^{-1} (900 to 195 nm) in the visible and near-ultraviolet. This leaves a gap from 0.9 to 2 μm which is filled by some spectrophotometers.

Questions

1. A diffraction grating is 10.40 cm wide with 600.0 grooves per millimetre and is blazed at 45.00°. What is the wavelength of radiation diffracted at this same angle in the first, fourth, and ninth orders? What is the resolving power in these orders? What is the resolution in terms of wavelength (nm), wavenumber (cm^{-1}), and frequency (GHz) in the ninth order at 300.0 nm?

2. At 18°C the refractive index of fused quartz varies with wavelength as follows:

λ/nm	n	λ/nm	n
185.47	1.5744	274.87	1.4962
193.58	1.5600	303.41	1.4859
202.55	1.5473	340.37	1.4787
214.44	1.5339	396.85	1.4706
226.50	1.5231	404.66	1.4697
250.33	1.5075	434.05	1.4670

Plot λ against n and hence obtain the resolving power of a fused quartz prism, with a base length of 3.40 cm, at 200, 250, 300, and 350 nm. What is the resolution, in nanometres, at these wavelengths? How would the resolving power and resolution be affected, quantitatively, by using two such prisms in tandem?

Bibliography

Bousquet, P. (1971). *Spectroscopy and Its Instrumentation*, Adam Hilger, London.

Harrison, G. R., Lord, R. C., and Loofbourow, J. R. (1948). *Practical Spectroscopy*, Prentice-Hall, Englewood, New Jersey.

Hecht, H., and Zajac, A. (1974). *Optics*, Addison-Wesley, Reading, Massachusetts.

Jenkins, F. A., and White, H. E. (1957). *Fundamentals of Optics*, McGraw-Hill, New York.

Longhurst, R. S. (1957). *Geometrical and Physical Optics*, Longman, London.

Sawyer, R. A. (1963). *Experimental Spectroscopy*, Dover, New York.

CHAPTER 4

Molecular Symmetry

The theory of molecular symmetry provides a satisfying and unifying thread which extends throughout spectroscopy and valence theory. Although it is possible to understand atoms and diatomic molecules without this theory, when it comes to understanding, say, spectroscopic selection rules in polyatomic molecules, molecular symmetry presents a small barrier which must be surmounted. However, for those not needing to progress so far this chapter may be bypassed without too much hindrance.

The application of symmetry arguments to atoms and molecules has its origins in group theory developed by mathematicians in the early nineteenth century, but it was not until the 1920s and 1930s that it was applied to atoms and molecules. It is because of this historical development that the teaching of the subject has often been preceded by a detailed treatment of matrix algebra. However, it is possible to progress quite a long way in understanding molecular symmetry without any such mathematical knowledge and it is such a treatment that is adopted here.

4.1 Elements of Symmetry

If we compare the symmetry of a circle, a square, and a rectangle it is intuitively obvious that the degree of symmetry decreases in the order given. What about the degrees of symmetry of a parallelogram and an isosceles triangle? Similarly, there is a decrease in the degree of symmetry along the series of molecules ethylene, 1,1-difluoroethylene, and fluoroethylene, shown in Figure 4.1. However, *cis*- and *trans*-1,2-difluoroethylene, also shown in the figure, present a similar problem to the parallelogram and isosceles triangle. In fact it will be apparent by the end of this section that *cis*-1,2-difluoroethylene and an isosceles triangle have the same symmetry as also do *trans*-1,2-difluoroethylene and a parallelogram.

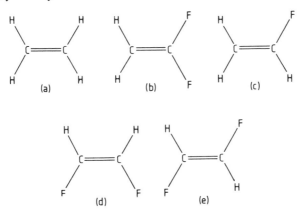

Figure 4.1 (a) Ethylene, (b) 1,1-difluoroethylene, (c) fluoroethylene, (d) *cis*-1,2-difluoroethylene, (e) *trans*-1,2-difluoroethylene.

These simple examples serve to show that instinctive ideas about symmetry are not going to get us very far. We must put symmetry classification on a much firmer footing if it is to be useful. In order to do this we need to define only five types of elements of symmetry—and one of these is almost trivial. In discussing these we refer only to the free molecule, realized in the gas phase at low pressure, and not, for example, to crystals which have additional elements of symmetry relating the positions of different molecules within the unit cell. We shall use, therefore, the Schönflies notation rather than the Hermann-Mauguin notation favoured in crystallography.

In discussing molecular symmetry it is essential that the molecular shape is accurately known, commonly by spectroscopic methods or X-ray, electron, or neutron diffraction.

4.1.1 *n*-Fold Axis of Symmetry, C_n

If a molecule has an *n*-fold axis of symmetry, for which the Schönflies symbol is C_n, rotation of the molecule by $2\pi/n$ radians, where $n = 1,2,3, \ldots \infty$, about the axis produces a configuration which, to a stationary observer, is indistinguishable from the initial one. Figure 4.2(a) to (e) illustrates a C_2 axis in H_2O, a C_3 axis in CH_3F, a C_4 axis in $XeOF_4$, a C_6 axis in C_6H_6, and a C_∞ axis (rotation by *any* angle produces an indistinguishable configuration) in HCN.

Corresponding to every symmetry element is a symmetry operation which is given the same symbol as the element. For example, C_n also indicates the actual operation of rotation of the molecule by $2\pi/n$ radians about the axis.

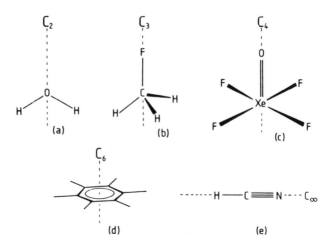

Figure 4.2 Examples of (a) C_2, (b) C_3, (c) C_4, (d) C_6, and (e) C_∞ axes.

4.1.2 Plane of Symmetry, σ

If a molecule has a plane of symmetry, for which the symbol is σ, reflection of all the nuclei through the plane to an equal distance on the opposite side produces a configuration indistinguishable from the initial one. Figure 4.3(a) shows the two planes of symmetry, $\sigma_v(xz)$ and $\sigma_v(yz)$, of H_2O using conventional axis notation. Just as the yz plane, the plane of the molecule, is a plane of symmetry so any planar molecule has at least one plane of symmetry. The subscript 'v' stands for 'vertical' and implies that the plane is vertical with respect to the highest-fold axis, C_2 in this case, which defines the vertical direction.

In the planar molecule BF_3, in Figure 4.3(b), the C_3 axis through B and perpendicular to the figure is the highest-fold axis, and therefore, the three planes of symmetry, perpendicular to the figure and through each of the B–F bonds, are labelled σ_v. On the other hand, the plane of the molecule is also a plane of symmetry and is labelled σ_h, where 'h' stands for 'horizontal' with respect to C_3.

In a molecule such as naphthalene, in Figure 4.3(c), there is no unique highest-fold axis and the three planes do not have subscripts.

A third subscript 'd', which stands for 'dihedral', is sometimes useful. Its use is illustrated by the allene molecule in Figure 4.3(d) in which the axes labelled C_2' are at 90° to each other and at 45° to the plane of the figure[†]. These are called dihedral axes which are, in general, C_2 axes at equal angles to each

---

[†] A molecular model is a great help in visualizing not only these particular axes but also all elements of symmetry.

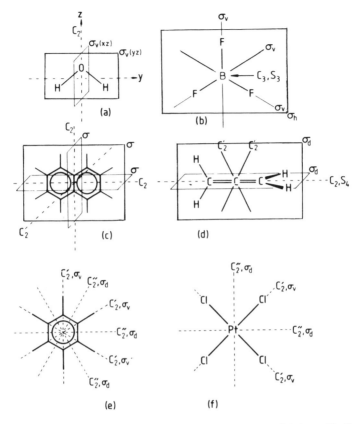

Figure 4.3 Planes and axes of symmetry in (a) H_2O, (b) BF_3, (c) naphthalene, (d) allene, (e) benzene, and (f) $[PtCl_4]^{2-}$.

other and perpendicular to the main axis. In allene, the main axis is the $C=C=C$ axis which is not only a C_2 axis but an S_4 axis (see Section 4.1.4). The σ_d planes are those which bisect the angles between dihedral axes. In molecules such as benzene and the square planar $[PtCl_4]^{2-}$ there is a choice of σ_v and σ_d labels for the planes perpendicular to the plane of the molecule. Figure 4.3(e) to (f) shows that, conventionally, the planes bisecting bond angles are usually labelled σ_d and those through bonds are labelled σ_v.

The symmetry operation σ is the operation of reflecting the nuclei across the plane.

4.1.3 Centre of Inversion, i

If a molecule has a centre of inversion (or centre of symmetry), i, reflection of each nucleus through the centre of the molecule to an equal distance on the

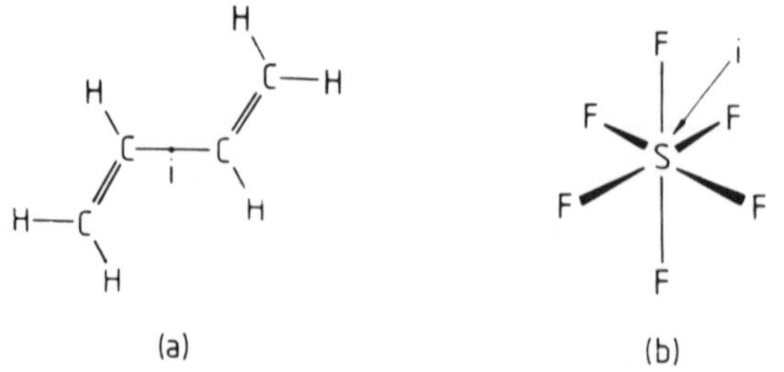

Figure 4.4 Inversion centre in (a) s-*trans*-buta-1,3-diene, (b) sulphur hexafluoride.

opposite side of the centre produces a configuration indistinguishable from the initial one. Figure 4.4 shows s-*trans*-buta-1,3-diene (the 's' refers to *trans* about a nominally single bond) and sulphur hexafluoride, both of which have inversion centres.

The symmetry operation i is the operation of inversion through the inversion centre.

4.1.4 n-Fold Rotation–Reflection Axis of Symmetry, S_n

For a molecule having an n-fold rotation–reflection axis of symmetry S_n rotation by $2\pi/n$ radians about the axis, followed by reflection through a plane perpendicular to the axis and through the centre of the molecule, produces a configuration indistinguishable from the initial one. Figure 4.3(d) shows that the $C=C=C$ axis in allene is an S_4 axis. The plane across which reflection takes place may or may not be a plane of symmetry; in allene it is not, but in BF_3 in Figure 4.3(b) the plane involved in the S_3 operation is the σ_h plane. This example also illustrates the point that, if there is a σ_h plane perpendicular to the highest-fold C_n axis, this axis must be an S_n axis also. We say that C_n and σ_h *generate* S_n, or

$$\sigma_h \times C_n = S_n \tag{4.1}$$

This equation can be interpreted also as implying that, if we carry out a C_n operation followed by a σ_h operation, the result is the same as carrying out an S_n operation. (The convention is to write $A \times B$ to mean carrying out operation B first and A second: in the case of C_n and σ_h the order does not matter but we shall come across examples where it does.)

From the definition of S_n it follows that $\sigma = S_1$ and $i = \infty S_2$; since σ and i are taken as separate symmetry elements the symbols S_1 and S_2 are never used.

4.1.5   The Identity Element of Symmetry, I (or E)

All molecules possess the identity element of symmetry, for which the symbol is I (some authors use E but this may cause confusion with the E symmetry

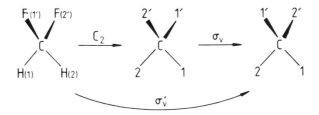

Figure 4.5 Illustration that, in CH_2F_2, C_2 and σ_v generate σ_v'.

species—see Section 4.3.2). The symmetry operation I consists of doing nothing to the molecule, so that it may seem too trivial to be of importance but it is a necessary element required by the rules of group theory. Since the C_1 operation is a rotation by 2π radians, $C_1 = I$ and the C_1 symbol is not used.

4.1.6 Generation of Elements

Equation (4.1) illustrates how the elements C_n and σ_h generate S_n. Figure 4.5 shows how C_2 and σ_v, in difluoromethane CH_2F_2, generate σ_v'; that is

$$\sigma_v \times C_2 = \sigma_v' \tag{4.2}$$

where σ_v is taken to be the plane containing CH_2 and σ_v' that containing CF_2. Similarly,

$$\sigma_v' \times C_2 = \sigma_v \tag{4.3}$$

$$\sigma_v' \times \sigma_v = C_2$$

From a C_n element we can generate other elements by raising it to the powers $1,2,3,\ldots,(n-1)$. For example, if there is a C_3 element there must also be C_3^2, where

$$C_3^2 = C_3 \times C_3 \tag{4.4}$$

and the C_3^2 operation is a rotation, which we take to be clockwise, by $2 \times (2\pi/3)$ radians. Similarly, if there is a C_6 element there must necessarily be $C_6^2(=C_3)$, $C_6^3(=C_2)$, $C_6^4(=C_3^2)$, and C_6^5. The C_6^5 operation is equivalent to an anticlockwise rotation by $2\pi/6$, an operation which is given the symbol C_6^{-1}. Similarly, C_3^2 is equivalent to C_3^{-1} and, in general,

$$C_n^{n-1} = C_n^{-1} \tag{4.5}$$

From an S_n element, also, we can generate S_n to the powers $1, 2, 3, \ldots, (n-1)$. Figure 4.6, for example, illustrates the S_4^2 and S_4^3 operations in allene and shows

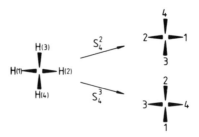

Figure 4.6 The S_4^2 and S_4^3 operations in allene.

that

$$S_4^2 = C_2 \qquad (4.6)$$
$$S_4^3 = S_4^{-1}$$

where S_4^{-1} implies an anticlockwise rotation by $2\pi/4$ followed by a reflection (note that a reflection is the same as its inverse, σ^{-1}).

4.1.7 Symmetry Conditions for Molecular Chirality

A chiral molecule is one which exists in two forms, known as enantiomers. Each of the enantiomers is optically active which means that they can rotate the plane of plane-polarized light. The enantiomer which rotates the plane to the right (clockwise) has been called the d- (or dextro-) form and the one which rotates it to the left (anticlockwise) the l- (or laevo-) form. Nowadays it is more usual to refer to the d- and l-forms as the $(+)$- and $(-)$-forms, respectively.

Very often a sample of a chiral molecule exists as an equimolar mixture of $(+)$- and $(-)$-enantiomers. Such a mixture will not rotate the plane of plane-polarized light and is called a racemic mixture with a prefix $(\pm)$ to indicate this. In the case of some chiral molecules the $(+)$- and $(-)$-components of a racemic mixture can be separated, or they can be synthesized separately, but there are many cases of chiral molecules which are known only in the form of racemic mixtures.

There are interesting examples of enantiomers which not only are found separately but have different chemical properties when reacting with some reagent which is itself an enantiomer. For example $(+)$-glucose is metabolized by animals and can be fermented by yeasts but $(-)$-glucose has neither of these properties. The enantiomer $(+)$-carvone smells of caraway while $(-)$-carvone smells of spearmint.

A useful way of describing the difference between $(+)$- and $(-)$-enantiomers is that one is the mirror image of the other. In other words neither enantiomer is superimposable on its mirror image.

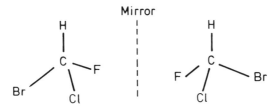

Figure 4.7 The CHFClBr molecule and its mirror image.

Figure 4.7 shows that any substituted methane, in which all four groups attached to the central carbon atom are different, as for example in CHFClBr, forms enantiomers. You can either use your imagination or construct models of these enantiomers to show that you can superimpose the carbon atoms and any two of the other atoms, such as H and F, but the remaining two atoms, Cl and Br, cannot be superimposed.

In the early days following the discovery of chirality it was thought that only molecules of the type CWXYZ, multiply substituted methanes, were important in this respect and it was said that a molecule with 'an asymmetric carbon atom' forms enantiomers. Nowadays this definition is totally inadequate for two reasons. The first is that the existence of enantiomers is not confined to molecules with a central carbon atom (it is not even confined to organic molecules), and the second is that, knowing what we do about the various possible elements of symmetry, the phrase 'asymmetric carbon atom' has no real meaning.

One useful rule that does survive is that:

If a molecule is not superimposable on its mirror image then it is a chiral molecule.

This rule is applicable to any molecule, whether or not it contains a central carbon atom, or indeed any carbon atom at all.

In considering whether a molecule is superimposable on its mirror image you may sense that the symmetry properties of the molecule should be able to give this information. This is in fact the case and the symmetry-related rule for chirality is a very simple one:

A molecule is chiral if it does not have any S_n symmetry element with any value of n.

We have seen in Section 4.1.4 that $S_1 = \sigma$ and that $S_2 = i$, so we can immediately exclude from chirality any molecule having a plane of symmetry or a centre of inversion. The condition that a chiral molecule may not have a plane of symmetry or a centre of inversion is sufficient in nearly all cases to decide whether a molecule is chiral. We have to go to a rather unusual molecule, such as the tetrafluorospiropentane shown in Figure 4.8, to find a case where there is no σ or i element of symmetry but there is a higher-fold S_n element. In

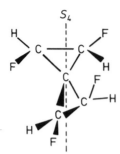

Figure 4.8 A chiral isomer of tetrafluorospiropentane.

this molecule the two three-membered carbon rings are mutually perpendicular and the pairs of fluorine atoms on each end of the molecule are *trans* to each other. There is an S_4 axis, as shown in Figure 4.8, but no σ or i element, and therefore the molecule is not chiral.

In the case of CHFClBr in Figure 4.7 there is no element of symmetry at all, except the identity I, and the molecule must be chiral.

In Section 4.2.1 it will be pointed out that hydrogen peroxide (Figure 4.11a) has only one symmetry element, a C_2 axis, and is therefore a chiral molecule although the enantiomers have never been separated. The complex ion [Co(ethylenediamine)$_3$]$^{3+}$, discussed in Section 4.2.4 and shown in Figure 4.11(f), is also chiral, having only a C_3 axis and three C_2 axes.

In organic chemistry there are many important molecules which contain two or more groups each of which, in isolation, would be chiral. A simple example is that of 2,3-difluorobutane, shown in Figure 4.9. The molecule can be regarded as a substituted ethane and we assume that, as in ethane itself, the stable structure is one in which one CHFCH$_3$ group is staggered relative to the other.

The five possible staggered structures of 2,3-difluorobutane are shown in Figure 4.9. These involve all possibilities of the two identical groups being *trans* or *gauche* to each other. In the *trans* positions they are as far away from each other, and in the *gauche* positions as close to each other, as they can get. The structures in Figure 4.9(a) and (b) are interconvertible by rotation about the central carbon–carbon bond and are known as conformers or rotamers. The same is true of the structures in Figure 4.9(c), (d) and (e).

The 'all *trans*' structure in Figure 4.9(a) is not chiral as it has an inversion centre i. This is called a *meso* structure. Although each CHFCH$_3$ group rotates the plane of plane-polarized light in one direction the other group rotates it an equal amount in the opposite direction. The result is that there is no rotation of the plane of polarization and, as the presence of an inversion centre tells us, the molecule is achiral.

All the other four structures in Figure 4.9 are chiral. The 'all *gauche*' structure in Figure 4.9(b) has no symmetry at all while each of the structures in Figure 4.9(c), (d) and (e) has a C_2 axis only.

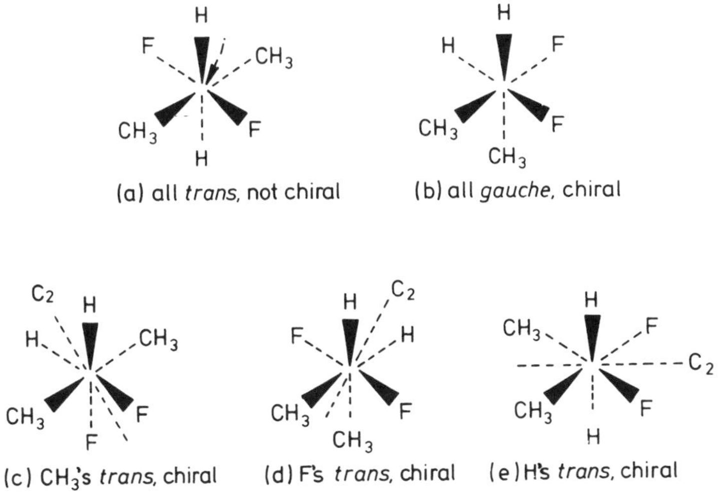

(a) all *trans*, not chiral   (b) all *gauche*, chiral

(c) CH₃'s *trans*, chiral   (d) F's *trans*, chiral   (e) H's *trans*, chiral

Figure 4.9 Five possible staggered structures of 2,3-difluorobutane.

The trisubstituted ammonia (tertiary amine) in Figure 4.10, in which all the substituents are different, has no symmetry element and is, therefore, chiral but there is an important proviso. The ammonia molecule itself is pyramidal and can invert, like an umbrella turning inside out. It does this so rapidly (in about 10^{-11} s, see Section 6.2.4.4a) that it can, for the present purposes, be regarded as effectively planar. The rate of inversion of the molecule in Figure 4.10 depends strongly on the masses of the groups R_1, R_2 and R_3. The heavier they are the more likely is inversion to be such a slow process that there is no feasible interconversion between the enantiomers. It is also possible that some of these tertiary amines are not chiral because they have a planar configuration about the nitrogen atom.

4.2 Point Groups

All the elements of symmetry which any molecule may have constitute a point group. Examples are the I, C_2, $\sigma_v(xz)$, $\sigma_v(yz)$ elements of H_2O and the I, S_4, S_4^{-1}, C_2, C_2', C_2', σ_d, σ_d elements of allene.

Point groups should be distinguished from space groups. Point groups are so called because, when all the operations of the group are carried out, at least

Figure 4.10 A general tertiary amine.

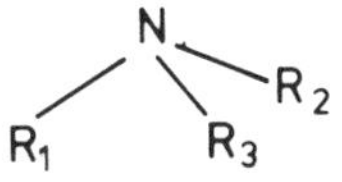

one point is unaffected: in the case of allene it is the point at the centre of the molecule, while in H_2O any point on the C_2 axis is unaffected. Space groups are appropriate to the symmetry properties of regular arrangements of molecules in space, such as are found in crystals, and we shall not be concerned with them here.

Many point groups are necessary to cover all possible molecules and it is convenient to collect together those which have certain types of elements in common. It is also convenient in defining point groups not to list all the elements of the group. In theory it is necessary to list only the generating elements from which all the other elements may be generated. In practice it is more helpful to give rather more than the generating elements and this is what we shall do here. Of course, all point groups contain the identity element and this will not be included in the point group definitions.

4.2.1 C_n Point Groups

A C_n point group contains a C_n axis of symmetry. By implication it does not contain σ, i, or S_n elements. However, it must contain C_n^2, C_n^3, ..., C_n^{n-1}.

Examples of molecules belonging to a C_n point group, for which $n = 1,2,3,...$, are not very common, but hydrogen peroxide (H_2O_2), shown in Figure 4.11(a), belongs to the C_2 point group; the only symmetry element, apart from I, is the C_2 axis bisecting the $118°$ angle between the two O–O–H planes. Bromo-

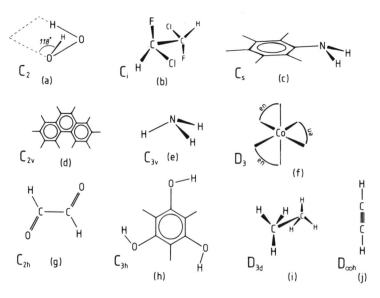

Figure 4.11 Examples of molecules belonging to various point groups.

chlorofluoromethane (CHBrClF), shown in Figure 4.7, has no symmetry element, apart from I, and belongs to the C_1 point group.

4.2.2 S_n Point Groups

An S_n point group contains an S_n axis of symmetry. The group must contain also $S_n^2, S_n^3, \ldots, S_n^{n-1}$.

Examples are rare except for the S_2 point group. This point group has only an S_2 axis but, since $S_2 = i$, it has only a centre of inversion and the symbol usually used for this point group is C_i. The isomer of the molecule ClFHC–CHFCl in which all pairs of identical H, F, or Cl atoms are *trans* to each other, shown in Figure 4.11(b), belongs to the C_i point group.

4.2.3 C_{nv} Point Groups

A C_{nv} point group contains a C_n axis of symmetry and n σ planes of symmetry, all of which contain the C_n axis. It also contains other elements which may be generated from these.

Many molecules belong to the C_{1v} point group. Apart from $I (= C_1)$ they have only a plane of symmetry. Fluoroethylene (Figure 4.1c) and aniline (Figure 4.11c), which has a pyramidal configuration about the nitrogen atom and a plane of symmetry bisecting the HNH angle, are examples. Instead of C_{1v} the symbol C_s is usually used.

H_2O (Figure 4.3a), difluoromethane (Figure 4.5), 1,1-difluoroethylene (Figure 4.1b), *cis*-1,2-difluoroethylene (Figure 4.1d), and phenanthrene (Figure 4.11d) have a C_2 axis and two σ_v planes and therefore belong to the C_{2v} point group.

NH_3 (Figure 4.11e) and methyl fluoride (Figure 4.2b) have a C_3 axis and three σ_v planes and belong to the C_{3v} point group. $XeOF_4$ (Figure 4.2c) belongs to the C_{4v} point group.

$C_{\infty v}$ is an important point group since all linear molecules without a σ_h plane belong to it. HCN (Figure 4.2e) has a C_∞ axis and an infinite number of σ_v planes, all containing the C_∞ axis and at any angle to the page, and therefore belongs to the $C_{\infty v}$ point group.

4.2.4 D_n Point Groups

A D_n point group contains a C_n axis and n C_2 axes. The C_2 axes are perpendicular to C_n and at equal angles to each other. It also contains other elements which may be generated from these.

Molecules belonging to D_n point groups are not common. They can be visualized as being formed by two identical C_{nv} fragments connected back to back in such a way that one fragment is staggered with respect to the other by any angle other than $m\pi/n$, where m is an integer.

The complex ion $[Co(ethylenediamine)_3]^{3+}$ shown in Figure 4.11(f), in which 'en' is the $H_2NCH_2CH_2NH_2$ group[†], belongs to the D_3 point group and can be thought of as consisting of two staggered C_{3v} fragments.

4.2.5 C_{nh} Point Groups

A C_{nh} point group contains a C_n axis and a σ_h plane, perpendicular to C_n. For n even the point group contains a centre of inversion i. It also contains other elements which may be generated from these.

The point group C_{1h} contains only a plane of symmetry, in addition to I. It is therefore the same as C_{1v} and is usually given the symbol C_s.

Examples of molecules having a C_2 axis, a σ_h plane, and a centre of inversion i, and therefore belonging to the C_{2h} point group, are *trans*-1,2-difluoroethylene (Figure 4.1e), *s-trans*-buta-1,3-diene (Figure 4.4a), and *s-trans*-glyoxal (Figure 4.11g). 1,3,5-Trihydroxybenzene (Figure 4.11h) belongs to the C_{3h} point group.

4.2.6 D_{nd} Point Groups

A D_{nd} point group contains a C_n axis, an S_{2n} axis, n C_2 axes perpendicular to C_n and at equal angles to each other, and n σ_d planes bisecting the angles between the C_2 axes. For n odd the point group contains a centre of inversion i. It also contains other elements which may be generated from these.

The point group D_{1d} is the same as C_{2v}. Molecules belonging to other D_{nd} point groups can be visualized as consisting of two identical fragments of C_{nv} symmetry back to back with one staggered at an angle of π/n to the other.

Allene (Figure 4.3d) belongs to the D_{2d} point group and ethane (Figure 4.11i), which has a staggered configuration in which each of the C–H bonds at one end of the molecule bisects an HCH angle at the other, belongs to the D_{3d} point group.

Apart from these two important examples molecules belonging to D_{nd} point groups are rare.

4.2.7 D_{nh} Point Groups

A D_{nh} point group contains a C_n axis, n C_2 axes perpendicular to C_n and at equal angles to each other, a σ_h plane, and n other σ planes. For n even the point group contains a centre of inversion i. It also contains other elements which may be generated from these.

A D_{nh} point group is related to the corresponding C_{nv} point group by the inclusion of a σ_h plane.

[†] The puckered nature of this ligand has been neglected.

The point group D_{1h} is the same as C_{2v}. Ethylene (Figure 4.1a) and naphthalene (Figure 4.3c) belong to the D_{2h} point group in which, because of the equivalence of the three mutually perpendicular C_2 axes, no subscripts are used for the planes of symmetry.

BF_3 (Figure 4.3b) belongs to D_{3h}, $[PtCl_4]^{2-}$ (Figure 4.3f) to D_{4h}, and benzene (Figure 4.3e) to the D_{6h} point group.

The $D_{\infty h}$ point group is derived from $C_{\infty v}$ by the inclusion of σ_h: therefore all linear molecules with a plane of symmetry perpendicular to the C_∞ axis belong to $D_{\infty h}$. Acetylene (Figure 4.1j), for example, and all homonuclear diatomic molecules belong to this point group.

4.2.8 T_d Point Group

The T_d point group contains four C_3 axes, three C_2 axes, and six σ_d planes. It also contains elements generated from these.

This is the point group to which all regular tetrahedral molecules, such as methane (Figure 4.12a), silane (SiH_4), and nickel tetracarbonyl ($Ni(CO)_4$), belong.

The C_3 axes, in the case of methane, are the directions of each of the C–H bonds. The σ_d planes are the six planes of all the possible CH_2 fragments. The C_2 axes are not quite so easy to see but Figure 4.12(a) shows that the line of intersection of any two mutually perpendicular σ_d planes is a C_2 axis.

4.2.9 O_h Point Group

The O_h point group contains three C_4 axes, four C_3 axes, six C_2 axes, three σ_h planes, six σ_d planes, and a centre of inversion i. It also contains elements generated from these.

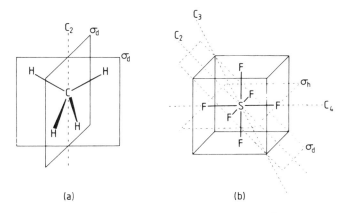

(a) (b)

Figure 4.12 (a) Some C_2 and σ_d elements in methane. (b) Some of the symmetry elements in sulphur hexafluoride.

This is the point group to which all regular octahedral molecules, such as SF_6 (Figure 4.12b) and $[Fe(CN)_6]^{3-}$, belong.

The main symmetry elements in SF_6 can be shown, as in Figure 4.12(b), by considering the sulphur atom at the centre of a cube and a fluorine atom at the centre of each face. The three C_4 axes are the three F–S–F directions, the four C_3 axes are the body diagonals of the cube, the six C_2 axes join the mid-points of diagonally opposite edges, the three σ_h planes are each halfway between opposite faces, and the six σ_d planes join diagonally opposite edges of the cube.

4.2.10 K_h Point Group

The K_h point group contains an infinite number of C_∞ axes and a centre of inversion i. It also contains elements generated from these.

This is the point group to which a sphere, and therefore all atoms, belong[†].

The point groups discussed here are all those that one is likely to use, but there are a few very uncommon ones which have not been included: they are to be found in several of the books on molecular symmetry mentioned in the bibliography at the end of this chapter.

4.3 Point Group Character Tables

Point groups may be divided into two important sets—non-degenerate and degenerate. A degenerate point group contains a C_n axis, with $n > 2$, or an S_4 axis. A molecule belonging to such a point group may have degenerate properties, e.g. different electronic or vibrational wave functions with identical associated energies, but a molecule belonging to a non-degenerate point group cannot have degenerate properties.

In Sections 4.1 and 4.2 we have seen how the symmetry properties of the equilibrium nuclear configuration of any molecule may be classified and the molecule assigned to a point group. However, molecules may have properties, such as electronic and vibrational wave functions, which do not preserve all the elements of symmetry and it is the symmetry classification of such properties with which character tables are concerned. We shall consider in detail three such tables, those of C_{2v}, an example of a non-degenerate point group; C_{3v}, an example of a degenerate point group; and $C_{\infty v}$, an example of an infinite group, which has an infinite number of elements.

[†] Although an atom with partially filled orbitals may not be spherically symmetrical, the electronic wave function is classified according to the K_h point group.

4.3.1 C_{2v} Character Table

A property such as a vibrational wave function of, say, H_2O may or may not preserve an element of symmetry. If it preserves the element, carrying out the corresponding symmetry operation, for example σ_v, has no effect on the wave function, which we write as

$$\psi_v \overset{\sigma_v}{\rightarrow} (+1)\psi_v \tag{4.7}$$

and we say that ψ_v is symmetric to σ_v. The only other possibility, in a non-degenerate point group, is that the wave function may be changed in sign by carrying out the operation

$$\psi_v \overset{\sigma_v}{\rightarrow} (-1)\psi_v \tag{4.8}$$

in which case we say that ψ_v is antisymmetric to σ_v. The $+1$ of equation (4.7) and the -1 of equation (4.8) are known as the character of, in this case, ψ_v with respect to σ_v.

We have seen that any two of the C_2, $\sigma_v(xz)$, $\sigma'_v(yz)$ elements may be regarded as generating elements. There are four possible combinations of $+1$ or -1 characters with respect to these generating elements, $+1$ and $+1$, $+1$ and -1, -1 and $+1$, -1 and -1, with respect to C_2 and $\sigma_v(xz)$. These combinations are entered in columns 3 and 4 of the C_{2v} character table in Table A.11 in the Appendix. The character with respect to I must always be $+1$ and, just as $\sigma'_v(yz)$ is generated from C_2 and $\sigma_v(xz)$, the character with respect to $\sigma'_v(yz)$ is the product of characters with respect to C_2 and $\sigma_v(xz)$. Each of the four rows of characters is called an irreducible representation of the group and, for convenience, each is represented by a symmetry species A_1, A_2, B_1, or B_2. The A_1

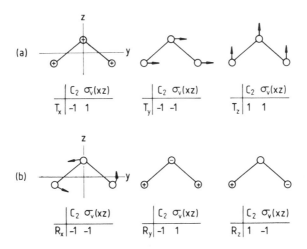

Figure 4.13 (a) Translations and (b) rotations in the H_2O molecule.

species is said to be totally symmetric since all the characters are $+1$: the other three species are non-totally symmetric.

The symmetry species labels are conventional: A and B indicate symmetry or antisymmetry respectively to C_2 and the subscripts 1 and 2 indicate symmetry or antisymmetry respectively to $\sigma_v(xz)$.

In the sixth column of the character table is indicated the symmetry species of translations (T) of the molecule along and rotations (R) about the cartesian axes. In Figure 4.13 vectors attached to the nuclei of H_2O represent these motions which are assigned to symmetry species by their behaviour under the operations C_2 and $\sigma_v(xz)$. Figure 4.13(a) shows that

$$\Gamma(T_x) = B_1; \qquad \Gamma(T_y) = B_2; \qquad \Gamma(T_z) = A_1 \qquad (4.9)$$

and Figure 4.13(b) shows that

$$\Gamma(R_x) = B_2; \qquad \Gamma(R_y) = B_1; \qquad \Gamma(R_z) = A_2 \qquad (4.10)$$

The symbol Γ, in general, stands for 'representation of ... '. Here it is an irreducible representation or symmetry species. In Sections 6.2.2.1 and 7.3.2, where we derive the infrared vibrational and electronic selection rules for polyatomic molecules, we shall require the symmetry species of the translations.

In the final column of the character table are given the assignments to symmetry species of α_{xx}, α_{yy}, α_{zz}, α_{xy}, α_{yz}, and α_{xz}. These are the components of the symmetric polarizability tensor α which is important in vibrational Raman spectroscopy to be discussed in Section 6.2.2.2.

An N-atomic non-linear molecule has $3N - 6$ normal modes of vibration. This follows from each atom having three degrees of freedom due to the need to specify three coordinates, say x, y, and z, to define the position of each. Of the total of $3N$ degrees of freedom for the molecule, three represent the translation of the molecule as a whole along the x, y, or z axis and another three represent rotation of the molecule about each of these axes. The remaining $3N - 6$ degrees of freedom represent motions of the nuclei relative to each other, namely the vibrations. For a linear molecule there are $3N - 5$ normal modes of vibration because there is no degree of freedom corresponding to rotation about the internuclear axis—there is no moment of inertia about this axis.

The H_2O molecule, therefore, has three normal vibrations which are illustrated in Figure 4.14 in which the vectors attached to the nuclei indicate the directions and relative magnitudes of the motions. Using the C_{2v} character table the wave functions ψ_v for each can easily be assigned to symmetry species. The characters of the three vibrations under the operations C_2 and $\sigma_v(xz)$ are respectively $+1$ and $+1$ for v_1, $+1$ and $+1$ for v_2, and -1 and -1 for v_3. Therefore

$$\Gamma(\psi_{v(1)}) = A_1; \qquad \Gamma(\psi_{v(2)}) = A_1; \qquad \Gamma(\psi_{v(3)}) = B_2 \qquad (4.11)$$

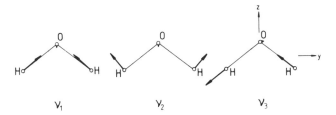

Figure 4.14 The normal vibrations of the H_2O molecule.

It is important to realize that the assignment of symmetry species to molecular properties depends on the labelling of axes. The axis labels used for the T's, R's, and α's in the last two columns of a character table are always the same but the choice of axis labels for the particular molecule under consideration is not easily standardized. Mulliken (see the bibliography) has suggested conventions for axis labelling in many cases and these conventions are widely used. For a planar C_{2v} molecule the convention is that the C_2 axis is taken to be the z axis and the x axis is perpendicular to the plane. This is the convention used here for H_2O. The importance of using a convention can be illustrated by the fact that if, say, we were to exchange the x and y axis labels then $\Gamma(\psi_{v(3)})$ would be B_1 rather than B_2. However, for a non-planar C_{2v} molecule, such as CH_2F_2 (Figure 4.5), although it is natural to take the C_2 axis to be the z axis, the choice of x and y axes is arbitrary. This example serves to demonstrate how important it is to indicate the axis notation being used.

There will be many occasions when we shall need to multiply symmetry species or, in the language of group theory, to obtain their direct product. For example, if H_2O is vibrationally excited simultaneously with one quantum each of v_1 and v_3, the symmetry species of the wave function for this vibrational combination state is

$$\Gamma(\psi_v) = A_1 \times B_2 = B_2 \qquad (4.12)$$

In order to obtain the direct product of two species we multiply the characters under each symmetry element using the rules

$$(+1) \times (+1) = 1; \qquad (+1) \times (-1) = -1; \qquad (-1) \times (-1) = 1 \quad (4.13)$$

In this way the result in equation (4.12) is obtained. If H_2O is excited with, say, two quanta of v_3 then

$$\Gamma(\psi_v) = B_2 \times B_2 = A_1 \qquad (4.14)$$

The results of multiplication in equations (4.12) and (4.14) that (a) the direct product of any species with the totally symmetric species leaves it unchanged, and (b) the direct product of any species with itself gives the totally symmetric species, are quite general for all non-degenerate point groups.

Using the rules for forming direct products it can be shown also that, in the C_{2v} point group,

$$A_2 \times B_1 = B_2; \qquad A_2 \times B_2 = B_1 \tag{4.15}$$

The character tables for all important point groups, degenerate and non-degenerate, are given in the Appendix.

4.3.2 C_{3v} Character Table

Inspection of this character table, given in Table A.12 in the Appendix, shows two obvious differences from a character table for any non-degenerate point group. The first is the grouping together of all elements of the same class, namely C_3 and C_3^2 as $2C_3$, and σ_v, σ_v', and σ_v'' as $3\sigma_v$.

Two elements P and Q are said to belong to the same class if there exists a third element R such that

$$P = R^{-1} \times Q \times R \tag{4.16}$$

Figure 4.15 shows that, in the C_{3v} point group,

$$C_3 = \sigma_v^{-1} \times C_3^2 \times \sigma_v \tag{4.17}$$

and therefore, that C_3 and C_3^2 belong to the same class. Elements belonging to the same class have the same characters and, as is the case in non-degenerate point groups for which each element is in a separate class, the number of symmetry species is equal to the number of classes.

The second difference is the appearance of a doubly degenerate E symmetry species whose characters are not always either the $+1$ or -1 that we have encountered in non-degenerate point groups.

The $+1$ and -1 characters of the A_1 and A_2 species have the same significance as in a non-degenerate point group. The characters of the E species may be understood by using the normal vibrations of NH_3, shown in Figure

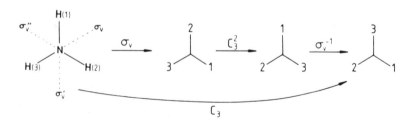

Figure 4.15 Illustration that, in the C_{3v} point group, C_3 and C_3^2 belong to the same class.

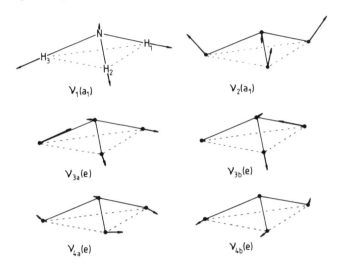

$V_1(a_1)$ $V_2(a_1)$

$V_{3a}(e)$ $V_{3b}(e)$

$V_{4a}(e)$ $V_{4b}(e)$

Figure 4.16   Normal vibrations of the NH_3 molecule.

4.16, as examples[†]. Vibrations v_1 and v_2 are clearly a_1[‡]. The two vibrations v_{3a} and v_{3b} are degenerate: although it is not obvious, it requires the same energy to excite one quantum of either of them and, clearly, they have different wave functions. Similarly, v_{4a} and v_{4b} are degenerate.

The symmetry properties of a fundamental vibrational wave function ψ_v are the same as those of the corresponding normal coordinate Q. For example, when the C_3 operation is carried out on Q_1, the normal coordinate for v_1, it is transformed into Q'_1, where

$$Q_1 \overset{C_3}{\to} Q'_1 = (+1)Q_1 \qquad (4.18)$$

[†] Mulliken has also recommended a vibrational numbering convention. The vibrations are grouped according to their symmetry species taken in the order in the character tables presented by Herzberg (see the bibliography). For each symmetry species the vibrations are sorted in order of decreasing wavenumber. The totally symmetric species is always the first and the totally symmetric vibration with the highest wavenumber is v_1, the next one v_2, and so on. Exceptionally, though, the doubly degenerate bending vibration of a linear triatomic, such as CO_2 or HCN, is *always* labelled v_2. Also, in unsymmetrical linear triatomics, such as HCN, there are two stretching vibrations belonging to the same symmetry species (σ^+): the *lower* wavenumber one is labelled v_1 and the *higher*, v_2.

[‡] Lower case letters are recommended for the symmetry species of a vibration (and for an electronic orbital) whereas upper case letters are recommended for the symmetry species of the corresponding wave functions.

On the other hand, when a symmetry operation is carried out on a degenerate normal coordinate it does not simply remain the same or change sign: in general, it is transformed into a linear combination of the two degenerate normal coordinates. Thus, on carrying out a symmetry operation S,

$$Q_{3a} \xrightarrow{S} Q'_{3a} = d_{aa} Q_{3a} + d_{ab} Q_{3b} \tag{4.19}$$
$$Q_{3b} \xrightarrow{S} Q'_{3b} = d_{ba} Q_{3a} + d_{bb} Q_{3b}$$

In matrix notation this can be written

$$\begin{pmatrix} Q'_{3a} \\ Q'_{3b} \end{pmatrix} = \begin{pmatrix} d_{aa} & d_{ab} \\ d_{ba} & d_{bb} \end{pmatrix} \begin{pmatrix} Q_{3a} \\ Q_{3b} \end{pmatrix} \tag{4.20}$$

The quantity $d_{aa} + d_{bb}$ is called the trace of the matrix which is then the character of the property, here a normal coordinate, under the symmetry operation S.

The character of the E species under the I operation is obtained from

$$Q_{3a} \xrightarrow{I} Q'_{3a} = 1 \times Q_{3a} + 0 \times Q_{3b} \tag{4.21}$$
$$Q_{3b} \xrightarrow{I} Q'_{3b} = 0 \times Q_{3a} + 1 \times Q_{3b}$$

or

$$\begin{pmatrix} Q'_{3a} \\ Q'_{3b} \end{pmatrix} = \begin{pmatrix} 1 & 0 \\ 0 & 1 \end{pmatrix} \begin{pmatrix} Q_{3a} \\ Q_{3b} \end{pmatrix} \tag{4.22}$$

The trace of the matrix is 2 which is the character of E under I.

Of the two components of ν_3 one, ν_{3a}, is symmetric to reflection across the σ_v plane bisecting the angle between H_1 and H_2, and the other is anti-symmetric giving

$$\begin{pmatrix} Q'_{3a} \\ Q'_{3b} \end{pmatrix} = \begin{pmatrix} 1 & 0 \\ 0 & -1 \end{pmatrix} \begin{pmatrix} Q_{3a} \\ Q_{3b} \end{pmatrix} \tag{4.23}$$

Therefore the character of E under σ_v is 0. Because of the equivalence of all the σ_v's the character must be the same under σ'_v and σ''_v.

When the operation is a rotation by an angle ϕ about a C_n axis, in this case by an angle of $2\pi/3$ radians about the C_3 axis, the resulting coordinate transformation, a result which will not be derived here, is given by

$$\begin{pmatrix} Q'_{3a} \\ Q'_{3b} \end{pmatrix} = \begin{pmatrix} \cos\phi & \sin\phi \\ -\sin\phi & \cos\phi \end{pmatrix} \begin{pmatrix} Q_{3a} \\ Q_{3b} \end{pmatrix} \tag{4.24}$$

$$= \begin{pmatrix} -\frac{1}{2} & \sqrt{\frac{3}{2}} \\ -\sqrt{\frac{3}{2}} & -\frac{1}{2} \end{pmatrix} \begin{pmatrix} Q_{3a} \\ Q_{3b} \end{pmatrix}$$

The trace of the matrix is -1 which is the character of E under C_3.

Except for the multiplication of E by E we follow the rules for forming direct products used in non-degenerate point groups: the characters under the various symmetry operations are obtained by multiplying the characters of the species being multiplied, giving

$$A_1 \times A_2 = A_2; \quad A_2 \times A_2 = A_1; \quad A_1 \times E = E; \quad A_2 \times E = E \quad (4.25)$$

In multiplying E by E we use, again, examples of the vibrations of NH_3. The result depends on whether we require $\Gamma(\psi_v)$ when (a) one quantum of each of two different e vibrations is excited, i.e. a combination level, or (b) two quanta of the same e vibration are excited, i.e. an overtone level. In case (a), such as for the combination $v_3 + v_4$, the product is written $E \times E$ and the result is obtained by first squaring the characters under each operation, giving

$$
\begin{array}{c|ccc}
 & I & 2C_3 & 3\sigma_v \\
\hline
E \times E & 4 & 1 & 0
\end{array}
\qquad (4.26)
$$

The characters 4, 1, 0 form a reducible representation in the C_{3v} point group and we require to reduce it to a set of irreducible representations, the sum of whose characters under each operation is equal to that of the reducible representation. We can express this algebraically as

$$\chi_C(k) \times \chi_D(k) = \chi_F(k) + \chi_G(k) + \cdots \qquad (4.27)$$

where χ is a character of any operation k and the result of multiplying the degenerate species C and D is

$$C \times D = F + G + \cdots \qquad (4.28)$$

The reduction of the $E \times E$ representation, like all such reductions, gives a unique set of irreducible representations which is

$$E \times E = A_1 + A_2 + E \qquad (4.29)$$

We can see from Table A.12 in the Appendix that the sum of the characters of A_1, A_2, and E under I, C_3, and σ_v gives the reducible representation of equation (4.26).

In case (b), where two quanta of the same e vibration are excited, such as $2v_3$, the product is written $(E)^2$ where

$$(E)^2 = A_1 + E \qquad (4.30)$$

which is called the symmetric part of $E \times E$, being the part which is symmetric to particle exchange. We arrive at the result of equation (4.30) by first obtaining the product $E \times E$. Then one of the species of the $E \times E$ product is forbidden. In degenerate point groups in general this is an A species and, where possible, one which is non-totally symmetric; in this case it is the A_2 species which is forbidden and this is the antisymmetric part of $E \times E$.

Tables, for all degenerate point groups, giving the symmetry species of

Tables, for all degenerate point groups, giving the symmetry species of vibrational combination states resulting from the excitation of one quantum of each of two different degenerate vibrations and of vibrational overtone states resulting from the excitation of two quanta of the same degenerate vibration are given in the books by Herzberg and by Hollas referred to in the bibliography.

4.3.3 $C_{\infty v}$ Character Table

In this character table, given in Table A.16 in the Appendix, there are an infinite number of classes since rotation about the C_∞ axis may be by *any* angle ϕ and each C_∞^ϕ element belongs to a different class. However, $C_\infty^{-\phi}$, an anticlockwise rotation by ϕ, belongs to the same class as C_∞^ϕ. Since there is an infinite number of classes there is an infinite number of symmetry species. Labels for these are $A_1, A_2, E_1, E_2, \ldots, E_\infty$ if we follow conventions used in other character tables. Unfortunately, before symmetry species labels had been adopted generally, another convention had grown up and was being used, particularly in electronic spectroscopy of diatomic molecules. Electronic states were given the labels Σ, Π, Δ, $\Phi, \ldots$ corresponding to the value of an electronic orbital angular momentum quantum number Λ (Section 7.2.2) which can take the values 0, 1, 2, 3, $\ldots$. It is mainly this system of labelling which is used in the $C_{\infty v}$, as well as the $D_{\infty h}$, point group, although both systems are given in Table A.16 in the Appendix.

Multiplication of symmetry species is carried out using the usual rules so that, for example,

$$\Sigma^+ \times \Sigma^- = \Sigma^-; \quad \Sigma^- \times \Pi = \Pi; \quad \Sigma^+ \times \Delta = \Delta \tag{4.31}$$

For the product $\Pi \times \Pi$, the reducible representation is

$$
\begin{array}{c|ccc}
 & I & 2C_\infty^\phi & \infty\sigma_v \\
\hline
\Pi \times \Pi & 4 & 4\cos^2\phi & 0 \\
 & & (=2+2\cos 2\phi) &
\end{array}
\tag{4.32}
$$

which reduces to give

$$\Pi \times \Pi = \Sigma^+ + \Sigma^- + \Delta \tag{4.33}$$

Use of the same rule as that for deriving $(E)^2$ from $E \times E$ in the C_{3v} point group gives

$$(\Pi)^2 = \Sigma^+ + \Delta \tag{4.34}$$

4.4 Symmetry and Dipole Moments

The dipole moment (strictly, the electric dipole moment) of a molecule is a measure of the charge asymmetry and is usually denoted by the symbol μ.

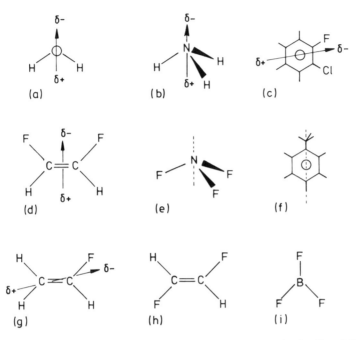

Figure 4.17 Molecules (a)–(g) have a permanent dipole moment but molecules (h) and (i) do not.

In the examples shown in Figure 4.17(a–g) all the molecules clearly have a charge asymmetry and, therefore, a non-zero dipole moment. Since a dipole moment has magnitude and direction it is a vector quantity and, if we wish to emphasize this, we use the vector symbol μ while, if we are concerned only with the magnitude, we use the symbol μ.

In H_2O and NH_3, shown in Figure 4.17(a) and (b), the direction of the dipole moment is along the C_2 or C_3 axis, respectively. In both molecules there are lone pairs of electrons directed away from the O–H or N–H bonds so that the negative end of the dipole is as shown in each case.

Charge asymmetry can be associated with a particular bond in a molecule and gives rise to what is called a bond dipole moment or, simply, bond moment. One use of bond moments is that they can be transferred, to a fair degree of approximation, from one molecule to another. In this way the dipole moment of a molecule can sometimes be estimated from a vector sum of bond moments. For example, the dipole moment of 2-chlorofluorobenzene, shown in Figure 4.17(c), can be estimated from the vector sum of the C–F and C–Cl bond moments found from, say, CH_3F and CH_3Cl respectively.

Bond moments for bonds involving a hydrogen atom are relatively small, so that the dipole moments of H_2O and NH_3 are dominated by the effects of the lone pairs of electrons. The dipole moment of *cis*-1,2-difluoroethylene, shown in Figure 4.17(d), is dominated by the large bond moments of the C–F bonds and the negative ends are directed towards the fluorine atoms, because of the high electronegativity of fluorine. The dipole moment is directed along the C_2 axis with the negative end as shown.

In NF_3 and $C_6H_5CH_3$ (toluene), shown in Figure 4.17(e) and (f), respectively, charge asymmetry confers a dipole moment on both molecules in the directions shown by the dotted lines but it is not obvious which is the negative end. In NF_3, the effect of the electronegative fluorine atoms tends to be counterbalanced by the lone pair of electrons on the nitrogen atom and, in $C_6H_5CH_3$, all the C–H bond moments are very small and their vector sum is not a reliable indicator of which is the negative end of the dipole.

The magnitude μ of a dipole moment is usually quoted in Debye units. Dipole moments are products of charges and their separations so that, for charges of $-q$ and $+q$ separated by a distance r,

$$\mu = qr \tag{4.35}$$

For a molecule with many nuclei and electrons with charges q_i

$$\mu = \sum_i q_i r_i \tag{4.36}$$

Equations (4.35) and (4.36) show that the SI unit of dipole moment is the product of the SI units of charge and distance which are the coulomb and the metre, respectively. However, the unit of C m is a rather cumbersome one and the old unit of the Debye (D) is still commonly used for the reason that molecular dipole moments are conveniently of the order of 1 D. The units are related by

$$1\ D \simeq 3.335\ 64 \times 10^{-30}\ C\ m \tag{4.37}$$

Quantitatively, using the Debye unit, the dipole moments of NH_3 (1.47 D), NF_3 (0.2 D) and $C_6H_5CH_3$ (0.36 D) confirm the simple arguments which we have made about their magnitudes.

It is important to realize that, although various types of experiments can give the value of a dipole moment, none is capable of determining the direction in which the negative or positive end lies. In molecules such as NF_3 and $C_6H_5CH_3$, in which the dipole moment is so small that simple arguments will not predict the direction, only accurate valence theory calculations can provide the answer. These calculations themselves are reliable only for fairly small molecules (see Section 5.2.3 for the case of CO).

The large C–F bond moment in fluoroethylene, in Figure 4.17(g), dominates the dipole moment which must lie in the molecular plane but its direction in that plane will depend on the effects of the other parts of the molecule with the result that it will not lie exactly along the C–F bond.

In molecules such as *trans*-1,2-difluoroethylene and BF_3, illustrated in Figure

4.17(h) and (i) respectively, it may be obvious at this stage that their dipole moments are zero but we shall see later that symmetry arguments confirm that this is so.

Looking at all the examples in Figure 4.17 suggests that molecular symmetry can be used to formulate a rule which will tell us whether any molecule has a non-zero dipole moment.

The dipole moment vector $\boldsymbol{\mu}$ must be totally symmetric, and therefore symmetric to all operations of the point group to which the molecule belongs: otherwise the direction of the dipole moment could be reversed by carrying out a symmetry operation, and this clearly cannot happen. The vector $\boldsymbol{\mu}$ has components μ_x, μ_y and μ_z along the cartesian axes of the molecule. In the examples of NH_3 and NF_3, in Figure 4.17(b) and (e), if the C_3 axis is the z-axis, $\mu_z \neq 0$ but $\mu_x = \mu_y = 0$. Similarly in H_2O and cis-1,2-difluoroethylene, in Figure 4.17(a) and (d), if the C_2 axis is the z-axis $\mu_z \neq 0$ but $\mu_x = \mu_y = 0$. In the case of fluoroethylene, in Figure 4.17(g), if the molecular plane is the xy-plane, then $\mu_x \neq 0$ and $\mu_y \neq 0$ but $\mu_z = 0$. In a molecule such as CHFClBr, in Figure 4.7, the dipole moment is not confined by symmetry in any way so that $\mu_x \neq 0$, $\mu_y \neq 0$ and $\mu_z \neq 0$.

If we compare the vectors representing a translation of, say, the H_2O molecule along the z-axis, as illustrated in Figure 4.13(a), with the dipole moment vector, which is also along the z-axis and is shown in Figure 4.17(a), it is clear that they have the same symmetry species, i.e. $\Gamma(\mu_z) = \Gamma(T_z)$ and, in general,

$$\Gamma(\mu_x) = \Gamma(T_x)$$

$$\Gamma(\mu_y) = \Gamma(T_y) \tag{4.38}$$

$$\Gamma(\mu_z) = \Gamma(T_z)$$

for all molecules. Because $\boldsymbol{\mu}$ must be totally symmetric, a molecule has a permanent dipole moment if any of the components μ_x, μ_y or μ_z is totally symmetric. With the relationships in equation (4.38) we can formulate the general rule:

A molecule has a permanent dipole moment if any of the translational symmetry species of the point group to which the molecule belongs is totally symmetric.

Then, if we look through all the point group character tables in the Appendix to see if any of the translational symmetry species is totally symmetric, it is apparent that molecules belonging to only the following point groups have a permanent dipole moment:

(a) C_1, in which $\Gamma(T_x) = \Gamma(T_y) = \Gamma(T_z) = A$;
(b) C_s, in which $\Gamma(T_x) = \Gamma(T_y) = A'$;
(c) C_n, in which $\Gamma(T_z) = A$;
(d) C_{nv}, in which $\Gamma(T_z) = A_1$ (or Σ^+ in $C_{\infty v}$).

In the C_1, C_s, C_n and C_{nv} point groups the totally symmetric symmetry species is A, A', A and A_1 (or Σ^+), respectively. For example, CHFClBr (Figure 4.7) belongs to the C_1 point group: therefore $\mu \neq 0$ and, since all three translations are totally symmetric, the dipole moment vector is not confined by symmetry to any particular direction. Fluoroethylene (Figure 4.17g) belongs to the C_s point group and therefore has a non-zero dipole moment which, because both $\Gamma(T_x)$ and $\Gamma(T_y)$ are totally symmetric, is confined to the xy-plane. Hydrogen peroxide (H_2O_2, Figure 4.11a) belongs to the C_2 point group and has a dipole moment directed along the z-axis (the C_2 axis). The NF_3 molecule (Figure 4.17e) belongs to the C_{3v} point group and has a dipole moment along the z-axis (the C_3 axis).

The BF_3 molecule, shown in Figure 4.17(i), is now seen to have $\mu = 0$ because it belongs to the D_{3h} point group for which none of the translational symmetry species is totally symmetric. Alternatively, we can show that $\mu = 0$ by using the concept of bond moments. If the B–F bond moment is μ_{BF} and we resolve the three bond moments along, say, the direction of one of the B–F bonds we get

$$\mu = \mu_{BF} - 2\mu_{BF} \cos 60° = 0 \tag{4.39}$$

If it seems obvious on looking at the BF_3 molecule in Figure 4.17(i) that it has no dipole moment then what we are doing in coming to that conclusion is a rapid, mental resolution of bond moments.

The molecule *trans*-1,2-difluoroethylene, in Figure 4.17(h), belongs to the C_{2h} point group in which none of the translational symmetry species is totally symmetric: therefore the molecule has no dipole moment. Arguments using bond moments would reach the same conclusion.

Although symmetry properties can tell us whether a molecule has a permanent dipole moment, they cannot tell us anything about the magnitude of a non-zero dipole moment. This can be determined most accurately from the microwave or millimetre wave spectrum of the molecule concerned (see Section 5.2.3).

Questions

1. The following molecules, in going from the ground to an excited electronic state, undergo a change of geometry which involves a change of point group:
   (a) ammonia (pyramidal ground state, planar excited state);
   (b) acetylene (linear ground state, *trans* bent excited state);
   (c) fluoroacetylene (linear ground state, *trans* bent excited state);
   (d) formaldehyde (planar ground state, pyramidal excited state).
   List the symmetry elements and point groups of these molecules in both electronic states.
2. List all the symmetry elements of the following molecules, assign each to a point group, and state whether they form enantiomers: (a) lactic acid, (b) *trans*-[Co(ethylenediamine)$_2$Cl$_2$]$^+$, (c) *cis*-[Co(ethylenediamine)$_2$Cl$_2$]$^+$,

(d) cyclopropane, (e) *trans*-1,2-dichlorocyclopropane, (f) 1-chloro-3-fluoro-allene.

3. Formaldehyde has six normal vibrations which can be represented as follows:

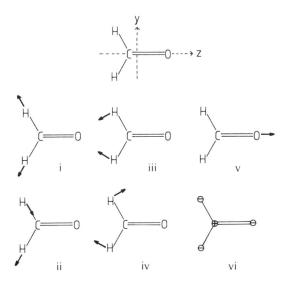

Using the conventional axis notation assign these vibrations to symmetry species of the appropriate point group.

4. Assign the allene molecule to a point group and use the character table to form the direct products $A_2 \times B_1$, $B_1 \times B_2$, $B_2 \times E$, and $E \times E$. Show how the symmetry species of the point group to which 1,1-difluoroallene belongs correlate with those of allene.

5. List the symmetry elements of each of the following molecules: (a) 1,2,3-trifluorobenzene, (b) 1,2,4-trifluorobenzene, (c) 1,3,5-trifluorobenzene, (d) 1,2,4,5-tetrafluorobenzene, (e) hexafluorobenzene, (f) 1,4-dibromo-2,5-difluorobenzene.

A molecule has a permanent dipole moment if any of the symmetry species of the translations T_x and/or T_y and/or T_z is totally symmetric. Using the appropriate character table apply this principle to each of the molecules and indicate the direction of any non-zero dipole moment.

Bibliography

Cotton, F. A. (1971). *Chemical Applications of Group Theory*, Wiley, New York.
Herzberg, G. (1945). *Infrared and Raman Spectra*, Van Nostrand, New York.
Hollas, J. M. (1972). *Symmetry in Molecules*, Chapman and Hall, London.
Jaffé, H. H., and Orchin, M. (1977). *Symmetry in Chemistry*, Krieger, New York.
Mulliken, R. S. (1955). *J. Chem. Phys.*, **23**, 1997.
Schonland, D. (1965). *Molecular Symmetry*, Van Nostrand, London.

CHAPTER 5 —————————————————————

Rotational Spectroscopy

5.1 Linear, Symmetric Rotor, Spherical Rotor, and Asymmetric Rotor Molecules

For the purposes of studying the rotational spectra of molecules it is essential to classify them according to their principal moments of inertia.

The moment of inertia I of any molecule about any axis through the centre of gravity is given by

$$I = \sum_i m_i r_i^2 \tag{5.1}$$

where m_i and r_i are the mass and distance of atom i from the axis. There is one of these axes, conventionally labelled the c axis, about which the moment of inertia has its maximum value and a second axis, labelled the a axis, about which the moment of inertia has its minimum value. It can be shown that the a and c axes must be mutually perpendicular. These, together with a third axis, the b axis which is perpendicular to the other two, are called the principal axes of inertia and the corresponding moments of inertia I_a, I_b, and I_c are the principal moments of inertia. In general, according to convention,

$$I_c \geqq I_b \geqq I_a \tag{5.2}$$

For a linear molecule, such as HCN in Figure 5.1(a),

$$I_c = I_b > I_a = 0 \tag{5.3}$$

where the b and c axes may be in any direction perpendicular to the internuclear a axis. Considering the nuclei as point masses on the a axis, it is clear that I_a must be zero, since all the r_i in equation (5.1) are zero.

For a symmetric rotor, or symmetric top as it is sometimes called, two of the principal moments of inertia are equal and the third is non-zero. If

$$I_c = I_b > I_a \tag{5.4}$$

the molecule is a prolate symmetric rotor. Figure 5.1(b) shows the example of

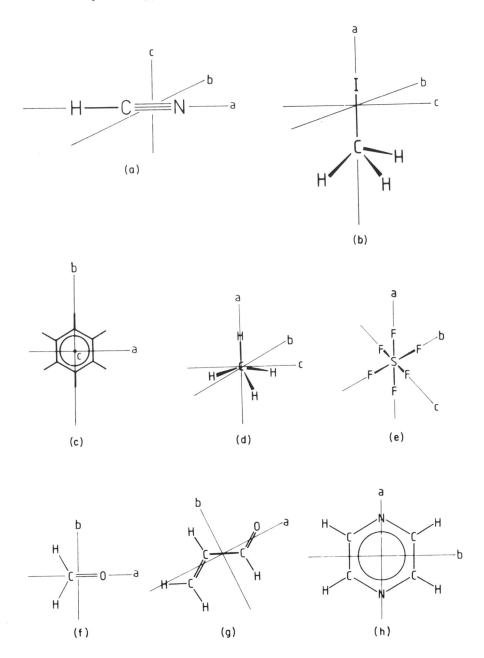

Figure 5.1 Principal inertial axes of (a) HCN, (b) methyl iodide, (c) benzene, (d) methane, (e) sulphur hexafluoride, (f) formaldehyde, (g) *s-trans*-acrolein, and (h) pyrazine.

methyl iodide. Since the heavy iodine nucleus makes no contribution to I_a, which is therefore relatively small, the molecule is a prolate symmetric rotor. The benzene molecule, shown in Figure 5.1(c), is an oblate symmetric rotor for which

$$I_c > I_b = I_a \tag{5.5}$$

As for I_c and I_b in methyl iodide, it is not immediately obvious that I_b and I_a in benzene are equal, but simple trigonometry will prove this.

A symmetric rotor must have either a C_n axis with $n > 2$ (see Section 4.1.1) or an S_4 axis (see Section 4.1.4). Methyl iodide has a C_3 axis and benzene a C_6 axis and, therefore, are symmetric rotors while allene, shown in Figure 4.3(d), is also a symmetric rotor since it has an S_4 axis which is the a axis: allene is a prolate symmetric rotor.

A spherical rotor has all three principal moments of inertia equal

$$I_c = I_b = I_a \tag{5.6}$$

as is the case for methane and sulphur hexafluoride, shown in Figure 5.1(d) and (e). In fact all molecules belonging to either the T_d or O_h point group (see Section 4.2.8 and 4.2.9) are spherical rotors.

An asymmetric rotor has all principal moments of inertia unequal

$$I_c \neq I_b \neq I_a \tag{5.7}$$

An example of an asymmetric rotor, a category which includes the majority of molecules, is formaldehyde in Figure 5.1(f). However, for many asymmetric rotors, either

$$I_c \simeq I_b > I_a \tag{5.8}$$

and the molecule is known as a prolate near-symmetric rotor, or

$$I_c > I_b \simeq I_a \tag{5.9}$$

when it is known as an oblate near-symmetric rotor. An example of the former is s-trans-acrolein (Figure 5.1(g)) and of the latter is pyrazine (Figure 5.1(h)).

5.2 Rotational Infrared, Millimetre Wave, and Microwave Spectra

5.2.1 Diatomic and Linear Polyatomic Molecules

5.2.1.1 Transition frequencies or wavenumbers

In Section 1.3.5 the expression

$$E_r = \frac{h^2}{8\pi^2 I} J(J + 1) \tag{5.10}$$

for the rotational energy levels E_r of a diatomic molecule, in the rigid rotor approximation, was introduced. In this equation, I is the moment of inertia ($=\mu r^2$ where the reduced mass $\mu = m_1 m_2/(m_1 + m_2)$) and the rotational quantum number $J = 0, 1, 2, \ldots$. The same expression applies also to any linear polyatomic molecule but, because I is likely to be larger than for a diatomic molecule, the energy levels of Figure 1.12 tend to be more closely spaced.

In practice, what is measured experimentally is not energy but frequency, in the millimetre wave and microwave regions, or wavenumber, in the far-infrared. So we convert the energy levels of equation (5.10) to what are known as term values $F(J)$ having dimensions of either frequency, by dividing by h, or wavenumber, by dividing by hc, giving

$$F(J) = \frac{E_r}{h} = \frac{h}{8\pi^2 I} J(J + 1) = BJ(J + 1) \tag{5.11}$$

or

$$F(J) = \frac{E_r}{hc} = \frac{h}{8\pi^2 cI} J(J + 1) = BJ(J + 1) \tag{5.12}$$

The use of the symbols $F(J)$ and B for quantities which may have dimensions of frequency or wavenumber is unfortunate, but the symbolism is used so commonly that there seems little prospect of change. In equations (5.11) and (5.12) the quantity B is known as the rotational constant. Its determination by spectroscopic means results in determination of internuclear distances and represents a very powerful structural technique.

Figure 5.2 shows a set of rotational energy levels, or, strictly, term values, for the CO molecule.

The transition intensity is proportional to the square of the transition moment which is given by

$$\boldsymbol{R}_r = \int \psi_r'^* \boldsymbol{\mu} \psi_r'' \, d\tau \tag{5.13}$$

analogous to equation (2.13). The rotational selection rules constitute the conditions for which the intensity, and therefore $\boldsymbol{R}_r$, is non-zero and are:

1. The molecule must have a permanent dipole moment ($\mu \neq 0$).
2. $\Delta J = \pm 1$.
3. $\Delta M_J = 0, \pm 1$, a rule which is important only if the molecule is in an electric or magnetic field (see equation 1.61).

Rule 1 shows that transitions are allowed in heteronuclear diatomic molecules such as CO, NO, HF, and even $^1H^2H$ (for which $\mu = 5.9 \times 10^{-4} D^\dagger$ com-

———————————

† 1 D = 3.335 64 $\times$ 10^{-30} C m.

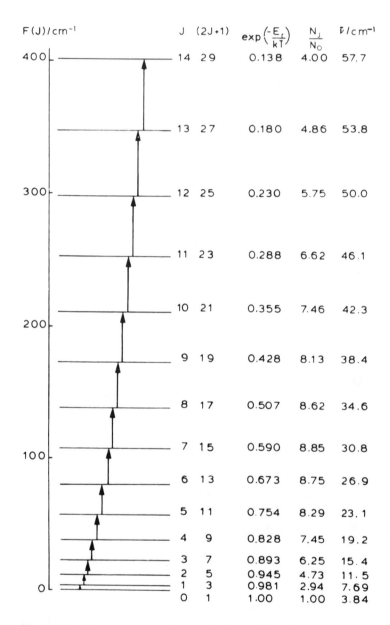

Figure 5.2 Rotational term values, relative populations and transition wavenumbers for CO.

pared to, say, HF for which $\mu = 1.82$ D), but not H_2, Cl_2, and N_2. Similarly 'unsymmetrical' linear polyatomic molecules (more specifically those having no centre of inversion—Section 4.1.3), such as $O{=}C{=}S$, $H{-}C{\equiv}N$, and even $^1H{-}C{\equiv}C{-}{^2}H$ ($\mu \simeq 0.012$ D), have allowed rotational transitions while the 'symmetrical' ones (those having a centre of inversion), such as $S{=}C{=}S$ and $H{-}C{\equiv}C{-}H$, do not.

So far as rule 2 is concerned, since ΔJ is conventionally taken to refer to $J'-J''$ where J' is the quantum number of the upper state and J'' that of the lower state of the transition, $\Delta J = -1$ has no physical meaning (although it emerges from the quantum mechanics). It is commonly, but incorrectly, thought that $\Delta J = +1$ and -1 refer to absorption and emission respectively: in fact $\Delta J = +1$ applies to *both*. Transition wavenumbers or frequencies are given by

$$\tilde{v} \text{ (or } v) = F(J + 1) - F(J) = 2B(J + 1) \quad (5.14)$$

where J is used conventionally instead of J''. The allowed transitions are shown in Figure 5.2 and are spaced, according to equation (5.14), 2B apart. The $J = 1{-}0$ transition (a transition is written conventionally as $J'-J''$) occurs at $2B$. Whether the transitions fall in the microwave, millimetre wave, or far-infrared region depends on the values of B and J. The spectrum of CO spans the millimetre wave and far-infrared regions. Part of the far-infrared spectrum, from 15 to 40 cm^{-1} with $J'' = 3$ to 9, is shown in Figure 5.3, and Table 5.1 lists the

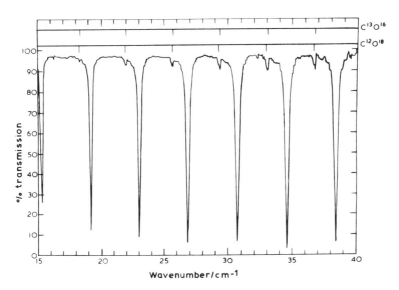

Figure 5.3 Far-infrared spectrum of CO showing transitions with $J'' = 3$ to 9. (Reproduced, with permission, from Fleming, J. W., and Chamberlain, J., *Infrared Phys.*, **14**, 277, 1974. Copyright (1974) Pergamon Press.)

Table 5.1 Frequencies and wavenumbers of rotational transitions of CO observed in the millimetre wave region.

$\tilde{v}/\mathrm{cm}^{-1}$	J''	J'	v/GHz	$\Delta v_{J''}^{J''+1}/\mathrm{GHz}$
3.845 033 19	0	1	115.271 195	115.271 195
7.689 919 07	1	2	230.537 974	115.266 779
11.534 509 6	2	3	345.795 900	115.257 926
15.378 662	3	4	461.040 68	115.244 78
19.222 223	4	5	576.267 75	115.227 07
23.065 043	5	6	691.472 60	115.204 85

characteristically highly accurate frequencies and wavenumbers of the transitions with $J'' = 0$ to 6 observed in the millimetre wave region.

It is apparent from Figure 5.3 and the last column of Table 5.1 that the spacing of adjacent transitions is fairly constant and, as equation (5.14) shows, the spacing is equal to $2B$.

Most linear polyatomic molecules have smaller B values and therefore more transitions tend to occur in the millimetre wave or microwave regions. Figure 5.4 shows a part of the microwave spectrum of cyanodiacetylene $(H-C\equiv C-C\equiv C-C\equiv N)$ which has such a small B value (1331.331 MHz) that six transitions with $J'' = 9$ to 14 lie in the 26.5 to 40.0 GHz region.

From Figure 5.3 it is apparent that transition intensities are not equal and, from Table 5.1, that the transition spacings show a slight decrease as J'' increases. We will now consider the reasons for these observations.

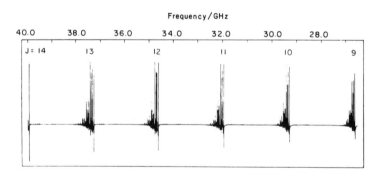

Figure 5.4 Part of the microwave spectrum of cyanodiacetylene. (The many 'satellite' transitions in each group are due to the molecule being in not only the zero-point vibrational state but also a multitude of excited vibrational states.) (Reproduced, with permission, from Alexander, A. J., Kroto, H. W., and Walton, D. R. M., *J. Mol. Spectrosc.*, **62**, 175, 1967.)

5.2.1.2 Intensities

Apart from depending on the numerical value of the square of the transition moment of equation (5.13) which varies relatively little with J, intensities depend on the population of the lower state of a transition. The population N_J of the Jth level relative to N_0 is obtained from Boltzmann's distribution law. Equation (2.11) gives

$$\frac{N_J}{N_0} = (2J + 1) \exp\left(-\frac{E_r}{kT}\right) \tag{5.15}$$

where $(2J + 1)$ is the degeneracy of the Jth level. This degeneracy arises from the fact that, in the absence of an electric or magnetic field, $(2J + 1)$ levels, resulting from the number of values that M_J (see equation 1.61) can take, are all of the same energy: in other words they are degenerate. Values of $(2J + 1)$, $\exp(-E_r/kT)$, and N_J/N_0 are given for CO in Figure 5.2 and illustrate the point that there are two opposing factors in N_J/N_0. The $(2J + 1)$ factor increases with J while the $\exp(-E_r/kT)$ factor decreases rapidly so that N_J/N_0 increases at low J until, at higher J, the exponential factor wins and N_J/N_0 approaches zero at high J. The population therefore shows a maximum at a value of $J = J_{max}$ corresponding to

$$\frac{d(N_J/N_0)}{dJ} = 0 \tag{5.16}$$

which gives

$$J_{max} = \left(\frac{kT}{2hB}\right)^{1/2} - \frac{1}{2} \tag{5.17}$$

for B having dimensions of frequency. Figure 5.2 shows that $J_{max} = 7$ for CO. In fact, Figure 5.3 shows that the observed maximum intensity is at about $J'' = 8$ because there are other, less important, factors that we have neglected.

5.2.1.3 Centrifugal distortion

The small decrease in transition spacings with increasing J, as shown for example in Table 5.1, compared to the constant spacings we expected is due to our original approximation in Section 1.3.5 of treating the molecule as a rigid rotor being not quite valid. In fact the bond is not rigid but is more accurately represented by a spring connecting the nuclei, as was apparent when we considered vibrational motion in Section 1.3.6. So it is not surprising that, as the speed of rotation increases, i.e. as J increases, the nuclei tend to be thrown outwards by centrifugal forces. The spring stretches, r increases, and, therefore,

B decreases. Originally this J-dependent decrease of B was included by changing the term values of equations (5.11) and (5.12) to

$$F(J) = B[1 - uJ(J + 1)]J(J + 1) \tag{5.18}$$

where u is an additional constant, but it is usually written in the form

$$F(J) = BJ(J + 1) - DJ^2(J + 1)^2 \tag{5.19}$$

where D is the centrifugal distortion constant and is always positive for diatomic molecules. The transition wavenumbers or frequencies are now modified from equation (5.14) to

$$\tilde{\nu} \,(\text{or } \nu) = F(J + 1) - F(J) = 2B(J + 1) - 4D(J + 1)^3 \tag{5.20}$$

The centrifugal distortion constant depends on the stiffness of the bond and it is not surprising that it can be related to the vibration wavenumber ω, in the harmonic approximation (see Section 1.3.6), by

$$D = \frac{4B^3}{\omega^2} \tag{5.21}$$

5.2.1.4 Diatomic molecules in excited vibrational states

There is a stack of rotational levels, with term values like those given by equation (5.19), associated with not only the zero-point vibrational level but all the other vibrational levels shown, for example, in Figure 1.13. However, the Boltzmann equation (equation 2.11), together with the vibrational energy level expression (equation 1.69), gives the ratio of the population N_v of the vth vibrational level to N_0, that of the zero-point level, as

$$\frac{N_v}{N_0} = \exp\left(- \frac{hcv\omega}{kT}\right) \tag{5.22}$$

where ω is the vibration wavenumber and v the vibrational quantum number. Since this ratio is, for example, only 0.10 for $v = 1$ and $\omega = 470 \text{ cm}^{-1}$ we can see that rotational transitions in excited vibrational states are generally very weak unless either the molecule is particularly heavy, leading to a relatively small value of ω, or the temperature is high; however the negative exponential nature of equation (5.22) means that increasing the temperature has only quite a small effect on populations.

The rotational constants B and D are both slightly vibrationally dependent so that the term values of equation (5.19) should be written

$$F_v(J) = B_v J(J + 1) - D_v J^2(J + 1)^2 \tag{5.23}$$

and equation (5.20) becomes

$$\tilde{\nu} \,(\text{or } \nu) = 2B_v(J + 1) - 4D_v(J + 1)^3 \tag{5.24}$$

The vibrational dependence of B is given, to a good approximation, by

$$B_v = B_e - \alpha(v + \tfrac{1}{2}) \tag{5.25}$$

where B_e refers to the hypothetical equilibrium state of the molecule at the bottom of the potential energy curve of Figure 1.13 and α is a vibration–rotation interaction constant. In order to obtain B_e, and hence the equilibrium bond length r_e, B_v must be obtained in at least two vibrational states. If there is insufficient population of the $v = 1$ level to obtain B_1 then rotational spectroscopy can give only B_0.

The vibrational dependence of the centrifugal distortion constant D_v is too small to concern us further.

5.2.2 Symmetric Rotor Molecules

In a diatomic or linear polyatomic molecule the rotational angular momentum vector $\textbf{\textit{P}}$ lies along the axis of rotation, as shown in Figure 5.5(a), which is analogous to Figure 1.5 for electronic orbital angular momentum. In a prolate symmetric rotor, such as methyl iodide shown in Figure 5.5(b), $\textbf{\textit{P}}$ need not be perpendicular to the a axis. In general, it takes up any direction in space and the molecule rotates around $\textbf{\textit{P}}$. The component of $\textbf{\textit{P}}$ along the a axis is P_a which can take only the values $K\hbar$, where K is a second rotational quantum number. The rotational term values are given by

$$F(J, K) = BJ(J + 1) + (A - B)K^2 \tag{5.26}$$

neglecting centrifugal distortion and the vibrational dependence of the rotational constants A and B. These are related to I_a and I_b by

$$A = \frac{h}{8\pi^2 I_a}; \qquad B = \frac{h}{8\pi^2 I_b} \tag{5.27}$$

when they have dimensions of frequency. The quantum number K may take values $K = 0, 1, 2, \ldots, J$. The fact that K cannot be greater than J follows from

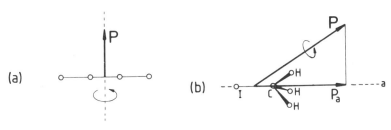

Figure 5.5 The rotational angular momentum vector $\textbf{\textit{P}}$ for (a) a linear molecule and (b) the prolate symmetric rotor CH_3I where P_a is the component along the a axis.

the fact that P_a cannot be greater than the magnitude of P. All levels with $K > 0$ are doubly degenerate which can be thought of, classically, as being due to the clockwise or anticlockwise rotation about the a axis resulting in the same angular momentum. For $K = 0$ there is no angular momentum about the a axis and, therefore, no K-degeneracy.

For an oblate symmetric rotor, such as NH_3, the rotational term values are given by

$$F(J, K) = BJ(J + 1) + (C - B)K^2 \tag{5.28}$$

analogous to equation (5.26), where

$$C = \frac{h}{8\pi^2 I_c} \tag{5.29}$$

when it has dimensions of frequency.

The rotational energy levels for a prolate and an oblate symmetric rotor are shown schematically in Figure 5.6. Although these present a much more complex picture than those for a linear molecule the fact that the selection rules

$$\Delta J = \pm 1; \qquad \Delta K = 0 \tag{5.30}$$

include $\Delta K = 0$ results in the expression for the transition frequencies or wavenumbers

$$v \text{ (or } \tilde{v}) = F(J + 1, K) - F(J, K) = 2B(J + 1) \tag{5.31}$$

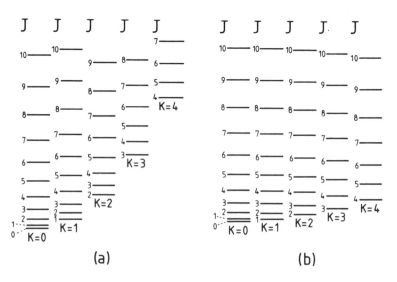

Figure 5.6 Rotational energy levels for (a) a prolate and (b) an oblate symmetric rotor.

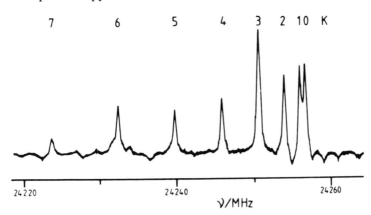

Figure 5.7 Eight components, with $K = 0$ to 7 and separated by centrifugal distortion, of the $J = 8 - 7$ microwave transition of SiH$_3$NCS.

This is the same as equation (5.14) for a diatomic or linear polyatomic molecule and, again, the transitions show an equal spacing of $2B$. The requirement that the molecule must have a permanent dipole moment applies to symmetric rotors also.

When the effects of centrifugal distortion are included the term values of a prolate symmetric rotor are given by

$$F(J, K) = BJ(J + 1) + (A - B)K^2 - D_J J^2 (J + 1)^2 - D_{JK} J(J + 1)K^2 - D_K K^4$$

$$(5.32)$$

where there are now three centrifugal distortion constants D_J, D_{JK}, and D_K. There is an analogous equation for an oblate symmetric rotor and, for both types, the transition frequencies or wavenumbers are given by

$$\nu \text{ (or } \tilde{\nu}) = F(J + 1, K) - F(J, K) = 2(B_v - D_{JK} K^2)(J + 1) - 4D_J(J + 1)^3 \quad (5.33)$$

The term $-2D_{JK}K^2(J + 1)$ has the effect of separating the $(J + 1)$ components of each $(J + 1) \leftarrow J$ transition with different values of K. This is illustrated in Figure 5.7 which shows the eight components of the $J = 8 - 7$ transition of silyl isothiocyanate (H$_3$Si—N=C=S) which has a linear SiNCS chain and is a prolate symmetric rotor.

5.2.3 Stark Effect in Diatomic, Linear, and Symmetric Rotor Molecules

As we saw in equation (1.61), space quantization of rotational angular momentum of a diatomic or linear polyatomic molecule is expressed by

$$(P_J)_z = M_J \hbar \quad (5.34)$$

where $M_J = J, J - 1, \ldots, -J$. Under normal conditions these $(2J + 1)$ components of each J level remain degenerate but, in the presence of an electric field $\mathscr{E}$, the degeneracy is partly removed: each level is split into $(J + 1)$ components according to the value of $|M_J| = 0, 1, 2, \ldots, J$. This splitting in an electric field is known as the Stark effect. The energy levels E_r of equation (5.10) are modified to $E_r + E_\mathscr{E}$, where

$$E_\mathscr{E} = \frac{\mu^2 \mathscr{E}^2 [J(J + 1) - 3M_J^2]}{2hBJ(J + 1)(2J - 1)(2J + 3)} \tag{5.35}$$

In this rather formidable equation there are two points to notice:

1. It involves M_J^2, and therefore the energy is independent of the sign of M_J.
2. It involves the molecular dipole moment μ.

Because of point 2, rotational microwave and millimetre wave spectroscopy are powerful techniques for determining dipole moments. However, the direction of the dipole moment cannot be determined. In the case of O=C=S, for which $\mu = 0.715\,21 \pm 0.000\,20$ D $[(2.3857 \pm 0.0007) \times 10^{-30}\,\text{C m}]$, a simple electronegativity argument leads to the correct conclusion—that the oxygen end of the molecule is the negative end of the dipole. However, in CO, the value of $\mu = 0.112$ D $(3.74 \times 10^{-31}\,\text{C m})$ is so small that only accurate electronic structure calculations can be relied upon to conclude correctly that the carbon end is the negative one.

For a symmetric rotor the modification $E_\mathscr{E}$ to the rotational energy levels in an electric field $\mathscr{E}$ is given by

$$E_\mathscr{E} = -\frac{\mu \mathscr{E} K M_J}{J(J + 1)}\tag{5.36}$$

$$+ \frac{\mu^2 \mathscr{E}^2}{2hB} \left\{ \frac{(J^2 - K^2)(J^2 - M_J^2)}{J^3(2J - 1)(2J + 1)} - \frac{[(J + 1)^2 - K^2][(J + 1)^2 - M_J^2]}{(J + 1)^3(2J + 1)(2J + 3)} \right\}$$

an even more formidable equation but, again, involving M_J^2 and the dipole moment μ. Using the Stark effect, the dipole moment of, for example, CH_3F has been found to be 1.857 ± 0.001 D $[(6.194 \pm 0.004) \times 10^{-30}\,\text{C m}]$.

5.2.4 Asymmetric Rotor Molecules

Although these molecules form much the largest group we shall take up the smallest space in considering their rotational spectra. The reason for this is that there are no closed formulae for their rotational term values. Instead these term values can be determined accurately only by a matrix diagonalization for each value of J, which remains a good quantum number. The selection rule $\Delta J = 0, \pm 1$ applies and the molecule must have a permanent dipole moment.

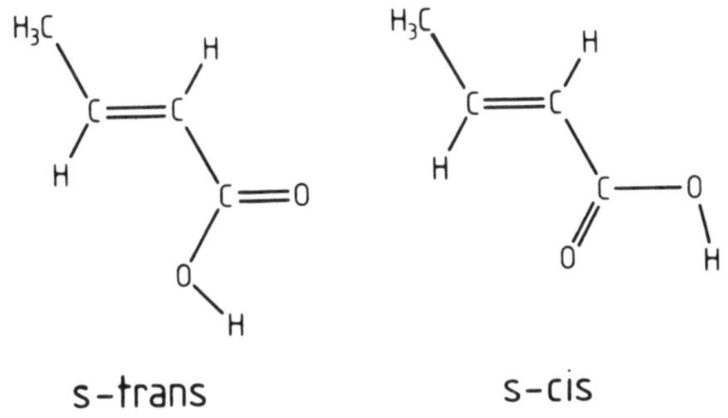

Figure 5.8 The *s-trans* and *s-cis* isomers of crotonic acid.

At a simple level the rotational transitions of near-symmetric rotors (see equations 5.8 and 5.9) are easier to understand. For a prolate or oblate near-symmetric rotor the rotational term values are given, approximately, by

$$F(J, K) \simeq \bar{B}J(J + 1) + (A - \bar{B})K^2 \tag{5.37}$$

and

$$F(J, K) \simeq \bar{B}J(J + 1) + (C - \bar{B})K^2 \tag{5.38}$$

respectively, where $\bar{B} = \frac{1}{2}(B + C)$ for a prolate rotor and $\frac{1}{2}(A + B)$ for an oblate rotor. Centrifugal distortion has been neglected. Because the molecules concerned are only approximately symmetric rotors K is not strictly a good quantum number *i.e.* does not take integral values.

Examples of prolate near-symmetric rotors are the *s-trans* and *s-cis* isomers of crotonic acid shown in Figure 5.8, the *a* axis straddling a chain of the heavier atoms in both species. The rotational term values for both isomers are given approximately by equation (5.37) but, because A and $\bar{B}$ are different for each of them, their rotational transitions are not quite coincident. Figure 5.9 shows a part of a low resolution microwave spectrum of crotonic acid in which the weaker series of lines is due to the less abundant *s-cis* isomer and the stronger series is due to the more abundant *s-trans* isomer.

Dipole moments of asymmetric rotors or, strictly, their components along the various inertial axes, may be determined using the Stark effect.

5.2.5 Spherical Rotor Molecules

We tend to think of a spherical rotor molecule, such as methane (see Figure 4.12a), as having no permanent dipole moment and, therefore, no infrared, millimetre wave, or microwave rotational spectrum. However, rotation about any of the C_3 axes, i.e. any of the four axes in methane containing a C–H bond, results in a centrifugal distortion in which the other three hydrogen atoms are

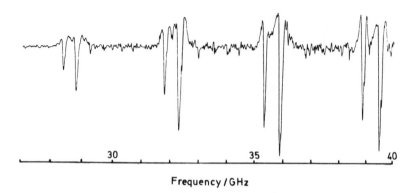

Frequency / GHz

Figure 5.9 Part of the microwave spectrum of crotonic acid. (Reproduced, with permission, from Scharpen, L. H., and Laurie, V. W., *Analyt. Chem.*, **44**, 378R, 1972. Copyright (1972) American Chemical Society.)

thrown outwards slightly from the axis. This converts the molecule into a symmetric rotor and gives it a small dipole moment resulting in a very weak rotational spectrum.

Part of the far-infrared rotational spectrum of silane (SiH_4) is shown in Figure 5.10. It was obtained using a Michelson interferometer (see Section 3.3.3.2), an absorbing path of 10.6 m, and a pressure of 4.03 atm (4.08×10^5 Pa), these conditions indicating how very weak the spectrum is. The dipole moment has been estimated from the intensities of the transitions to be 8.3×10^{-6} D (2.7×10^{-35} C m).

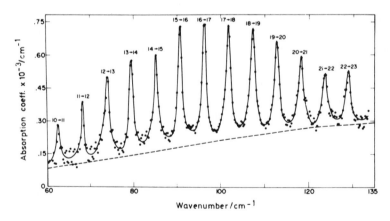

Figure 5.10 Part of the far-infrared spectrum of silane. (Reproduced from Rosenberg, A., and Ozier, I., *Can. J. Phys.*, **52**, 575, 1974.)

Neglecting centrifugal distortion the rotational term values for a spherical rotor are given by

$$F(J) = BJ(J + 1) \tag{5.39}$$

This is an identical expression to that for a diatomic or linear polyatomic molecule (equation 5.11 and 5.12) and, as the rotational selection rule is the same, namely $\Delta J = \pm 1$, the transition wavenumbers or frequencies are given by

$$v \text{ (or } \tilde{v}) = F(J + 1) - F(J) = 2B(J + 1) \tag{5.40}$$

and adjacent transitions are separated by $2B$.

All regular tetrahedral molecules, which belong to the T_d point group (Section 4.2.8), may show such a rotational spectrum. However, those spherical rotors which are regular octahedral molecules and which belong to the O_h point group (Section 4.2.9) do not show any such spectrum. The reason for this is that when, for example, SF_6 (see Figure 5.1e) rotates about a C_4 axis (any of the F–S–F axes) no dipole moment is produced when the other four fluorine atoms are thrown outwards.

5.2.6 Interstellar Molecules Detected by their Radiofrequency, Microwave, or Millimetre Wave Spectra

Radiotelescopes are used to scan the universe for radiation in the radio-frequency region of the spectrum (see Figure 3.1). As illustrated in Figure 5.11 such a telescope consists of a parabolic reflecting dish which focuses all parallel rays reaching it onto a radiofrequency detector supported at the focus of the paraboloid. The surface of such a dish must be constructed accurately but only sufficiently so that the irregularities are small compared to the wavelength of the radiation which is of the order of 0.5 m.

One of the uses of such a telescope is in detecting atomic hydrogen which is found in large quantities, but with varying density, throughout the universe and which has an emission line with a wavelength of 21 cm. The emission occurs

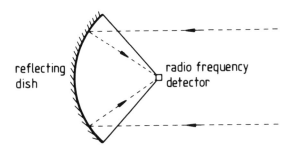

Figure 5.11 A radiotelescope.

between closely spaced sub levels into which the $n = 1$ level (see Figure 7.8) is split. In 1963 the first molecule, OH, was detected with such a telescope. A transition with a wavelength of about 18 cm was observed in absorption and is electronic in character: it is a transition between components of Λ-doublets which are due to the ground electronic state being $^2\Pi$ and split into two components (see Section 7.2.6.2).

Both emission and absorption processes rely on the background radiation which is present throughout the universe and which has a wavelength distribution characteristic of a black body and a temperature of about 2.7 K. This radiation is a consequence of the big bang with which the universe supposedly started its life.

Since 1963 spectra of many molecules have been detected, mainly in emission but some in absorption. Telescopes have been constructed with more accurately engineered paraboloids in order to extend observations into the millimetre wave and microwave regions.

The regions of space where molecules have been detected are the nebulae which are found not only in our own galaxy but in other galaxies as well. In our galaxy the nebulae are found in the Milky Way which appears as a hazy band of light due to its containing millions of stars. Associated with the luminous clouds comprising the nebulae are dark clouds of interstellar dust and gas. The presence of the dust particles is indicated by the fact that visible starlight passing through such a cloud is reddened due to preferential scattering, which is proportional to λ^{-4}, of the blue light by the dust particles. The nature of the dust particles is not known but they are about 0.2 μm in diameter. Since new stars are formed by gravitational collapse in the region of the nebulae, which must contain the raw material from which the new stars are formed, the detection of molecules in these regions is of the greatest importance. Molecules have been detected in several nebulae but the large cloud known as Sagittarius B2, which is close to the centre of our galaxy, has proved a particularly fruitful source.

The first polyatomic molecule was detected in 1968 using a telescope having a dish 6.3 m in diameter at Hat Creek, California (USA), designed to operate in the millimetre wave region. Emission lines were found in the 1.25 cm wavelength region due to NH_3. The transitions are not rotational but are between the very closely spaced $v_2 = 0$ and $v_2 = 1$ levels of the inversion vibration v_2 (see Section 6.2.4.4a).

Table 5.2 lists some of the molecules which have been detected. It is interesting to note that some of them, such as the linear triatomics C_2H, HCO^+ and N_2H^+, were found in the interstellar medium before they were searched for and found in the laboratory. In all molecules, except OH and NH_3, the transitions observed are rotational in nature.

Identification of a molecule known in the laboratory is usually unambiguous because of the uniqueness of the highly precise transition frequencies. However

Table 5.2 Interstellar molecules detected by their radiofrequency or millimetre wave spectra.

Diatomics	OH, CO, CN, CS, SiO, SO, SiS, NO, NS, CH, CH$^+$
Triatomics	H$_2$O, HCN, HNC, OCS, H$_2$S, N$_2$H$^+$, SO$_2$, HNO, C$_2$H, HCO, HCO$^+$, HCS$^+$, H$_2$D$^+$
Tetratomics	NH$_3$, H$_2$CO, HNCO, H$_2$CS, HNCS, N≡C—C≡C, H$_3$O$^+$, C$_3$H (linear), C$_3$H (cyclic)
5-Atomics	N≡C—C≡C—H, HCOOH, CH$_2$=NH, H—C≡C—C≡C, NH$_2$CN
6-Atomics	CH$_3$OH, CH$_3$CN, NH$_2$CHO, CH$_3$SH
7-Atomics	CH$_3$—C≡C—H, CH$_3$CHO, CH$_3$NH$_2$, CH$_2$=CHCN, N≡C—C≡C—C≡C—H
8-Atomics	HCOOCH$_3$, CH$_3$—C≡C—C≡N
9-Atomics	CH$_3$OCH$_3$, CH$_3$CH$_2$OH, N≡C—C≡C—C≡C—C≡C—H
11-Atomics	N≡C—C≡C—C≡C—C≡C—C≡C—H

before frequencies detected in the interstellar medium can be compared with laboratory frequencies they must be corrected for the Doppler effect (see Section 2.3.2) due to the motion of the clouds. In Sagittarius B2 the molecules are found to be travelling fairly uniformly with a velocity of 60 km s^{-1} relative to a local standard of rest, which is taken to be certain stars close to the sun. In other clouds, there is a wider range of molecular velocities.

Figure 5.12 shows the $J = 1 - 0$ transition of the linear molecule cyanodiacetylene (H—C≡C—C≡C—C≡N) observed in emission in Sagittarius B2 (Figure 5.4 shows part of the absorption spectrum in the laboratory). The three hyperfine components into which the transition is split are due to interaction between the rotational angular momentum and the nuclear spin of the ^{14}N

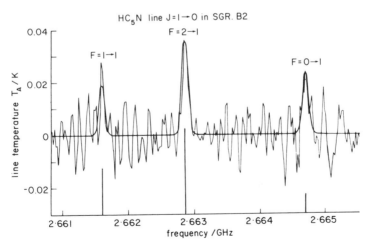

Figure 5.12 The $J = 1 - 0$ transition of cyanodiacetylene observed in emission in Sagittarius B2. (Reproduced, with permission, from Broton, N. W., MacLeod, J. M., Oka, T., Avery, L. W., Brooks, J. W., McGee, R. X., and Newton, L. M., *Astrophys. J.*, **209**, L143, 1976, published by the University of Chicago Press; © 1976 The American Astronomical Society.)

nucleus for which $I = 1$ (see Table 1.3). The vertical scale is a measure of the change of the temperature of the antenna due to the received signal.

Table 5.2 shows that quite large molecules, of which the cyanopolyacetylenes form a remarkable group, have been detected. The presence of such sizeable molecules in the interstellar medium came as a considerable surprise. Previously it was supposed that the ultraviolet radiation present throughout all galaxies would photodecompose most of the molecules, and particularly the larger ones. It seems likely that the dust particles play an important part not only in the formation of the molecules but also in preventing their decomposition.

In considering the molecules in Table 5.2 it should be remembered that the method of detection filters out any molecules with zero dipole moment. There is known to be large quantities of H_2 and, no doubt, there are such molecules as C_2, N_2, O_2, H—C≡C—H, and polyacetylenes to be found in the clouds but these escape detection by radiofrequency, millimetre wave, or microwave spectroscopy.

It is also apparent from Table 5.2 that, to date, H_2D^+ and C_3H are the only cyclic molecules which have been found although, from the atoms commonly present, such molecules as pyrrole or pyridine might have been anticipated.

5.3 Rotational Raman Spectroscopy

When electromagnetic radiation falls on an atomic or molecular sample it may be absorbed if the energy of the radiation corresponds to the separation of two energy levels of the atoms or molecules. If it does not, the radiation will be either transmitted or scattered by the sample. Of the scattered radiation most is of unchanged wavelength λ and is the Rayleigh scattering. It was Lord Rayleigh in 1871 who showed that the intensity I_s of scattered light is related to λ by

$$I_s \propto \lambda^{-4} \tag{5.41}$$

For this reason blue radiation from the sun is scattered preferentially by particles in the atmosphere and the result is that a cloudless sky appears blue.

It was predicted in 1923 by Smekal and shown experimentally in 1928 by Raman and Krishnan that a small amount of radiation scattered by a gas, liquid, or solid is of increased or decreased wavelength (or wavenumber). This is called the Raman effect and the scattered radiation with decreased or increased wavenumber is referred to as Stokes or anti-Stokes Raman scattering respectively.

5.3.1 Experimental Methods

The incident radiation should be highly monochromatic for the Raman effect to be observed clearly and, because Raman scattering is so weak, the incident radiation should be very intense. This is particularly important when, as in rotational Raman spectroscopy, the sample is in the gas phase.

In outline the method used is to pass the monochromatic radiation through the gaseous sample and disperse and detect the scattered radiation. Usually this radiation is collected in directions normal to the incident radiation in order to avoid this incident radiation passing to the detector.

Until the advent of lasers, the most intense monochromatic sources available were atomic emission sources from which an intense, discrete line in the visible or near-ultraviolet region was isolated by optical filtering if necessary. The most-often-used source of this kind was the mercury discharge lamp operating at the vapour pressure of mercury. Three of the most intense lines are at 253.7 nm (near-ultraviolet), 404.7 nm, and 435.7 nm (both in the visible region). Although the line width is typically small the narrowest has a width of about 0.2 cm^{-1}, which places a limit on the resolution which can be achieved.

Lasers (see Chapter 9) are sources of intense, monochromatic radiation which are ideal for Raman spectroscopy and have entirely replaced atomic emission sources. They are more convenient to use, have higher intensity, and are more highly monochromatic: e.g. the line width at half-intensity of 632.8 nm (red) radiation from a helium–neon laser can be less than 0.05 cm^{-1}.

Multiple-reflection mirrors for efficiently collecting the weak Raman scattering of gases were used commonly when excitation was by mercury discharge lamps. With a laser source two parallel mirrors, M_1 and M_2 in Figure 5.13, may be used for passing the highly parallel beam many times across the gas cell. A system of four concave mirrors, M_3 to M_6, efficiently collects the Raman scattered radiation which is then dispersed and detected by a spectrograph (photographic detection) or spectrometer (photoelectric detection).

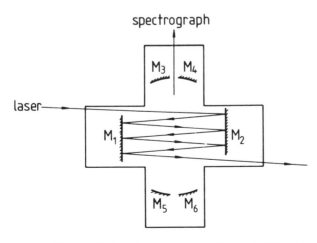

Figure 5.13 Multiple-reflection gas cell for Raman spectroscopy.

Until the mid 1970s only lasers operating in the visible region of the spectrum, such as the helium–neon laser (Section 9.2.5) at 632.8 nm and the argon ion laser (Section 9.2.6) at 514.5 nm, were used as sources of monochromatic radiation for Raman spectroscopy. However, many molecules are coloured and therefore absorb and fluoresce in the visible region. This fluorescence tends to mask the much weaker Raman scattering thereby excluding many molecules from investigation by the Raman effect. Raman spectra are obtained, very often, with the sample in the liquid or solid phase when even a small amount of coloured impurity may produce sufficient fluorescence to interfere with the very weak Raman scattering of the main component of the sample.

Using a laser operating in the infrared overcomes the problem of fluorescence, which normally occurs following the absorption of only visible or ultraviolet radiation (see Section 7.2.5.2). However, with an ordinary spectrometer having a diffraction grating as the dispersing element, the advantage of using an infrared laser is more than counteracted by the fact that the intensity of Raman scattering decreases as the fourth power of the wavelength, as equation (5.41) indicates, making detection extremely difficult.

It was not until the development of Fourier transform infrared (FTIR) spectrometers (see Section 3.3.3.2) that the possibility of using an infrared laser routinely was opened up. The intensity advantage of an infrared interferometer, with which a single spectrum can be obtained very rapidly and then many spectra co-added, coupled with the development of more sensitive Ge and InGaAs semiconductor infrared detectors, more than compensate for the loss of scattering intensity in the infrared region.

The infrared laser which is most often used in this technique of Fourier transform Raman, or FT–Raman, spectroscopy is the Nd–YAG laser (see Section 9.2.3) operating at a wavelength of 1064 nm.

In FT–Raman spectroscopy the radiation emerging from the sample contains not only the Raman scattering but also the extremely intense laser radiation used to produce it. If this were allowed to contribute to the interferogram, before Fourier transformation, the corresponding cosine wave would over-whelm those due to the Raman scattering. To avoid this, a sharp cut-off (interference) filter is inserted after the sample cell to remove 1064 nm (and lower wavelength) radiation.

An FT–Raman spectrometer is often simply an FTIR spectrometer adapted to accommodate the laser source, the filters to remove the laser radiation and a variety of infrared detectors.

5.3.2 Theory of Rotational Raman Scattering

Electronic, vibrational, and rotational transitions may be involved in Raman scattering but, in this chapter, we consider only rotational transitions.

The property of the sample which determines the degree of scattering is the

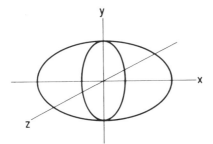

Figure 5.14 The polarizability ellipsoid.

polarizability α. When the incident radiation is in the visible or near-ultraviolet region, as it usually is in Raman spectroscopy, the polarizability is a measure of the degree to which the electrons in the molecule can be displaced relative to the nuclei. In general the polarizability of a molecule is an anisotropic property which means that, at equal distances from the centre of the molecule, α may have different magnitudes when measured in different directions. A surface drawn so that the distance from the origin to a point on the surface has a length $\alpha^{-1/2}$, where α is the polarizability *in that direction*, forms an ellipsoid. This has elliptical cross-sections in the xy and yz planes, as shown in Figure 5.14, and the lengths of the axes in the x, y and z directions are, in general, unequal. Like other anisotropic properties, such as the moment of inertia of a molecule (Section 5.1) and the electrical conductivity of a crystal, polarizability is a tensor property. The tensor α can be expressed in the form of a matrix:

$$\alpha = \begin{pmatrix} \alpha_{xx} & \alpha_{xy} & \alpha_{xz} \\ \alpha_{yx} & \alpha_{yy} & \alpha_{yz} \\ \alpha_{zx} & \alpha_{zy} & \alpha_{zz} \end{pmatrix} \tag{5.42}$$

where the diagonal elements α_{xx}, α_{yy}, and α_{zz} are the values of α along the x, y, and z axes of the molecule. The matrix is symmetrical in the sense that $\alpha_{yx} = \alpha_{xy}$, $\alpha_{zx} = \alpha_{xz}$, and $\alpha_{zy} = \alpha_{yz}$ so that there are, in general, six different components of α, namely α_{xx}, α_{yy}, α_{zz}, α_{xy}, α_{xz}, and α_{yz}. Each of these can be assigned to one of the symmetry species of the point group to which the molecule belongs. These assignments are indicated in the right-hand column of each character table given in the Appendix and will be required when we consider vibrational Raman spectra in Section 6.2.2.2.

When monochromatic radiation falls on a molecular sample in the gas phase, and is not absorbed by it, the oscillating electric field E (see equation 2.1) of the radiation induces in the molecule an electric dipole μ which is related to E by the polarizability

$$\mu = \alpha E \tag{5.43}$$

where μ and E are vector quantities. The magnitude E of the vector can be written

$$E = A \sin 2\pi c \tilde{v} t \tag{5.44}$$

where A is the amplitude and $\tilde{v}$ the wavenumber of the monochromatic radiation. The magnitude of the polarizability varies during rotation and a simple classical treatment will serve to illustrate the rotational Raman effect.

The polarizability ellipsoid rotates with the molecule at a frequency v_{rot}, say, and the radiation 'sees' the polarizability changing at *twice* the frequency of rotation since, as can be seen from Figure 5.14, the ellipsoid appears the same for a rotation by π radians about any of the cartesian axes. The variation of α with rotation is given by

$$\alpha = \alpha_{0,r} + \alpha_{1,r} \sin 2\pi c (2\tilde{v}_{rot}) t \tag{5.45}$$

where $\alpha_{0,r}$ is the average polarizability and $\alpha_{1,r}$ is the amplitude of the change of polarizability during rotation. Substitution of equations (5.45) and (5.44) into equation (5.43) gives the magnitude of the induced dipole moment as

$$\mu = \alpha_{0,r} A \sin 2\pi c \tilde{v} t - \tfrac{1}{2}\alpha_{1,r} A \cos 2\pi c (\tilde{v} + 2\tilde{v}_{rot}) t \tag{5.46}$$
$$+ \tfrac{1}{2}\alpha_{1,r} A \cos 2\pi c (\tilde{v} - 2\tilde{v}_{rot}) t$$

All three terms in this equation represent scattering of the radiation. The first term corresponds to Rayleigh scattering of unchanged wavenumber $\tilde{v}$ and the second and third terms correspond to anti-Stokes and Stokes Raman scattering with wavenumbers of $(\tilde{v} + 2\tilde{v}_{rot})$ and $(\tilde{v} - 2\tilde{v}_{rot})$ respectively.

Although in a classical system $\tilde{v}_{rot}$ can take any value, in a quantum mechanical system it can take only certain values and we shall now see what these are for diatomic and linear polyatomic molecules.

5.3.3 Rotational Raman Spectra of Diatomic and Linear Polyatomic Molecules

In a diatomic or linear polyatomic molecule rotational Raman scattering obeys the selection rule

$$\Delta J = 0, \pm 2 \tag{5.47}$$

but the $\Delta J = 0$ transitions are unimportant as they correspond to the intense Rayleigh scattering. In addition the molecule must have an anisotropic polarizability which means that α must not be the same in all directions. This is not a very stringent requirement since all molecules except spherical rotors, whose polarizability is in the form of a sphere, have this property. As a result all diatomics and linear polyatomics which are symmetrical or unsymmetrical, i.e. whether or not they have an inversion centre i, show a rotational Raman spectrum.

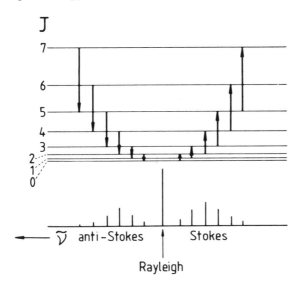

Figure 5.15 Rotational Raman spectrum of a diatomic or linear polyatomic molecule.

The resulting spectrum is illustrated in Figure 5.15 while Figure 5.16 shows in detail the processes involved in the first Stokes and anti-Stokes transitions and in the Rayleigh scattering.

Molecules initially in the $J = 0$ state encounter intense, monochromatic radiation of wavenumber $\tilde{v}$. Provided the energy $hc\tilde{v}$ does not correspond to the difference in energy between $J = 0$ and any other state (electronic, vibrational, or rotational) of the molecule it is not absorbed but produces an induced dipole in the molecule as expressed by equation (5.43). The molecule is said to

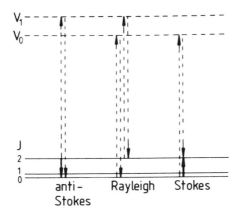

Figure 5.16 Raman and Rayleigh scattering process involving virtual states V_0 and V_1.

be in a virtual state which, in the case shown in Figure 5.16, is V_0. When scattering occurs the molecule may return, according to the selection rules, to $J = 0$ (Rayleigh) or $J = 2$ (Stokes). Similarly a molecule initially in the $J = 2$ state goes to the virtual state V_1 and returns to $J = 2$ (Rayleigh), $J = 4$ (Stokes) or $J = 0$ (anti-Stokes). The overall transitions, $J = 2$ to 0 and 0 to 2, are indicated by solid lines in Figure 5.16, while Figure 5.15 shows more of these overall transitions.

Conventionally we take ΔJ to mean J (upper) $- J$ (lower) so we need consider only $\Delta J = +2$. The magnitude of a Raman displacement from the exciting radiation $\tilde{v}$ is given by

$$|\Delta \tilde{v}| = F(J + 2) - F(J) \tag{5.48}$$

where $\Delta \tilde{v} = \tilde{v} - \tilde{v}_L$ ($\tilde{v}_L$ is the wavenumber of the exciting laser radiation) is positive for anti-Stokes and negative for Stokes lines. When centrifugal distortion is neglected the rotational term values $F(J)$ are given by equation (5.12), and equation (5.48) gives

$$|\Delta \tilde{v}| = 4B_0 J + 6B_0 \tag{5.49}$$

for molecules in the zero-point vibrational state. The spectrum shows two sets of equally spaced lines with a spacing of $4B_0$ and a separation of $12B_0$ between the first Stokes and anti-Stokes lines.

When centrifugal distortion is taken into account the rotational term values are given by equation (5.19) and we have

$$|\Delta \tilde{v}| = (4B_0 - 6D_0)(J + \tfrac{3}{2}) - 8D_0(J + \tfrac{3}{2})^3 \tag{5.50}$$

Series of rotational transitions are referred to as branches and they are labelled with a letter according to the value of ΔJ as follows:

$$
\begin{array}{lccccccc}
\Delta J & \cdots & -2, & -1, & 0, & +1, & +2, & \cdots \\
\text{Branch} & \cdots & O, & P, & Q, & R, & S, & \cdots
\end{array}
\tag{5.51}
$$

so that the branches in Figure 5.15 are both S branches (although some authors refer to the anti-Stokes S branch as an O branch).

The intensities along the branches show a maximum because the populations of the initial levels of the transitions, given by equation (5.15), show similar behaviour.

Figure 5.17 shows the rotational Raman spectrum of $^{15}N_2$ obtained with 476.5 nm radiation from an argon ion laser. From this spectrum a very accurate value for B_0 of $1.857\,672 \pm 0.000\,027\ \text{cm}^{-1}$ has been obtained from which a value for the bond length r_0 of $1.099\,985 \pm 0.000\,010$ Å results. Such accuracy is typical of high resolution rotational Raman spectroscopy.

A feature of the $^{15}N_2$ spectrum is an intensity alternation of 1:3 for the J value of the initial level of the transition even:odd. This is an effect due to the nuclear spin of the ^{15}N nuclei which will now be discussed in some detail.

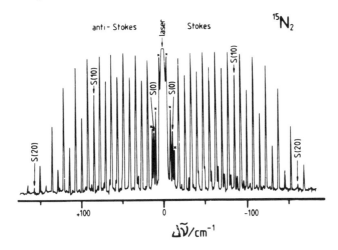

Figure 5.17 Rotational Raman spectrum of $^{15}N_2$. (The lines marked with a cross are grating 'ghosts' and not part of the spectrum.)

5.3.4 Nuclear Spin Statistical Weights

When nuclear spin is included the total wave function ψ for a molecule is modified from that of equation (1.58) to

$$\psi = \psi_e \psi_v \psi_r \psi_{ns} \qquad (5.52)$$

where ψ_e, ψ_v, ψ_r are the electronic, vibrational, and rotational wave functions and ψ_{ns} is the nuclear spin wave function. We shall be concerned here mainly with the symmetry properties of ψ_{ns} and ψ_r.

For a symmetrical $(D_{\infty h})$ diatomic or linear polyatomic molecule with two, or any even number, of identical nuclei having the nuclear spin quantum number (see equation 1.47) $I = n + \frac{1}{2}$, where n is zero or an integer, exchange of any two which are equidistant from the centre of the molecule results in a change of sign of ψ which is then said to be antisymmetric to nuclear exchange. In addition the nuclei are said to be Fermi particles (or fermions) and obey Fermi–Dirac statistics. On the other hand, if $I = n$, ψ is symmetric to nuclear exchange and the nuclei are said to be Bose particles (or bosons) and obey Bose–Einstein statistics.

We will now consider the consequences of these rules in the simple case of $^1H_2$. In this molecule both ψ_v, whatever the value of v, and ψ_e, in the ground electronic state, are symmetric to nuclear exchange: so we need consider only the behaviour of $\psi_r \psi_{ns}$. Since $I = \frac{1}{2}$ for ^1H, ψ and therefore $\psi_r \psi_{ns}$ must be antisymmetric to nuclear exchange. It can be shown that, for even values of the rotational quantum number J, ψ_r is symmetric (s) to exchange and, for odd values of J, ψ_r is antisymmetric (a) to exchange, as shown in Figure 5.18.

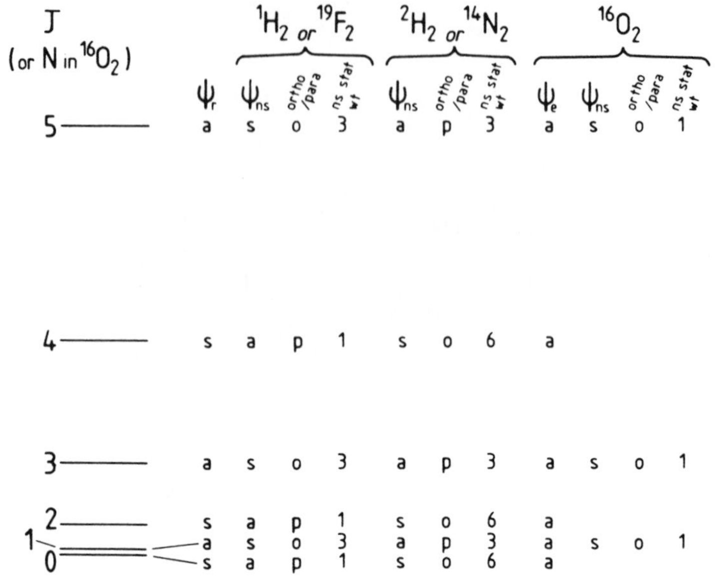

J (or N in $^{16}O_2$)	$^1H_2$ or $^{19}F_2$				$^2H_2$ or $^{14}N_2$			$^{16}O_2$			
	ψ_r	ψ_{ns}	ortho/para	ns stat wt	ψ_{ns}	ortho/para	ns stat wt	ψ_e	ψ_{ns}	ortho/para	ns stat wt
5	a	s	o	3	a	p	3	a	s	o	1
4	s	a	p	1	s	o	6	a			
3	a	s	o	3	a	p	3	a	s	o	1
2	s	a	p	1	s	o	6	a			
1	a	s	o	3	a	p	3	a	s	o	1
0	s	a	p	1	s	o	6	a			

Figure 5.18 Nuclear spin statistical weights of rotational states of various diatomic molecules.

Equation (1.48) shows that, for $I = \frac{1}{2}$, space quantization of nuclear spin angular momentum results in the quantum number M_I taking the values $\frac{1}{2}$ or $-\frac{1}{2}$. The nuclear spin wave function ψ_{ns} is usually written as α or β, corresponding to $M_I = \frac{1}{2}$ or $-\frac{1}{2}$, respectively, and both ^1H nuclei, labelled 1 and 2, can have either α or β spin wave functions. There are therefore four possible forms of ψ_{ns} for the molecule as a whole:

$$\psi_{ns} = \alpha(1)\alpha(2); \quad \beta(1)\beta(2); \quad \alpha(1)\beta(2); \quad \text{or} \quad \beta(1)\alpha(2) \tag{5.53}$$

Although the $\alpha(1)\alpha(2)$ and $\beta(1)\beta(2)$ wave functions are clearly symmetric to interchange of the 1 and 2 labels, the other two functions are neither symmetric nor antisymmetric. For this reason it is necessary to use instead the linear combinations $2^{-1/2}[\alpha(1)\beta(2) + \beta(1)\alpha(2)]$ and $2^{-1/2}[\alpha(1)\beta(2) - \beta(1)\alpha(2)]$, where $2^{-1/2}$ is a normalization constant. Then three of the four nuclear spin wave functions are seen to be symmetric (s) to nuclear exchange and one is antisymmetric (a):

$$(s) \quad \psi_{ns} = \begin{cases} \alpha(1)\alpha(2) \\ 2^{-1/2}[\alpha(1)\beta(2) + \beta(1)\alpha(2)] \\ \beta(1)\beta(2) \end{cases} \tag{5.54}$$

$$(a) \quad \psi_{ns} = 2^{-1/2}[\alpha(1)\beta(2) - \beta(1)\alpha(2)] \tag{5.55}$$

In general, for a homonuclear diatomic molecule there are $(2I + 1)(I + 1)$ symmetric and $(2I + 1)I$ antisymmetric nuclear spin wave functions; therefore

$$\frac{\text{Number of } (s) \text{ functions}}{\text{Number of } (a) \text{ functions}} = \frac{I + 1}{I} \tag{5.56}$$

In order that $\psi_r \psi_{ns}$ is always antisymmetric for $^1H_2$ the antisymmetric ψ_{ns} are associated with even J states and the symmetric ψ_{ns} with odd J states, as shown in Figure 5.18. Interchange between states with ψ_{ns} symmetric and antisymmetric is forbidden so that $^1H_2$ can be regarded as consisting of two distinct forms:

1. *para*-hydrogen with ψ_{ns} antisymmetric and with what is commonly referred to as antiparallel nuclear spins:
2. *ortho*-hydrogen with ψ_{ns} symmetric and parallel nuclear spins.

As indicated in Figure 5.18, *para*-$^1H_2$ can exist in only even J states and *ortho*-$^1H_2$ in odd J states. At temperatures at which there is appreciable population up to fairly high values of J there is roughly three times as much *ortho*- as there is *para*-$^1H_2$. However, at very low temperatures at which the population of all rotational levels other than $J = 0$ is small, $^1H_2$ is mostly in the *para* form.

All other homonuclear diatomic molecules with $I = \frac{1}{2}$ for each nucleus, such as $^{19}F_2$, also have *ortho* and *para* forms with odd and even J and nuclear spin statistical weights of 3 and 1 respectively, as shown in Figure 5.18.

If $I = 1$ for each nucleus, as in $^2H_2$ and $^{14}N_2$, the total wave function must be symmetric to nuclear exchange. There are nine nuclear spin wave functions of which six are symmetric and three antisymmetric to exchange. Figure 5.18 illustrates the fact that *ortho*-$^2H_2$ (or $^{14}N_2$) can have only even J and the *para* form only odd J, and that there is roughly twice as much of the *ortho* form as there is of the *para* form at normal temperatures: at low temperatures there is a larger excess of the *ortho* form.

The effect of low temperatures affecting the *ortho:para* ratio is more important for light molecules, such as $^1H_2$ and $^2H_2$, than for heavy ones, such as $^{19}F_2$ and $^{14}N_2$. The reason is that the separation of the $J = 0$ and $J = 1$ levels is smaller for a heavier molecule and a lower temperature is required before a significant deviation from the normal *ortho:para* ratio is observed.

For the symmetrical linear polyatomic molecule acetylene

$$^1H-^{12}C\equiv^{12}C-^1H$$

the situation is very similar to that for $^1H_2$ since $I = 0$ for ^{12}C. The main difference is that, since the rotational energy levels are much more closely spaced, a much lower temperature is necessary to produce acetylene predominantly in the *para* form.

For ^{16}O, I is zero and there are no antisymmetric nuclear spin wave functions for $^{16}O_2$. Since each ^{16}O nucleus is a boson the total wave function must be antisymmetric to nuclear exchange. In the case of $^{16}O_2$, with two electrons with unpaired spins in the ground state (see Section 7.2.1.1), ψ_e is antisymmetric, unlike the other molecules we have considered. Since $I = 0$ the nuclear spin wave function $\psi_{ns} = 1$, which is symmetric to the exchange of nuclei. Therefore, as Figure 5.18 shows, $^{16}O_2$ has only levels with the rotational quantum number having odd values and, in its rotational Raman spectrum, alternate lines with N'' even are missing. (In molecules such as $^{16}O_2$, which have a resultant electron spin angular momentum due to unpaired electrons, the quantum number N, rather than J, distinguishes the rotational levels.)

5.3.5 Rotational Raman Spectra of Symmetric and Asymmetric Rotor Molecules

For a symmetric rotor molecule the selection rules for the rotational Raman spectrum are

$$\Delta J = 0, \pm 1, \pm 2; \qquad \Delta K = 0 \qquad (5.57)$$

resulting in R and S branches for each value of K, in addition to the Rayleigh scattering.

For asymmetric rotors the selection rule in J is $\Delta J = 0, \pm 1, \pm 2$, but the fact that K is not a good quantum number results in the additional selection rules being too complex for discussion here.

5.4 Structure Determination from Rotational Constants

Measurement and assignment of the rotational spectrum of a diatomic or other linear molecule result in a value of the rotational constant. In general, this will be B_0, which relates to the zero-point vibrational state. If the rotational constant can be determined in one or more excited vibrational states, B_e, relating to the molecule in its unattainable equilibrium configuration, may be obtained using equation (5.25). For a diatomic molecule B_0 and B_e can be converted to moments of inertia, using equation (5.11) or equation (5.12), and thence to bond lengths r_0 and r_e, since $I = \mu r^2$. This begs the question of whether we refer to r_0 or r_e as the bond length.

Part of the answer is that, unless we require a high degree of accuracy, it does not matter very much. This is illustrated by the values for r_0 and r_e for $^{14}N_2$ and $^{15}N_2$ in Table 5.3. Nevertheless, an important difference between r_0 and r_e in general is that r_e is independent of the isotopic species whereas r_0 is not. Table 5.3 illustrates this point also. It is because of the isotope-independence of r_e that it is this which is used in discussing bond lengths at the highest degree of accuracy.

Table 5.3 Values of r_0 and r_e for N_2.

Molecule	r_0/Å	r_e/Å
$^{14}N_2$	1.100 105 ± 0.000 010	1.097 651 ± 0.000 030
$^{15}N_2$	1.099 985 ± 0.000 010	1.097 614 ± 0.000 030

The reason that r_e does not change with isotopic substitution is that it refers to the bond length at the minimum of the potential energy curve (see Figure 1.13) and this curve, whether it refers to the harmonic oscillator approximation (Section 1.3.6) or an anharmonic oscillator (to be discussed in Section 6.1.3.2), does not change with isotopic substitution. On the other hand, the vibrational energy levels within the potential energy curve, and therefore r_0, are affected by isotopic substitution: this is illustrated by the mass-dependence of the vibration frequency demonstrated by equation (1.68).

These arguments can be extended to linear and non-linear polyatomic molecules whose zero-point structure, in terms of bond lengths and angles, is isotope-dependent but whose equilibrium structure is not.

As in diatomic molecules the structure of greatest importance is the equilibrium structure, but one rotational constant can give, at most, only one structural parameter. In a non-linear but planar molecule the out-of-plane principal moment of inertia I_c is related to the other two by

$$I_c = I_a + I_b \tag{5.58}$$

so there are only two independent rotational constants.

In, for example, the planar asymmetric rotor molecule formaldehyde, ^1H_2CO, shown in Figure 5.1(f), it is possible by obtaining, say, A_v and B_v in the zero-point level and in the $v = 1$ level of all six vibrations to determine A_e and B_e. Two rotational constants are insufficient, however, to give the three structural parameters $r_e(CH)$, $r_e(CO)$, and $(\angle HCH)_e$ necessary for a complete equilibrium structure. It is at this stage that the importance of dealing with equilibrium, rather than zero-point, structure is apparent. Determination of A_e and B_e for ^2H_2CO, whose equilibrium structure is identical to that of ^1H_2CO, would allow a complete structure determination, three structural parameters being found from four rotational constants.

However, even for a small molecule such as H_2CO, determination of the rotational constants in the $v = 1$ levels of all the vibrations presents considerable difficulties. In larger molecules it may be possible to determine only A_0, B_0, and C_0. In such cases the simplest way to determine the structure is to ignore the differences from A_e, B_e, and C_e and make sufficient isotopic substitutions to give a complete, but approximate, structure, called the r_0 structure.

An improvement on the r_0 structure is the substitution structure, or r_s structure. This is obtained using the so-called Kraitchman equations which give

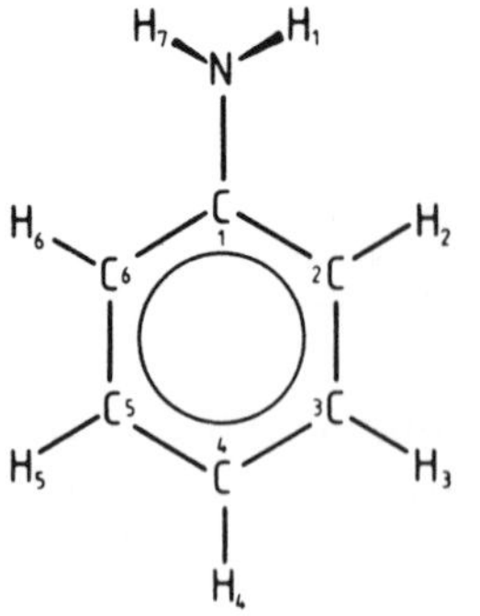

XY	$r_s(XY)/\text{Å}$	XYZ	$(\angle XYZ)_s/\text{deg}$
NH_1	1.001 ± 0.01	H_1NH_7	113.1 ± 2
C_1N	1.402 ± 0.002	$C_6C_1C_2$	119.4 ± 0.2
C_1C_2	1.397 ± 0.003	$C_1C_2C_3$	120.1 ± 0.2
C_2C_3	1.394 ± 0.004	$H_2C_2C_3$	120.1 ± 0.2
C_3C_4	1.396 ± 0.002	$C_2C_3C_4$	120.7 ± 0.1
C_2H_2	1.082 ± 0.004	$H_3C_3C_2$	119.4 ± 0.1
C_3H_3	1.083 ± 0.002	$C_3C_4C_5$	118.9 ± 0.1
C_4H_4	1.080 ± 0.002		

out-of-plane angle of NH_2 is $37.5 \pm 2°$

Figure 5.19 The r_s structure of aniline.

the coordinates of an atom, which has been isotopically substituted, in relation to the principal inertial axes of the molecule before substitution. The substitution structure is also approximate but is nearer to the equilibrium structure than the zero-point structure.

One of the largest molecules for which an r_s structure has been obtained is aniline, shown in Figure 5.19. The benzene ring shows small deviations from a regular hexagon in the angles but no meaningful deviations in the bond lengths. As might be expected, by comparison with the pyramidal NH_3 molecule, the plane of the NH_2-group is not coplanar with the rest of the molecule.

Questions

1. Use equation (5.16) to derive equation (5.17). Compare values of J_{max} and the transition frequencies involving the values of J_{max} for HCN ($B \simeq 44.316$ GHz) and $N{\equiv}C{-}(C{\equiv}C)_2{-}H$ ($B \simeq 1.313$ GHz).
2. Rearrange equation (5.20) into the form $y = mx + c$ so that m involves D only. Plot y against x using the data in Table 5.1 to obtain B and D, in hertz, for carbon monoxide (use a computer or calculator which will work to nine-figure accuracy).
3. By making measurements from Figure 5.3 determine the average separation of rotational transitions in carbon monoxide and hence estimate the bond length.
4. Make measurements of the transition wavenumbers in the rotational spectrum of silane from Figure 5.10 and hence determine the Si $-$H bond length.
5. Assuming reasonable bond lengths estimate the frequency of the $J = 15 - 14$ transition in the linear molecule $N{\equiv}C{-}(C{\equiv}C)_6{-}H$. In which region of the electromagnetic spectrum does it lie?

6. Starting from an expression for rotational term values show that

$$|\Delta\tilde{\nu}| = (4B_0 - 6D_0)(J + \tfrac{3}{2}) - 8D_0(J + \tfrac{3}{2})^3$$

for rotational Raman transitions of a diatomic or linear polyatomic molecule.

7. From the value for B_0 of $1.923\ 604 \pm 0.000\ 027\ \text{cm}^{-1}$ obtained from the rotational Raman spectrum of $^{14}N^{15}N$ calculate the bond length r_0. Why does it differ from r_0 for $^{14}N_2$? Would the values of r_e differ? Would there be an intensity alternation in the spectrum of $^{14}N^{15}N$? Would $^{14}N^{15}N$ show a rotational infrared spectrum?

8. The first three Stokes lines in the rotational Raman spectrum of $^{16}O_2$ are separated by 14.4, 25.8, and $37.4\ \text{cm}^{-1}$ from the exciting radiation. Using the rigid rotor approximation obtain an approximate value for r_0.

Bibliography

Carrington, A. (1974). *Microwave Spectroscopy of Free Radicals*, Academic Press, New York.

Gordy, W., and Cook, R. L. (1984). *Microwave Molecular Spectra*, 3rd ed., Wiley-Interscience, New York.

Herzberg, G. (1945). *Infrared and Raman Spectra*, Van Nostrand, New York.

Herzberg, G. (1950). *Spectra of Diatomic Molecules*, Van Nostrand, New York.

Kroto, H. W. (1975). *Molecular Rotation Spectra*, Wiley, London; (1992) Dover, New York.

Long, D. A. (1977). *Raman Spectroscopy*, McGraw-Hill, London.

Sugden, T. M., and Kenney, C. N. (1965). *Microwave Spectroscopy of Gases*, Van Nostrand, London.

Townes, C. H., and Schawlow, A. L. (1955). *Microwave Spectroscopy*, McGraw-Hill, New York.

Wollrab, J. E. (1967). *Rotational Spectra and Molecular Structure*, Academic Press, New York.

CHAPTER 6

Vibrational Spectroscopy

6.1 Diatomic Molecules

We have seen in Section 1.3.6 how the vibrational energy levels E_v of a diatomic molecule, treated by the harmonic oscillator approximation, are given by

$$E_v = hv(v + \tfrac{1}{2}) \tag{6.1}$$

where the vibrational quantum number $v = 0,1,2 \ldots$. The classical vibrational frequency v is related to the reduced mass $\mu [= m_1 m_2/(m_1 + m_2)]$ and the force constant k by

$$v = \frac{1}{2\pi} \left(\frac{k}{\mu}\right)^{1/2} \tag{6.2}$$

The force constant can be regarded as a measure of the strength of the spring in the ball-and-spring model for molecular vibration. Table 6.1 gives some typical values in units[†] of aJ Å^{-2} ($=10^2$ N m^{-1}). The pre-SI unit of k was commonly mdyn Å^{-1} and it is fortunate that 1 aJ Å^{-2} = 1 mdyn Å^{-1}. The values illustrate the increase of k with bond order. HCl, HF, Cl$_2$, and F$_2$ all have single bonds and relatively low values of k, although that of HF is somewhat higher. O$_2$, NO, CO, and N$_2$ have bond orders of 2, 2½, 3, and 3 respectively, which are reflected in their k values.

Physically the strength of the spring representing the bond is due to a subtle balance of nuclear repulsions, electron repulsions, and electron–nuclear attractions. None of these is affected by nuclear mass and, therefore, k is not affected by isotopic substitution.

Figure 1.13 shows the potential function, vibrational wave functions, and energy levels for a harmonic oscillator. Just as for rotation it is convenient to use term values instead of energy levels. Vibrational term values $G(v)$ invariably

---

[†] The prefix 'atto' stands for 10^{-18}.

Table 6.1 Force constants for some diatomic molecules.

Molecule	k/aJ Å^{-2}	Molecule	k/aJ Å^{-2}	Molecule	k/aJ Å^{-2}
HCl	5.16	F$_2$	4.45	CO	18.55
HF	9.64	O$_2$	11.41	N$_2$	22.41
Cl$_2$	3.20	NO	15.48		

have dimensions of wavenumber so we have, from equation (1.69),

$$\frac{E_v}{hc} = G(v) = \omega(v + \tfrac{1}{2}) \tag{6.3}$$

where ω is the vibration wavenumber (commonly but incorrectly known as the vibration frequency).

6.1.1 Infrared Spectra

The transition moment (equation 2.13) for a transition between lower and upper states with vibrational wave functions ψ_v'' and ψ_v' respectively is given by

$$R_v = \int \psi_v'^* \mu \psi_v'' \, dx \tag{6.4}$$

where x is $(r-r_e)$, the displacement of the internuclear distance from equilibrium. The dipole moment μ is zero for a homonuclear diatomic molecule resulting in $R_v = 0$ and all vibrational transitions being forbidden. For a heteronuclear diatomic molecule μ is non-zero and varies with x. This variation can be expressed as a Taylor series expansion

$$\mu = \mu_e + \left(\frac{d\mu}{dx}\right)_e x + \frac{1}{2!}\left(\frac{d^2\mu}{dx^2}\right)_e x^2 + \ldots \tag{6.5}$$

where the subscript 'e' refers to the equilibrium configuration. The transition moment of equation (6.4) now becomes

$$R_v = \mu_e \int \psi_v'^* \psi_v'' \, dx + \left(\frac{d\mu}{dx}\right)_e \int \psi_v'^* x \psi_v'' \, dx + \ldots \tag{6.6}$$

Since ψ_v' and ψ_v'' are eigenfunctions of the same hamiltonian, namely that in equation (1.65), they are orthogonal which means that, when $v' \neq v''$,

$$\int \psi_v'^* \psi_v'' \, dx = 0 \tag{6.7}$$

Equation (6.6) then becomes

$$R_v = \left(\frac{d\mu}{dx}\right)_e \int \psi_v'^* x \psi_v''\, dx + \ldots \tag{6.8}$$

The first term in this series is non-zero only if

$$\Delta v = \pm 1 \tag{6.9}$$

which constitutes the vibrational selection rule. Since Δv refers to v(upper) − v(lower) the selection rule is effectively $\Delta v = +1$. In the harmonic oscillator, where all level spacings are equal, all transitions obeying this selection rule are coincident at a wavenumber ω.

If the spectrum is observed in absorption, as it usually is, and at normal temperatures intensities of the transitions decrease rapidly as v'' increases, since the population N_v of the vth vibrational level is related to N_0 by the Boltzmann factor

$$\frac{N_v}{N_0} = \exp\left(-\frac{E_v}{kT}\right) \tag{6.10}$$

derived from equation (2.11).

Each vibrational transition observed in the gas phase gives rise to what is called a band in the spectrum. The word 'line' is reserved for describing a transition between rotational levels associated with the two vibrational levels giving rise to the fine structure of a band. However, in the solid or liquid phase, where this fine structure is not present, each vibrational transition is sometimes referred to as a line rather than a band.

All bands with $v'' \neq 0$ are referred to as hot bands because, as indicated by equation (6.10), the populations of the lower levels of such transitions, and therefore the transition intensities, increase with temperature.

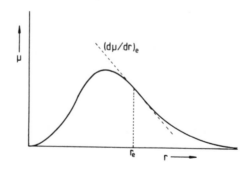

Figure 6.1 Variation of dipole moment with internuclear distance in a heteronuclear diatomic molecule.

The transition intensities are proportional also to $|R_v|^2$ and therefore, according to equation (6.8), to $(d\mu/dx)_e^2$. Figure 6.1 shows how the magnitude μ of the dipole moment varies with internuclear distance in a typical heteronuclear diatomic molecule. Obviously $\mu \to 0$ when $r \to 0$ and the nuclei coalesce. For neutral diatomics, $\mu \to 0$ when $r \to \infty$ because the molecule dissociates into neutral atoms. Therefore, between $r = 0$ and ∞ there must be a maximum value of μ. Figure 6.1 has been drawn with this maximum at $r < r_e$, giving a negative slope $d\mu/dr$ at r_e. If the maximum were at $r > r_e$ there would be a positive slope at r_e. It is possible that the maximum is at r_e, in which case $d\mu/dr = 0$ at r_e and the $\Delta v = 1$ transitions, although allowed, would have zero intensity.

It is an important general point that spectroscopic selection rules tell us only whether a transition *may* occur but tell us nothing about intensities which may be accidentally zero or very low.

6.1.2 Raman Spectra

In both heteronuclear and homonuclear diatomic molecules the polarizability α (see Section 5.3.2) varies during vibrational motion leading to a vibrational Raman effect. This variation can be visualized as being due to the polarizability ellipsoid (Figure 5.14) expanding and contracting as the bond length increases and decreases due to vibration. A classical treatment, analogous to that for rotation in Section 5.3.2, leads to a variation with time of the dipole moment μ, induced by irradiation of the sample with intense monochromatic radiation of wavenumber $\tilde{\nu}$, given by

$$\mu = \alpha_{0,v} A \sin 2\pi c \tilde{\nu} t - \tfrac{1}{2}\alpha_{1,v} A \cos 2\pi c(\tilde{\nu} + \omega)t \qquad (6.11)$$
$$+ \tfrac{1}{2}\alpha_{1,v} A \cos 2\pi c(\tilde{\nu} - \omega)t$$

In this equation $\alpha_{0,v}$ is the average polarizability during vibration, $\alpha_{1,v}$ is the amplitude of the change of polarizability due to vibration, A is the amplitude of the oscillating electric field of the incident radiation (equation 5.44), and ω is the vibration wavenumber. Equation (6.11) is similar to equation (5.46) for rotation except that the second and third terms in equation (6.11) correspond to Raman scattering with a wavenumber of $(\tilde{\nu} + \omega)$ and $(\tilde{\nu} - \omega)$, the anti-Stokes and Stokes Raman scattering respectively. Whereas for rotation α changes at twice the rotational frequency, during a complete vibrational cycle α goes through only one cycle and, therefore, changes at the same frequency as the frequency of vibration. This accounts for the absence of the factor of two in equation (6.11) compared to equation (5.46).

As for the dipole moment, the change of polarizability with vibrational displacement x can be expressed as a Taylor series

$$\alpha = \alpha_e + \left(\frac{d\alpha}{dx}\right)_e x + \frac{1}{2!}\left(\frac{d^2\alpha}{dx^2}\right)_e x^2 + \dots \qquad (6.12)$$

By analogy with equation (6.6) the vibrational Raman transition moment R_v is given by

$$R_v = \left(\frac{\mathrm{d}\alpha}{\mathrm{d}x}\right)_e A \int \psi_v'^* x \psi_v'' \, \mathrm{d}x + \dots \tag{6.13}$$

and the first term is non-zero only if

$$\Delta v = \pm 1 \tag{6.14}$$

which constitutes the vibrational Raman selection rule. This is the same as for infrared vibrational transitions but vibrational Raman spectroscopy has the advantage that transitions are allowed in homonuclear, as well as heteronuclear, diatomic molecules.

Intensities of Raman transitions are proportional to $|R_v|^2$ and therefore, from equation (6.13), to $(\mathrm{d}\alpha/\mathrm{d}x)_e^2$. Since α is a tensor property we cannot illustrate easily its variation with x: instead we use the mean polarizability $\bar{\alpha}$, where

$$\bar{\alpha} = \tfrac{1}{3}(\alpha_{xx} + \alpha_{yy} + \alpha_{zz}) \tag{6.15}$$

Figure 6.2 shows typically how $\bar{\alpha}$ varies with r; $\mathrm{d}\bar{\alpha}/\mathrm{d}r$ is usually positive and, unlike $\mathrm{d}\mu/\mathrm{d}r$ in Figure 6.1, varies little with r. For this reason vibrational Raman intensities are less sensitive than infrared intensities to the environment of the molecule, such as the solvent in a solution spectrum.

The mechanism for Stokes and anti-Stokes vibrational Raman transitions is analogous to that for rotational transitions illustrated in Figure 5.16. As shown in Figure 6.3, intense monochromatic radiation may take the molecule from the $v = 0$ state to a virtual state V_0. Then it may return to $v = 0$ in a Rayleigh scattering process or to $v = 1$ in a Stokes Raman transition. Alternatively it may go from the $v = 1$ state to the virtual state V_1 and return to $v = 1$ (Rayleigh) or to $v = 0$ (Raman anti-Stokes). However, in many molecules at normal

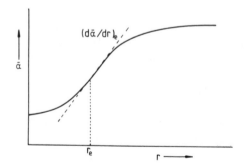

Figure 6.2   Variation of the mean polarizability with internuclear distance in a diatomic molecule.

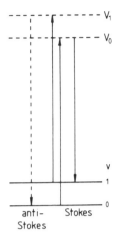

Figure 6.3 Stokes and anti-Stokes vibrational Raman scattering

temperatures the initial population of the $v = 1$ state is so low that anti-Stokes transitions may be too weak to be observed.

6.1.3 Anharmonicity

6.1.3.1 Electrical anharmonicity

Equations (6.5) and (6.12) contain terms in x to the second and higher powers. If the expressions for the dipole moment μ and the polarizability α were linear in x, then μ and α would be said to vary harmonically with x. The effect of higher terms is known as anharmonicity and, because this particular kind of anharmonicity is concerned with electrical properties of a molecule, it is referred to as electric anharmonicity. One effect of it is to cause the vibrational selection rule $\Delta v = \pm 1$ in infrared and Raman spectroscopy to be modified to $\Delta v = \pm 1$, $\pm 2, \pm 3, \ldots$. However, since electrical anharmonicity is usually small, the effect is to make only a very small contribution to the intensities of $\Delta v = \pm 2$, $\pm 3, \ldots$ transitions, which are known as vibrational overtones.

6.1.3.2 Mechanical anharmonicity

Just as the electrical behaviour of a real diatomic molecule is not accurately harmonic, neither is its mechanical behaviour. The potential function, vibrational energy levels, and wave functions shown in Figure 1.13 were derived by assuming that vibrational motion obeys Hooke's law, as expressed by equation (1.63), but this assumption is reasonable only when r is not very different from r_e, i.e. when x is small. At large values of r we know that the molecule dissociates: two neutral atoms are formed and, since they do not influence each other, the force constant is zero and r can then be increased to infinity with no further change of the potential energy V. Therefore the potential energy curve flattens out at $V = D_e$, where D_e is the dissociation energy measured relative to the equilibrium potential energy, as shown in Figure 6.4. As dissociation is approached the force constant $k \rightarrow 0$ and the bond gets weaker. The effect is to make the potential energy curve shallower than for a harmonic oscillator, when $r > r_e$, as the figure shows. At small values of r the positive charges on the two nuclei cause mutual repulsion which increasingly opposes their approaching each other. Consequently the potential energy curve is steeper than for a harmonic oscillator, as the figure also shows. The deviations found in the curve for a real molecule from that resulting from the harmonic oscillator approximation are due to mechanical anharmonicity.

A molecule may show both electrical and mechanical anharmonicity but the latter is generally much more important and it is usual to define a harmonic oscillator as one which is harmonic in the mechanical sense. It is possible, therefore, that a harmonic oscillator may show electrical anharmonicity.

One effect of mechanical anharmonicity is to modify the $\Delta v = \pm 1$ infrared and Raman selection rule to $\Delta v = \pm 1, \pm 2, \pm 3, \ldots$, but the overtone transitions with $\Delta v = \pm 2, \pm 3, \ldots$ are usually weak compared to those with $\Delta v = \pm 1$. Since electrical anharmonicity also has this effect both types of anharmonicity may contribute to overtone intensities.

However, unlike electrical anharmonicity, mechanical anharmonicity modifies the vibrational term values and wave functions. The harmonic oscillator term values of equation (6.3) are modified to a power series in $(v + \frac{1}{2})$:

$$G(v) = \omega_e(v + \tfrac{1}{2}) - \omega_e x_e(v + \tfrac{1}{2})^2 + \omega_e y_e(v + \tfrac{1}{2})^3 + \ldots \qquad (6.16)$$

where ω_e is the vibration wavenumber which a classical oscillator would have for an infinitesimal displacement from equilibrium. $\omega_e x_e, \omega_e y_e, \ldots$ are

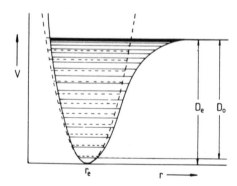

Figure 6.4 Potential energy curve and energy levels for a diatomic molecule behaving as an anharmonic oscillator compared with those for a harmonic oscillator (dashed curve).

anharmonic constants and are written in this way, rather than, say, x_e, y_e, ..., because earlier authors wrote equation (6.16) in the form

$$G(v) = \omega_e[(v + \tfrac{1}{2}) - x_e(v + \tfrac{1}{2})^2 + y_e(v + \tfrac{1}{2})^3 + \cdots] \tag{6.17}$$

The reason for the negative sign in the second term of the expansion is that the constant $\omega_e x_e$ has the same sign for all diatomic molecules and, when the negative sign is included, $\omega_e x_e$ is always positive. Further terms in the expansion may be positive or negative.

There is a rapid decrease in value along the series ω_e, $\omega_e x_e$, $\omega_e y_e$, ... For example, for $^1H^{35}Cl$, $\omega_e = 2990.946$ cm^{-1}, $\omega_e x_e = 52.8186$ cm^{-1}, $\omega_e y_e = 0.2244$ cm^{-1}, $\omega_e z_e = -0.0122$ cm^{-1}. The effect of the positive value of $\omega_e x_e$ is to close up the energy levels with increasing v. The corresponding energy levels are compared with those of a harmonic oscillator with constant separation in Figure 6.4. The anharmonic oscillator levels converge to the dissociation limit D_e above which there is a continuum of levels.

One effect of the modification to the harmonic oscillator term values is that, unlike the case of the harmonic oscillator, ω_e cannot be measured directly. Wavenumbers $\Delta G_{v+1/2}$ for $(v + 1)-v$ transitions[†] are given by

$$\Delta G_{v+1/2} = G(v + 1) - G(v) = \omega_e - \omega_e x_e(2v + 2) + \omega_e y_e(3v^2 + 6v + \tfrac{13}{4}) + \cdots \tag{6.18}$$

In order to determine, say, ω_e and $\omega_e x_e$, at least two transition wavenumbers, for example $G(1) - G(0) = \omega_0$ and $G(2) - G(1) = \omega_1$, must be obtained.

The dissociation energy D_e is given approximately by

$$D_e \simeq \frac{\omega_e^2}{4\omega_e x_e} \tag{6.19}$$

the approximation being due to the neglect of all anharmonic constants other than $\omega_e x_e$.

Experimentally the dissociation energy can be measured only relative to the zero-point level and is symbolized by D_0. It is clear from Figure 6.4 that

$$D_0 = \sum_v \Delta G_{v+1/2} \tag{6.20}$$

If all anharmonic constants except $\omega_e x_e$ are neglected, $\Delta G_{v+1/2}$ is a linear function of v (equation 6.18) and D_0 is the area under a plot of $\Delta G_{v+1/2}$ versus v, shown by a dashed line in Figure 6.5. In many cases only the first few ΔG values can be observed and a linear extrapolation to $\Delta G_{v+1/2} = 0$ has to be made. This is called a Birge–Sponer extrapolation and the area under the

---

[†] Note that the upper state quantum number of a transition is given first and the lower state quantum number second.

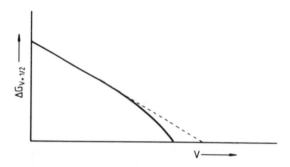

Figure 6.5 A Birge–Sponer extrapolation (dashed line) for determining D_0. The actual points lie on the solid line in a typical case.

extrapolated plot gives an approximate value for D_0. However, most plots deviate considerably from linearity at high v in the way shown in Figure 6.5, so that the value of D_0 is usually an overestimate.

Experimental values of $\Delta G_{v+1/2}$ for high values of v are not normally obtainable from infrared or Raman spectroscopy because of the low intensities of $\Delta v = \pm 2, \pm 3, \ldots$ transitions and low populations of excited vibrational energy levels. Information on higher vibrational levels is obtained mostly from electronic emission spectroscopy (Chapter 7).

The dissociation energy D_e is unaffected by isotopic substitution because the potential energy curve, and therefore the force constant, is not affected by the number of neutrons in the nucleus. On the other hand the vibrational energy levels are changed due to the mass dependence of ω (proportional to $\mu^{-1/2}$ where μ is the reduced mass) resulting in D_0 being isotope-dependent. D_0 is smaller than D_e by the zero-point energy. The term value $G(0)$, corresponding to the zero-point energy, is given by

$$G(0) = \tfrac{1}{2}\omega_e - \tfrac{1}{4}\omega_e x_e + \tfrac{1}{8}\omega_e y_e + \ldots \tag{6.21}$$

For example, ω_e for $^2H_2$ is less than for $^1H_2$ so that

$$D_0(^2H_2) > D_0(^1H_2) \tag{6.22}$$

An important consequence of the isotope-dependence of D_0 is that, if a chemical reaction involves bond dissociation in a rate-determining step, the rate of reaction is decreased by substitution of a heavier isotope at either end of the bond. Because of the relatively large effect on D_0 substitution of ^2H for ^1H is particularly effective in reducing the reaction rate.

Due to the effects of mechanical anharmonicity—to which we shall refer in future simply as anharmonicity since we encounter electrical anharmonicity much less frequently—the vibrational wave functions are also modified compared to those of a harmonic oscillator. Figure 6.6 shows some wave functions

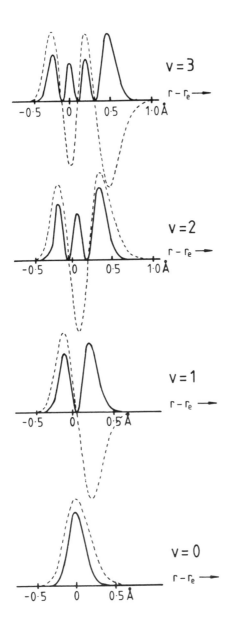

Figure 6.6 ψ_v (dashed lines) and $(\psi_v{}^* \psi_v)^2$ (solid lines) for $v = 0$ to 3 for an anharmonic oscillator.

and probability density functions $(\psi_v^*\psi_v)^2$ for an anharmonic oscillator. The asymmetry in ψ_v and $(\psi_v^*\psi_v)^2$, compared to the harmonic oscillator wave functions in Figure 1.13, increases their magnitude on the shallow side of the potential curve compared to the steep side.

In 1929 Morse suggested

$$V(x) = D_e[1 - \exp(-ax)]^2 \qquad (6.23)$$

as a useful potential function relating to the behaviour of an anharmonic oscillator. In the Morse function $x = r - r_e$, and a and D_e are constants characteristic of a particular electronic state of the molecule—here, the ground electronic state. For this function $V(x) \to D_e$ as $x \to \infty$, as it should do. On the other hand, as $r \to 0$, i.e. as $x \to -r_e$, $V(x)$ becomes large but *not* infinite. This deficiency in the Morse function is not as serious as it might seem since the region where $r \to 0$ is of little experimental importance. The vibrational term values which follow from the Morse function involve terms in only $(v + \frac{1}{2})$ and $(v + \frac{1}{2})^2$ but, although the function is limited in its quantitative use, its ease of manipulation accounts for its sustained popularity compared to more accurate, but much more complex, functions.

6.1.4 Vibration–Rotation Spectroscopy

6.1.4.1 Infrared spectra

We have seen in Section 5.2.1.4 that there is a stack of rotational energy levels associated with all vibrational levels. In rotational spectroscopy we observe transitions between rotational energy levels associated with the same vibrational level (usually $v = 0$). In vibration–rotation spectroscopy we observe transitions between stacks of rotational energy levels associated with two different vibrational levels. These transitions accompany all vibrational transitions but, whereas vibrational transitions may be observed even when the sample is in the liquid or solid phase, the rotational transitions may be observed only in the gas phase at low pressure and usually in an absorption process.

When a molecule has both vibrational and rotational energy the total term values S are given by the sum of the rotational term values $F_v(J)$, given in equation (5.23), and the vibrational term values $G(v)$, given in equation (6.16):

$$S = G(v) + F_v(J)$$
$$= \omega_e(v + \tfrac{1}{2}) - \omega_e x_e(v + \tfrac{1}{2})^2 + \cdots + B_v J(J + 1) - D_v J^2(J + 1)^2 \qquad (6.24)$$

Figure 6.7(a) illustrates the rotational energy levels associated with two vibrational levels v' (upper) and v'' (lower) between which a vibrational transition is allowed by the $\Delta v = \pm 1$ selection rule. The rotational selection rule governing

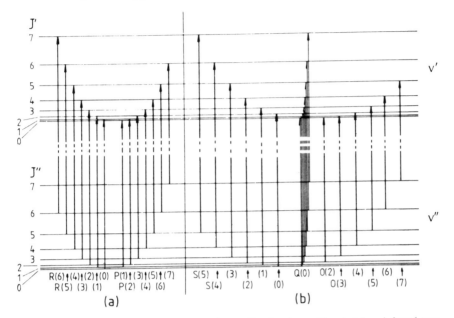

Figure 6.7 Rotational transitions accompanying a vibrational transition in (a) an infrared spectrum and (b) a Raman spectrum of a diatomic molecule.

transition between the two stacks of levels is

$$\Delta J = \pm 1 \tag{6.25}$$

giving an R branch ($\Delta J = +1$) and a P branch ($\Delta J = -1$). Each transition is labelled $R(J)$ or $P(J)$, where J is understood to represent J'', the J value of the lower state. The fact that $\Delta J = 0$ is forbidden means that the pure vibrational transition is not observed. The position at which it would occur is known as the band centre. Exceptions to this selection rule are molecules such as nitric oxide, which have an electronic angular momentum in the ground electronic state. The rotational selection rule for such a molecule is

$$\Delta J = 0, \pm 1 \tag{6.26}$$

and the $(J' = 0) - (J'' = 0)$ transition, the first line of the Q branch ($\Delta J = 0$), marks the band centre.

More usual is the kind of vibrational–rotation band shown in Figure 6.8. This spectrum was obtained with a grating spectrometer having a resolution of about 2 cm^{-1} and shows the $v = 1$–0 transition in ^{1}H^{35}Cl and ^{1}H^{37}Cl. The ^{35}Cl and ^{37}Cl isotopes occur naturally with a 3:1 abundance ratio. The band due to ^{1}H^{37}Cl is displaced to low wavenumber relative to that due to ^{1}H^{35}Cl because of the larger reduced mass (see equation 6.2).

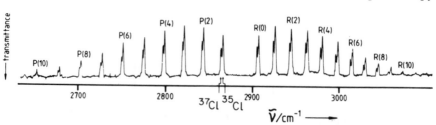

Figure 6.8 The $v = 1$–0 infrared spectrum of $^1H^{35}Cl$ and $^1H^{37}Cl$ showing the P- and R-branch rotational structure.

It is clear from Figure 6.8 that the band for each isotope is fairly symmetrical about the corresponding band centre and that there is approximately equal spacing between adjacent R-branch lines and between adjacent P-branch lines, with twice this spacing between the first R- and P-branch lines, $R(0)$ and $P(1)$. This spacing between $R(0)$ and $P(1)$ is called the zero gap and it is in this region where the band centre falls.

The approximate symmetry of the band is due to the fact that $B_1 \simeq B_0$, i.e. the vibration–rotation interaction constant (equation 5.25) is small. If we assume that $B_1 = B_0 = B$ and neglect centrifugal distortion the wavenumbers of the R-branch transitions, $\tilde{v}[R(J)]$, are given by

$$\tilde{v}[R(J)] = \tilde{v}_0 + B(J + 1)(J + 2) - BJ(J + 1) \qquad (6.27)$$
$$= \tilde{v}_0 + 2BJ + 2B$$

where $\tilde{v}_0$ is the wavenumber of the pure vibrational transition. Similarly the wavenumbers of the P-branch transitions, $\tilde{v}[P(J)]$, are given by

$$\tilde{v}[P(J)] = \tilde{v}_0 + B(J - 1)J - BJ(J + 1) \qquad (6.28)$$
$$= \tilde{v}_0 - 2BJ$$

From equations (6.27) and (6.28) it follows that the zero gap, $\tilde{v}[R(0)] - \tilde{v}[P(1)]$, is $4B$ and that the spacing is $2B$ between adjacent R-branch lines and also between adjacent P-branch lines, hence the approximate symmetry of the band.

A close look at Figure 6.8 reveals that the bands are not quite symmetrical but show a convergence in the R branch and a divergence in the P branch. This behaviour is due principally to the inequality of B_0 and B_1 and there is enough information in the band to be able to determine these two quantities separately. The method used is called the method of combination differences which employs a principle quite common in spectroscopy. The principle is that, if we wish to derive information about a series of lower states and a series of upper states, between which transitions are occurring, then differences in wavenumber between transitions with a common upper state are dependent on properties of the lower states only. Similarly differences in wavenumber between transitions with a common lower state are dependent on properties of the upper states only.

In the case of a vibration–rotation band it is clear from Figure 6.7(a) that, since $R(0)$ and $P(2)$ have a common upper state with $J' = 1$, then $\tilde{v}[R(0)] - \tilde{v}[P(2)]$ must be a function of B'' only. The transitions $R(1)$ and $P(3)$ have $J' = 2$ in common and, in general, $\tilde{v}[R(J - 1)] - \tilde{v}[P(J + 1)]$, usually written as $\Delta''_2 F(J)^\dagger$, is a function of B'' only. If we still neglect centrifugal distortion the function is given by

$$\begin{aligned} \Delta''_2 F(J) &= \tilde{v}[R(J - 1)] - \tilde{v}[P(J + 1)] \\ &= \tilde{v}_0 + B'J(J + 1) - B''(J - 1)J \\ &\quad - [\tilde{v}_0 + B'J(J + 1) - B''(J + 1)(J + 2)] \\ &= 4B''(J + \tfrac{1}{2}) \end{aligned}$$

$$(6.29)$$

After assignment and measurement of the wavenumbers of the rotational lines a graph of $\Delta''_2 F(J)$ versus $(J + \tfrac{1}{2})$ is a straight line of slope $4B''$.

Similarly, since all pairs of transitions $R(J)$ and $P(J)$ have common lower states, $\tilde{v}[R(J)] - \tilde{v}[P(J)]$ is a function of B' only and we have

$$\begin{aligned} \Delta'_2 F(J) &= \tilde{v}[R(J)] - \tilde{v}[P(J)] \\ &= \tilde{v}_0 + B'(J + 1)(J + 2) - B''J(J + 1) \\ &\quad - [\tilde{v}_0 + B'(J - 1)J - B''J(J + 1)] \\ &= 4B'(J + \tfrac{1}{2}) \end{aligned}$$

$$(6.30)$$

and a graph of $\Delta'_2 F(J)$ versus $(J + \tfrac{1}{2})$ is a straight line of slope $4B'$.

The band centre is not quite midway between $R(0)$ and $P(1)$ but its wavenumber $\tilde{v}_0$ can be obtained from

$$\begin{aligned} \tilde{v}_0 &= \tilde{v}[R(0)] - 2B' \\ &= \tilde{v}[P(1)] + 2B'' \end{aligned}$$

$$(6.31)$$

Any effects of centrifugal distortion will show up as slight curvature of the $\Delta_2 F(J)$ versus $(J + \tfrac{1}{2})$ graphs. If the term $-DJ^2(J + 1)^2$ is included, as in equation (6.24), in the rotational term value expression we get

$$\Delta''_2 F(J) = (4B'' - 6D'')(J + \tfrac{1}{2}) - 8D''(J + \tfrac{1}{2})^3 \qquad (6.32)$$

and

$$\Delta'_2 F(J) = (4B' - 6D')(J + \tfrac{1}{2}) - 8D'(J + \tfrac{1}{2})^3 \qquad (6.33)$$

A graph of $\Delta''_2 F(J)/(J + \tfrac{1}{2})$ versus $(J + \tfrac{1}{2})^2$ is a straight line of slope $8D''$ and intercept $4B''$ (or, strictly, $(4B'' - 6D'')$, but $6D''$ may be too small to affect the

---

† The reason for the subscript 2 in the $\Delta''_2 F(J)$ symbol is that these are the differences between rotational term values, in a particular vibrational state, with J differing by 2.

Table 6.2 Rotational and vibrational constants for $^1H^{35}Cl$.

$v = 0$	$v = 1$
$B_0 = 10.440\ 254\ cm^{-1}$	$B_1 = 10.136\ 228\ cm^{-1}$
$D_0 = 5.2828 \times 10^{-4}\ cm^{-1}$	$D_1 = 5.2157 \times 10^{-4}\ cm^{-1}$
ω_0 (for $v = 1$–0 transition) = 2885.9775 cm^{-1}	
$B_e = 10.593\ 42\ cm^{-1}$	
$\alpha_e = 0.307\ 18\ cm^{-1}$	

intercept). Similarly a graph of $\Delta'_2 F(J)/(J + \frac{1}{2})$ versus $(J + \frac{1}{2})^2$ is a straight line of slope $8D'$ and intercept $4B'$.

If B_v can be obtained for at least two vibrational levels, say B_0 and B_1, then B_e and the vibrational–rotation interaction constant α can be obtained from equation (5.25). Values for B_e and α, together with other constants, are given for $^1H^{35}Cl$ in Table 6.2.

The intensity distribution among rotational transitions in a vibration–rotation band is governed principally by the Boltzmann distribution of population among the initial states, giving

$$\frac{N_{J''}}{N_0} = (2J'' + 1) \exp\left[-\frac{hcB''J''(J'' + 1)}{kT} \right] \tag{6.34}$$

as for pure rotational transitions (equation 5.15).

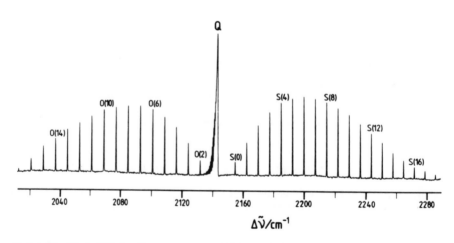

Figure 6.9 The 1–0 Stokes Raman spectrum of CO showing the O-, Q-, and S-branch rotational structure.

6.1.4.2 Raman spectra

The rotational selection rule for vibrational–rotation Raman transitions in diatomic molecules is

$$\Delta J = 0, \pm 2 \tag{6.35}$$

giving a $Q\,(\Delta J = 0)$, an $S\,(\Delta J = +2)$, and an $O\,(\Delta J = -2)$ branch, as shown in Figure 6.7(b).

Figure 6.9 shows the resulting rotational structure of the $v = 1$–0 Stokes Raman transition of CO. The approximate symmetry of the band is due, as in the infrared vibration–rotation spectrum, to the fact that $B_1 \simeq B_0$. If we assume that $B_1 = B_0 = B$ then the wavenumbers $\tilde{v}[S(J)]$, $\tilde{v}[O(J)]$, and $\tilde{v}[Q(J)]$ of the S-, O- and Q-branch lines respectively are given by

$$\tilde{v}[S(J)] = \tilde{v}_0 + B(J + 2)(J + 3) - BJ(J + 1) \tag{6.36}$$
$$= \tilde{v}_0 + 4BJ + 6B$$

$$\tilde{v}[O(J)] = \tilde{v}_0 + B(J - 2)(J - 1) - BJ(J + 1) \tag{6.37}$$
$$= \tilde{v}_0 - 4BJ + 2B$$

$$\tilde{v}[Q(J)] = \tilde{v}_0 \tag{6.38}$$

The Q-branch lines are coincident in this approximation. The separation of the first S-branch line $S(0)$ and the first O-branch line $O(2)$ is $12B$ and the separation of adjacent S-branch lines and of adjacent O-branch lines is $4B$.

More accurately, we can use the method of combination differences, while still neglecting centrifugal distortion, to obtain B'' and B'. Transitions having wavenumbers $\tilde{v}[S(J - 2)]$ and $\tilde{v}[O(J + 2)]$ have a common upper state so that the corresponding combination difference $\Delta_4'' F(J)$ is a function of B'' only:

$$\Delta_4'' F(J) = \tilde{v}[S(J - 2)] - \tilde{v}[O(J + 2)] = 8B''(J + \tfrac{1}{2}) \tag{6.39}$$

Similarly transitions having wavenumbers $\tilde{v}[S(J)]$ and $\tilde{v}[O(J)]$ have a common lower state and

$$\Delta_4' F(J) = \tilde{v}[S(J)] - \tilde{v}[O(J)] = 8B'(J + \tfrac{1}{2}) \tag{6.40}$$

Plotting a graph of $\Delta_4'' F(J)$ versus $(J + \tfrac{1}{2})$ and $\Delta_4' F(J)$ versus $(J + \tfrac{1}{2})$ gives straight lines with slopes $8B''$ and $8B'$ respectively.

Nuclear spin statistical weights have been discussed in Section 5.3.4 and the effects on the populations of the rotational levels in the $v = 0$ states of $^1H_2$, $^{19}F_2$, $^2H_2$, $^{14}N_2$, and $^{16}O_2$ illustrated as examples in Figure 5.18. The effect of these statistical weights in the vibration–rotation Raman spectra is to cause a J'' even:odd intensity alternation of 1:3 for $^1H_2$ and $^{19}F_2$, 6:3 for $^2H_2$ and $^{14}N_2$, and, for $^{16}O_2$, all transitions with J'' even are absent. It is for the purposes of Raman spectra of such homonuclear diatomics that the 's' or 'a' labels, indicating symmetry or antisymmetry of the rotational wave function ψ_r, with respect to the interchange of nuclei, are attached to Figure 5.18.

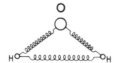

Figure 6.10 Ball-and-spring model of H_2O.

6.2 Polyatomic Molecules

6.2.1 Group Vibrations

An N-atomic molecule has $3N - 5$ normal modes of vibration if it is linear and $3N - 6$ if it is non-linear: these expressions were derived in Section 4.3.1.

Classically we can think of the vibrational motions of a molecule as being those of a set of balls representing the nuclei, of various masses, connected by Hooke's law springs representing the various forces acting between the nuclei. Such a model for the H_2O molecule is illustrated in Figure 6.10. The stronger forces between the bonded O and H nuclei are represented by strong springs which provide resistance to stretching the bonds. The weaker force between the non-bonded hydrogen nuclei is represented by a weaker spring which provides resistance to an increase or decrease of the HOH angle.

Even with this simple model it is clear that if one of the nuclei is given a sudden displacement it is very likely that the whole molecule will undergo a very complicated motion, a Lissajous motion, consisting of a mixture of angle-bending and bond-stretching. The Lissajous motion can always be broken down into a combination of the so-called normal vibrations of the system which, in the Lissajous motion, are superimposed in varying proportions.

A normal mode of vibration is one in which all the nuclei undergo harmonic motion, have the same frequency of oscillation, and move in phase but generally with different amplitudes. Examples of such normal modes[†] are v_1 to v_3 of H_2O, in Figure 4.14, and v_1 to v_{4b} of NH_3 in Figure 4.16. The arrows attached to the nuclei are vectors representing the relative amplitudes and directions of motion.

The form of the normal vibrations may be obtained from a knowledge of the bond lengths and angles and of the bond-stretching and angle-bending force constants, which are a measure of the strengths of the various springs in the ball-and-spring model. However, the calculations are complex and will not be considered here, but useful references are given in the bibliography.

In an approximation which is analogous to that which we have used for a diatomic molecule, each of the vibrations of a polyatomic molecule can be regarded as harmonic. Quantum mechanical treatment in the harmonic oscilla-

---

[†] For the vibrational numbering scheme see the footnote on page 87.

tor approximation shows that the vibrational term values $G(v_i)$ associated with each normal vibration i, all taken to be non-degenerate, are given by

$$G(v_i) = \omega_i(v_i + \tfrac{1}{2}) \qquad (6.41)$$

where ω_i is the classical vibration wavenumber and v_i the vibrational quantum number which can take the values $0,1,2,3,\ldots$. In general, for vibrations with a degree of degeneracy d_i, equation (6.41) becomes

$$G(v_i) = \omega_i\left(v_i + \frac{d_i}{2}\right) \qquad (6.42)$$

As for a diatomic molecule, the general harmonic oscillator selection rule for infrared and Raman vibrational transitions is

$$\Delta v_i = \pm 1 \qquad (6.43)$$

for each vibration, with $\Delta v_i = \pm 2, \pm 3, \ldots$ overtone transitions allowed, but generally weak, when account is taken of anharmonicity.

In addition there is the possibility of combination tones involving transitions to vibrationally excited states in which more than one normal vibration is excited. Fundamental, overtone, and combination tone transitions involving two vibrations v_i and v_j are illustrated in Figure 6.11.

For vibrational transitions to be allowed in the infrared spectrum there is an additional requirement that there must be an accompanying change of dipole moment and, in the Raman spectrum, a change of amplitude of the induced dipole moment (equation 5.43). These requirements necessitate further selection rules which depend on the symmetry properties of the molecule concerned and will be discussed in Section 6.2.2. However, in this discussion of group vibrations we shall be concerned primarily with molecules of low symmetry for which

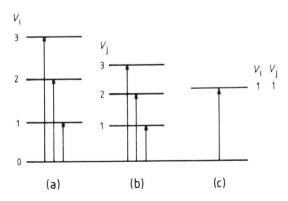

Figure 6.11 (a,b) Fundamental and overtone, and (c) combination tone transitions involving vibrations v_i and v_j.

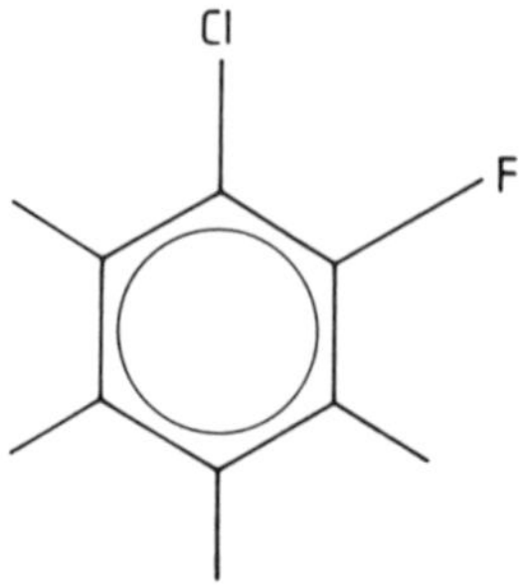

Figure 6.12 2-Chlorofluorobenzene.

these selection rules are fairly unrestrictive. For example, in 2-chlorofluoro-benzene having only one plane of symmetry (C_s point group), shown in Figure 6.12, all the thirty normal vibrations involve a change of dipole moment and of amplitude of the induced dipole moment. Therefore all $\Delta v_i = 1$ transitions are allowed in the infrared and Raman spectra. However, their intensities depend on the magnitudes of the change of dipole moment and of the change of the amplitude of the induced dipole moment: these may be so small for some vibrations that, although the transitions are allowed, they are too weak to be observed.

Although, in general, a normal mode of vibration involves movement of all the atoms in a molecule there are circumstances in which movement is more or less localized in a part of the molecule. For example, if the vibration involves the stretching or bending of a terminal —X—Y group, where X is heavy compared to Y, as in the —O—H group of ethyl alcohol (CH_3CH_2OH), the corresponding vibration wavenumbers are almost independent of the rest of the molecule to which —X—Y is attached. In ethyl alcohol the motions of the hydrogen atom of the OH group are approximately those which it would have if it were attached to an infinite mass by a bond whose force constants are typical of an OH bond. For this reason we speak of a typical wavenumber of an OH-stretching vibration, for which the symbol $v(O—H)$ is used[†], of 3590 to 3650 cm^{-1} in the absence of hydrogen bonding. The wavenumber range is small and reflects the relatively slight dependence on the part of the molecule in the immediate neighbourhood of the group. Such a typical wavenumber is called a group wavenumber or, incorrectly but commonly, a group frequency. Another group wavenumber of the OH group is the bending, or deformation, vibration which is typically in the 1050 to 1200 cm^{-1} range.

Other general circumstances in which normal vibrations tend to be localized in a particular group of atoms arise when there is a chain of atoms in which the

---

[†] It is common to use the symbols $v(X–Y)$, $\delta(X–Y)$, and $\gamma(X–Y)$, for stretching, in-plane bending, and out-of-plane bending respectively, in the X–Y group. In addition the word 'deformation' is often used to imply a bending motion.

force constant between two of them is very different from those between other atoms in the chain. For example, in the molecule $HC\equiv C-CH=CH_2$ the force constants in the $C-C$, $C=C$, and $C\equiv C$ bonds are quite dissimilar. It follows that the stretchings of the bonds are not strongly coupled and that each stretching vibration wavenumber is typical of the $C-C$, $C=C$, or $C\equiv C$ group.

Table 6.3 lists a number of group vibration wavenumbers for both bond-stretching and angle-bending vibrations.

Table 6.3 Typical bond-stretching and angle-bending group vibration wavenumbers $\tilde{\nu}$.

Bond-stretching		Bond-stretching	
Group	$\tilde{\nu}/\text{cm}^{-1}$	Group	$\tilde{\nu}/\text{cm}^{-1}$
$\equiv C-H$	3300	$-O-H$	3600†
$=C\big\langle{}^H$	3020	$\rangle N-H$	3350
except $O=C\big\langle{}^H$	2800	$\rangle\!\!\!=\!P=O$	1295
$\geqq C-H$	2960	$\rangle S=O$	1310
$-C\equiv C-$	2050	Angle-bending	
		$\equiv\!\overset{\frown}{C}\!-H$	700
$\rangle C=C\langle$	1650	$=C\overset{\frown}{\big\langle}{}^H_H$	1100
$\geqq C-C\leqq$	900		
		$-\overset{\frown}{\underset{H}{C}}{}^H_H$	1000
$\geqq Si-Si\leqq$	430		
$\rangle C=O$	1700	$\rangle C\overset{\frown}{\big\langle}{}^H_H$	1450
$-C\equiv N$	2100		
$\geqq C-F$	1100		
$\geqq C-Cl$	650	$C\overset{\frown}{\equiv}\overset{\frown}{C}-C$	300
$\geqq C-Br$	560		
$\geqq Cl-I$	500		

† May be reduced in a condensed phase by hydrogen bonding.

Not all parts of a molecule are characterized by group vibrations. Many normal modes involve strong coupling between stretching or bending motions of atoms in a straight chain, a branched chain, or a ring. Such vibrations are called skeletal vibrations and tend to be specific to a particular molecule. For this reason the region where skeletal vibrations mostly occur, from about 1300 cm^{-1} to low wavenumber, is sometimes called the fingerprint region. On the other hand, the region from 1500 to 3700 cm^{-1}, where many transferable group vibrations occur, is known as the functional group region.

In addition to the descriptions of group vibrations as stretch and bend (or deformation) the terms rock, twist, scissors, wag, torsion, ring breathing, and inversion (or umbrella) are used frequently: these motions are illustrated in Figure 6.13.

The use of group vibrations as an important tool in qualitative analysis of a molecular sample was, until the advent of laser Raman spectroscopy, confined largely to infrared spectroscopy. In the infrared spectrum the intensity of absorption due to a particular vibration depends on the change of dipole moment during the vibration, much as for a diatomic molecule (see Figure 6.1). For example, the stretching vibration of the strongly polar C=O bond gives a strong absorption band whereas that of the C=C bond gives a weak band. Indeed, if the C=C bond is in a symmetrical molecule such as $H_2C=CH_2$ there is no change of dipole moment at all and the vibration is infrared inactive. If the C=C bond is in, say, $HFC=CH_2$ there is a small change of dipole moment due to stretching of the bond but clearly not as large a change as that due to stretching of the C—F bond.

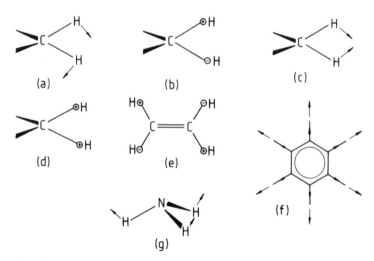

Figure 6.13 Illustration of (a) rocking, (b) twisting, (c) scissoring, and (d) wagging vibrations in a CH_2 group. Also shown are (e) the torsional vibration in ethylene, (f) the ring breathing vibration in benzene, and (g) the inversion, or umbrella, vibration in ammonia.

Just as group vibration wavenumbers are fairly constant from one molecule to another, so are their intensities. For example, if a molecule were being tested for the presence of a C—F bond there must be not only an infrared absorption band due to bond-stretching at about 1100 cm^{-1} but it must also be intense. A weak band in this region might be attributable to another normal mode.

Infrared spectra for purposes of qualitative analysis may be obtained from a pure liquid, the solid, embedded in a KBr disk, or a solution using a non-polar solvent: polar solvents may affect both group vibration wavenumbers and band intensities and may be involved in hydrogen-bonding with the solute. For example, the O—H stretching vibration of phenol in solution in hexane has a wavenumber of 3622 cm^{-1}, but this is reduced to 3344 cm^{-1} in diethylether because of the hydrogen-bonding illustrated in Figure 6.14.

The use of vibrational Raman spectroscopy in qualitative analysis has increased greatly since the introduction of lasers which have replaced mercury arcs as monochromatic sources. Although a laser Raman spectrometer is more expensive than a typical infrared spectrometer used for qualitative analysis, it does have the advantage that low and high wavenumber vibrations can be observed with equal ease whereas in the infrared a different, far-infrared, spectrometer may be required for observations below abut 400 cm^{-1}.

The observation of a vibrational band in the Raman spectrum depends on there being an accompanying change of the amplitude of the induced dipole moment and the intensity, as in the case of a diatomic molecule (see Figure 6.2), depends on the magnitude of this change. It seems that, in general, this change is less sensitive than the change of dipole moment to the environment of the vibrating group. As a result, group vibration intensities are more accurately transferable from one molecule to another, and from one phase or solvent to another, in the Raman than in the infrared spectrum.

Figure 6.15 shows the infrared spectrum of s-trans-crotonaldehyde, illustrated in Figure 6.16, and Figure 6.17 shows the laser Raman spectrum. The infrared spectrum is mostly of a solution in carbon tetrachloride but partly of a thin

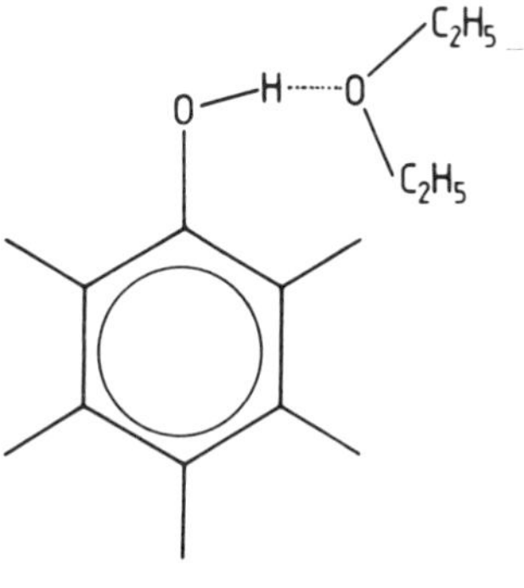

Figure 6.14 Hydrogen bonding between phenol and diethylether.

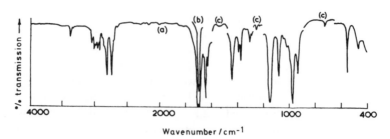

Figure 6.15 The infrared vibrational spectrum of crotonaldehyde. The parts marked (a), (b), and (c) refer to a 10 per cent (by volume) solution in CCl_4, a 1 per cent solution in CCl_4, and a thin liquid film respectively. (Reproduced, with permission, from Bowles, A. J., George, W. O., and Maddams, W. F., *J. Chem. Soc.* (B), 810, 1969.)

film of the pure liquid in the region where carbon tetrachloride itself absorbs. The Raman spectrum is of the pure liquid. Table 6.4 records the vibration wavenumbers of all twenty-seven normal modes, together with an approximate description of the vibrational motions. A comparison of this table with Table 6.3 shows that v_1, v_2, v_5, v_6, v_7 and v_{15} are all well-behaved group vibrations. A comparison of the infrared and Raman spectra shows many similarities of intensities but also some large differences. For example, v_{15}, the $C—CH_3$ stretching vibration, is strong in the infrared but very weak in the Raman whereas v_3, the CH_3 antisymmetric stretching vibration, is very strong in the Raman but weak in the infrared.

The Raman spectrum can be used to give additional information regarding the symmetry properties of vibrations. This information derives from the measurement of the depolarization ratio ρ for each Raman band. The quantity ρ is a measure of the degree to which the polarization properties of the incident radiation may be changed after scattering has occurred, and is often used to distinguish between totally symmetric and non-totally symmetric vibrations.

In addition to bands in the infrared and Raman spectra due to $\Delta v = 1$ transitions, combination and overtone bands may occur with appreciable intensity, particularly in the infrared. Care must be taken not to confuse such bands with weakly active fundamentals. Occasionally combinations and, more often, overtones may be used to aid identification of group vibrations.

Figure 6.16 *s-trans*-Crotonaldehyde.

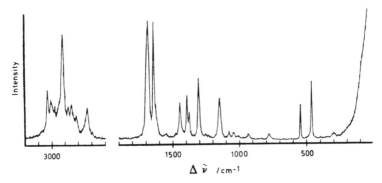

Figure 6.17 The laser Raman vibrational spectrum of liquid crotonaldehyde. (Reproduced, with permission, from Durig, J. R., Brown, S. C., Kalasinsky, V. F., and George, W. O., *Spectrochim. Acta*, **32A**, 807, 1976. Copyright (1976) Pergamon Press.)

Most of the discussion of group vibrations so far has been confined to organic molecules because it is to such molecules that their use is most commonly applied. In inorganic molecules and ions the approximations involved in the concept of group vibration wavenumbers are not so often valid. Nevertheless, vibrations involving stretching of M—H, M—C, M—X, and M=O bonds, where M is a metal atom and X a halogen atom, are useful group vibrations.

Inorganic complexes containing organic ligands exhibit at least some vibrations which are characteristic of the ligands.

Because of the presence of heavy atoms in many inorganic molecules there may be several low wavenumber vibrations. For this reason it is generally more important than for organic molecules to obtain the far-infrared or Raman spectrum.

6.2.2 Vibrational Selection Rules

6.2.2.1 Infrared spectra

In the process of absorption or emission of infrared radiation involving transitions between two vibrational states the interaction is usually between the molecule and the electric, rather than the magnetic, component of the electromagnetic radiation (see Section 2.1). For this reason infrared selection rules are sometimes referred to as electric dipole, or simply dipole, selection rules.

In the case of H_2O it is easy to see from the form of the normal modes shown in Figure 4.14 that all the vibrations v_1, v_2, and v_3 involve a change of dipole moment and are infrared active, that is $v = 1-0$ transitions in each vibration are allowed. The transitions may be labelled 1_0^1, 2_0^1, and 3_0^1 according to a useful, but not universal, convention for polyatomic molecules in which $N_{v''}^{v'}$ refers to a transition with lower and upper state vibrational quantum numbers v'' and v' respectively in vibration N.

Table 6.4 Fundamental vibration wavenumbers of crotonaldehyde obtained from the infra-red and Raman spectra.

Vibration[†]	Approximate description	$\tilde{\nu}/\text{cm}^{-1}$	
		Infrared	Raman
In plane			
ν_1	CH antisymmetric stretch on C=C	3042	3032
ν_2	CH symmetric stretch on C=C	3002	3006
ν_3	CH_3 antisymmetric stretch	2944	2949
ν_4	CH_3 symmetric stretch	2916	2918
ν_5	CH stretch on CHO	2727	2732
ν_6	C=O stretch	1693	1682
ν_7	C=C stretch	1641	1641
ν_8	CH_3 antisymmetric deformation	1444	1445
ν_9	CH rock (in-plane bend) on CHO	1389	1393
ν_{10}	CH_3 symmetric deformation	1375	1380
ν_{11}	CH symmetric deformation on C=C	1305	1306
ν_{12}	CH antisymmetric deformation on C=C	1253	1252
ν_{13}	CH_3 in-plane rock	1075	1080
ν_{14}	C—CHO stretch	1042	1046
ν_{15}	C—CH_3 stretch	931	931
ν_{16}	CH_3—C=C bend	542	545
ν_{17}	C=C—C bend	459	464
ν_{18}	C—C=O bend	216	230
Out-of-plane			
ν_{19}	CH_3 antisymmetric stretch	2982	2976
ν_{20}	CH_3 antisymmetric deformation	1444	1445
ν_{21}	CH_3 rock	1146	1149
ν_{22}	CH antisymmetric[‡] deformation on C=C	966	—
ν_{23}	CH symmetric[‡] deformation on C=C	—	780
ν_{24}	CH wag (out-of-plane bend) on CHO	727	—
ν_{25}	CH_3 bend	297	300
ν_{26}	CH_3 torsion	173	—
ν_{27}	CHO torsion	121	—

[†] See the footnote on page 87 for vibrational numbering convention.
[‡] To inversion of the two hydrogens through the centre of the C=C bond.

Acetylene, HC≡CH, being a linear, four-atomic molecule, has seven normal modes of vibration which are illustrated in Figure 6.18. The *trans* and *cis* bending vibrations, ν_4 and ν_5, are each doubly degenerate. This is rather like ν_3 and ν_4 of ammonia (Figure 4.16) except that in acetylene the degeneracy of the two components is more obvious. For example, in the ν_4 mode the molecule can bend in any two mutually perpendicular planes but the vectors representing the motion in one plane have no components in the other. The two motions clearly have different wave functions but equal energies.

Inspection of the normal modes shows that only ν_3 and ν_5 involve a change of dipole moment and are infrared active.

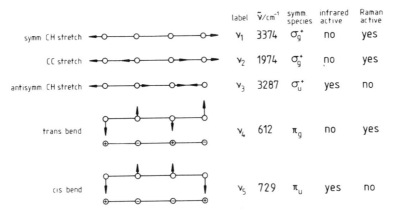

	label	$\tilde{v}/cm^{-1}$	symm. species	infrared active	Raman active
symm CH stretch	v_1	3374	σ_g^+	no	yes
CC stretch	v_2	1974	σ_g^+	no	yes
antisymm CH stretch	v_3	3287	σ_u^+	yes	no
trans bend	v_4	612	π_g	no	yes
cis bend	v_5	729	π_u	yes	no

Figure 6.18 Normal modes of vibration of acetylene.

Although we have been able to see on inspection which vibrational fundamentals of water and acetylene are infrared active, in general this is not the case. It is also not the case for vibrational overtone and combination tone transitions. To be able to obtain selection rules for all infrared vibrational transitions in any polyatomic molecule we must resort to symmetry arguments.

The vibrational transition intensity is proportional to $|R_v|^2$, the square of the vibrational transition moment R_v where

$$R_v = \int \psi_v'^* \mu \psi_v'' \, d\tau_v \qquad (6.44)$$

just as for a diatomic molecule (equation 6.4), where the integration is over vibrational coordinates. Clearly

$$R_v = 0 \text{ for a forbidden transition} \qquad (6.45)$$

$$R_v \neq 0 \text{ for an allowed transition} \qquad (6.46)$$

There are simple symmetry requirements for the integral of equation (6.44) to be non-zero and therefore for the transition to be allowed. If both vibrational states are non-degenerate, the requirement is that the symmetry species of the quantity to be integrated is totally symmetric: this can be written as

$$\Gamma(\psi_v') \times \Gamma(\mu) \times \Gamma(\psi_v'') = A \qquad (6.47)$$

where Γ stands for 'symmetry species of' and A denotes the totally symmetric species of any non-degenerate point group. If either, or both, of the vibrational states are degenerate then equation (6.47) is not quite satisfactory as it stands since the product of two degenerate species, for example $E \times E = A_1 + A_2 + E$ in equation (4.29) and $\Pi \times \Pi = \Sigma^+ + \Sigma^- + \Delta$ in equation (4.33), results in more than one species. So equation (6.47) is modified, for degenerate states, to

$$\Gamma(\psi'_v) \times \Gamma(\mu) \times \Gamma(\psi''_v) \supset A \qquad (6.48)$$

where the boolean symbol $\supset$ means 'contains'. For example, equation (4.29) shows that $E \times E$ contains A_1, that is

$$E \times E \supset A_1 \qquad (6.49)$$

The transition moment of equation (6.44) is a vector and has components

$$R_{v,x} = \int \psi'^{*}_v \mu_x \psi''_v \, d\tau_v; \qquad R_{v,y} = \int \psi'^{*}_v \mu_y \psi''_v \, d\tau_v; \qquad R_{v,z} = \int \psi'^{*}_v \mu_z \psi''_v \, d\tau_v \qquad (6.50)$$

along the x, y, and z axes and, since

$$|R_v|^2 = (R_{v,x})^2 + (R_{v,y})^2 + (R_{v,z})^2 \qquad (6.51)$$

the $v'-v''$ transition is allowed if any of $R_{v,x}$, $R_{v,y}$, or $R_{v,z}$ is non-zero.

Since the dipole moment is a vector in a particular direction it has the same symmetry species as a translation of the molecule in the same direction. Figure 6.19 shows this for H_2O in which the dipole moment and the translation in the same direction have the same symmetry species, the totally symmetric A_1 species. In general

$$\Gamma(\mu_x) = \Gamma(T_x); \qquad \Gamma(\mu_y) = \Gamma(T_y); \qquad \Gamma(\mu_z) = \Gamma(T_z) \qquad (6.52)$$

Then equation (6.47) becomes

$$\Gamma(\psi'_v) \times \Gamma(T_x) \times \Gamma(\psi''_v) = A$$

and/or

$$\Gamma(\psi'_v) \times \Gamma(T_y) \times \Gamma(\psi''_v) = A \qquad (6.53)$$

and/or

$$\Gamma(\psi'_v) \times \Gamma(T_z) \times \Gamma(\psi''_v) = A$$

where the 'and/or' implies that, for an allowed transition, one or more of the components of R_v may be non-zero.

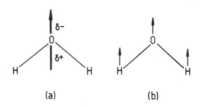

Figure 6.19 (a) the dipole moment vector for H_2O and (b) a translation in the same direction.

If the lower state of the transition is $v'' = 0$, as it very often is, $\Gamma(\psi_v'') = A$, i.e. the wave function is totally symmetric, and equations (6.53) become

$$\Gamma(\psi_v') \times \Gamma(T_x) = A$$

and/or

$$\Gamma(\psi_v') \times \Gamma(T_y) = A \tag{6.54}$$

and/or

$$\Gamma(\psi_v') \times \Gamma(T_z) = A$$

since multiplying something by the totally symmetric species leaves it unchanged. Also, if two species multiplied together give the totally symmetric species, they must be equal. Therefore, from equations (6.54), we have

$$\Gamma(\psi_v') = \Gamma(T_x) \text{ and/or } \Gamma(T_y) \text{ and/or } \Gamma(T_z) \tag{6.55}$$

which is the vibrational selection rule for transitions between the $v = 0$ and any other non-degenerate fundamental, overtone, or combination level. The analogous selection rule for transitions between the $v = 0$ and any degenerate level follows from equation (6.48) and is

$$\Gamma(\psi_v') \supset \Gamma(T_x) \text{ and/or } \Gamma(T_y) \text{ and/or } \Gamma(T_z) \tag{6.56}$$

The procedure for determining the selection rules for a particular molecule is:

1. Assign the molecule to a point group.
2. Look up the translational symmetric species in the relevant character table.
3. The allowed transitions, with $v'' = 0$ as the lower state, are

$$\Gamma(T_x) - A; \qquad \Gamma(T_y) - A; \qquad \Gamma(T_z) - A \tag{6.57}$$

where we indicate a transition between an upper state Y and a lower state X as $Y - X$.

Using the C_{2v} character table (Table A.11 in the Appendix) we can immediately write down the allowed transitions involving the zero-point level as

$$A_1 - A_1; \qquad B_1 - A_1; \qquad B_2 - A_1 \tag{6.58}$$

These are polarized along the z, x, and y axes respectively, since they involve μ_z, μ_x, and μ_y. This means that when, for example, a $B_1 - A_1$ transition occurs an oscillating electric dipole is set up along the x axis.

It follows from equation (6.58) that the 1_0^1, 2_0^1, and 3_0^1 transitions of H_2O are allowed since v_1, v_2, and v_3 are a_1, a_1, and b_2 vibrations[†] respectively, as

---

[†] Lower case letters are used here to label vibrations while upper case letters are used for the corresponding wave function but, frequently, upper case letters are used for both.

equation (4.11) shows. We had derived this result previously simply by observing that all three vibrations involve a changing dipole moment but the rules of equation (6.57) enable us to derive selection rules for overtone and combination transitions as well.

Equation (4.14) tells us that, if H_2O is vibrating with two quanta of v_3, then $\Gamma(\psi'_v)$ is A_1 and, in general, if it is vibrating with n quanta of a vibration with symmetry species S then

$$\Gamma(\psi'_v) = S^n \tag{6.59}$$

Figure 6.20 shows, for example, that the symmetry species of vibrational fundamental and overtone levels for v_3 alternate, being A_1 for v even and B_2 for v odd. It follows that the $3_0^1, 3_0^2, 3_0^3, \ldots$ transitions are allowed and polarized along the $y, z, y, \ldots$ axes (see Figure 4.13 for axis labelling).

Equation (4.12) gives the result that, if H_2O is vibrating with one quantum of each of v_1 and v_3, then $\Gamma(\psi'_v) = B_2$. The result is the same as if it were vibrating with one quantum of each of v_2 and v_3, as indicated in Figure 6.20. We deduce, therefore, that both the $1_0^1 3_0^1$ and $2_0^1 3_0^1$ combination transitions are allowed and polarized along the y axis.

The H_2O molecule has no a_2 or b_1 vibrations but selection rules for, say, CH_2F_2, which has vibrations of all symmetry species, could be applied in an analogous way.

In NH_3, belonging to the C_{3v} point group, there are degenerate E vibrations v_3 and v_4, shown in Figure 4.16. The transitions 1_0^1 and 2_0^1 are allowed and polarized along the $z(C_3)$ axis because both v_1 and v_2 are a_1 vibrations and Table A.12 in the Appendix shows that $\Gamma(T_z) = A_1$. Similarly 3_0^1 and 4_0^1 are allowed and polarized in the xy plane because $\Gamma(T_x, T_y) = E$.

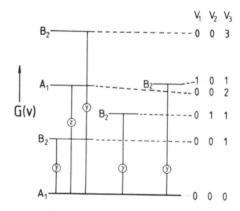

Figure 6.20 Symmetry species of some overtone and combination levels of H_2O together with directions of polarization of transition moments. The vibration wavenumbers are $\tilde{v}_1 = 3657.1$ cm^{-1}, $\tilde{v}_2 = 1594.8$ cm^{-1}, $\tilde{v}_3 = 3755.8$ cm^{-1}.

If the upper state is a combination or overtone level $\Gamma(\psi'_v)$ may not be a single symmetry species and we must use equation (6.56). For example, for the combination transition $3^1_0 4^1_0$ in NH_3 we have, according to equation (4.29),

$$\Gamma(\psi'_v) = E \times E = A_1 + A_2 + E \qquad (6.60)$$

Since $\Gamma(T_x, T_y) = E$ and $\Gamma(T_z) = A_1$ it is clear that $\Gamma(\psi'_v) \supset \Gamma(T_x, T_y)$ and $\Gamma(T_z)$ and the transition is allowed. More specifically, transitions involving two of the three components (A_1 and E) of the combination state are allowed but that involving the A_2 component is forbidden.

For the overtone transition 3^2_0 we have, according to equation (4.30),

$$\Gamma(\psi'_v) = (E)^2 = A_1 + E \qquad (6.61)$$

and transitions to both components are allowed.

Acetylene ($HC{\equiv}CH$) belongs to the $D_{\infty h}$ point group whose character table is given in Table A.37 in the Appendix, and its vibrations are illustrated in Figure 6.18. Since v_3 is a σ_u^+ vibration and $\Gamma(T_z) = \Sigma_u^+$, the 3^1_0 transition is allowed and the transition moment is polarized along the z axis. Similarly, since v_5 is a π_u vibration, the 5^1_0 transition is allowed with the transition moment in the xy plane.

The symmetry species of the $4^1 5^1$ combination state is given by

$$\Gamma(\psi'_v) = \Pi_g \times \Pi_u = \Sigma_u^+ + \Sigma_u^- + \Delta_u \qquad (6.62)$$

a result which can be obtained in a similar way to that of equation (4.33) for the $C_{\infty v}$ point group. Therefore, one component of the $4^1_0 5^1_0$ transition, namely the one involving the Σ_u^+ level, is allowed.

The symmetry species of the 5^2 state is given by

$$\Gamma(\psi'_v) = (\Pi_u)^2 = \Sigma_g^+ + \Delta_g \qquad (6.63)$$

a result similar to that in equation (4.34). Therefore the 5^2_0 transition is forbidden.

It is important to remember that selection rules, in general, tell us nothing about transition intensities other than their being zero or non-zero. It is possible that, even though it is non-zero, it may be so small that the transition escapes observation.

It should also be remembered that the selection rules derived here are relevant to the free molecule and may break down in the liquid or solid state. This is

Figure 6.21 The v_4 vibration of ethylene.

the case, for example, with the electric dipole forbidden 4_0^1 transition in ethylene, where v_4 is the a_u torsional vibration shown in Figure 6.21. It is not observed in the infrared spectrum of the gas but is observed weakly in the liquid and solid phases.

6.2.2.2 Raman spectra

Analogous to equation (6.47) for an allowed infrared transition is the requirement

$$\Gamma(\psi_v') \times \Gamma(\alpha_{ij}) \times \Gamma(\psi_v'') = A \qquad (6.64)$$

for a Raman vibrational transition to be allowed between non-degenerate states. Here α_{ij} represents any of the components of the polarizability tensor $\boldsymbol{\alpha}$ in equation (5.42). In general, there are six different components, α_{xx}, α_{yy}, α_{zz}, α_{xy}, α_{xz}, and α_{yz}, and their symmetry species are always given at the right-hand side of each character table, as is apparent from those in the Appendix. For transitions between vibrational states, at least one of which is degenerate, equation (6.64) is replaced by

$$\Gamma(\psi_v') \times \Gamma(\alpha_{ij}) \times \Gamma(\psi_v'') \supset A \qquad (6.65)$$

In the most usual cases, where the lower level is the zero-point level, $\Gamma(\psi_v'') = A$ and the requirement for a Raman vibrational transition becomes

$$\Gamma(\psi_v') = \Gamma(\alpha_{ij}) \qquad (6.66)$$

for degenerate or non-degenerate vibrations.

Inspection of the C_{2v} and C_{3v} character tables (Tables A.11 and A.12 in the Appendix) shows that all the fundamentals of H_2O (Figure 4.14) are allowed in the Raman spectrum since $\Gamma(\alpha_{xx}, \alpha_{yy}, \alpha_{zz}) = A_1$ and $\Gamma(\alpha_{yz}) = B_2$. Similarly, all those of NH_3 (Figure 4.16) are allowed since $\Gamma(\alpha_{xx}, \alpha_{yy}, \alpha_{zz}) = A_1$ and $\Gamma[(\alpha_{xx} - \alpha_{yy}, \alpha_{xy}) (\alpha_{xz}, \alpha_{yz})] = E$.

The $D_{\infty h}$ character table (Table A.37 in the Appendix) shows that, of the 1–0 bands of acetylene (Figure 6.18), only 1_0^1, 2_0^1, and 4_0^1 are allowed in the Raman spectrum.

The vibrations of acetylene provide an example of the so-called mutual exclusion rule. The rule states that, for a molecule with a centre of inversion, the fundamentals which are active in the Raman spectrum (g vibrations) are inactive in the infrared spectrum while those active in the infrared spectrum (u vibrations) are inactive in the Raman spectrum, i.e. the two spectra are mutually exclusive. However, there are some vibrations which are forbidden in both spectra, such as the a_u torsional vibration of ethylene in Figure 6.21: in the D_{2h} point group (Table A.32 in the Appendix) a_u is the species of neither a translation nor a component of the polarizability.

6.2.3 Vibration–Rotation Spectroscopy

Raman scattering is normally of such very low intensity that gas phase Raman spectroscopy is one of the more difficult techniques. This is particularly the case for vibration–rotation Raman spectroscopy since scattering involving vibrational transitions is much weaker than that involving rotational transitions which were described in Sections 5.3.3 and 5.3.5. For this reason we shall consider here only the more easily studied infrared vibration–rotation spectroscopy which must also be investigated in the gas phase.

As for diatomic molecules, there are stacks of rotational energy levels associated with all vibrational levels of a polyatomic molecule. The resulting term values S are given by the sum of the rotational and vibrational term values

$$S = F_{v_i} + G(v_i) \tag{6.67}$$

where i refers to a particular vibration. When each vibration is treated in the harmonic oscillator approximation the vibrational term values are given by

$$G(v_i) = \omega_i \left(v_i + \frac{d_i}{2} \right) \tag{6.68}$$

where ω_i is the classical vibration wavenumber, v_i the quantum number, and d_i the degree of degeneracy.

6.2.3.1 *Infrared spectra of linear molecules*

Linear molecules belong to either the $D_{\infty h}$ (with an inversion centre) or the $C_{\infty v}$ (without an inversion centre) point group. Using the vibrational selection rule in equation (6.56) and the $D_{\infty h}$ (Table A.37 in the Appendix) or $C_{\infty v}$ (Table A.16 in the Appendix) character table we can see that the vibrational selection rules for transitions from the zero-point level (Σ_g^+ in $D_{\infty h}$, Σ^+ in $C_{\infty v}$) allow transitions of the type

$$\Sigma_u^+ - \Sigma_g^+ \text{ and } \Pi_u - \Sigma_g^+ \tag{6.69}$$

in $D_{\infty h}$ and

$$\Sigma^+ - \Sigma^+ \text{ and } \Pi - \Sigma^+ \tag{6.70}$$

in $C_{\infty v}$.

For all types of Σ vibrational levels the stack of rotational levels associated with them is given by

$$F_v(J) = B_v J(J + 1) - D_v J^2(J + 1)^2 \tag{6.71}$$

just as for a diatomic molecule (equation 5.23). Two such stacks are shown in Figure 6.22 for a Σ_u^+ and Σ_g^+ vibrational level.

The rotational selection rule is

$$\Delta J = \pm 1 \tag{6.72}$$

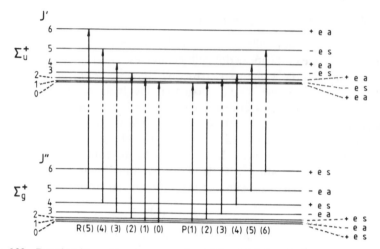

Figure 6.22 Rotational transitions accompanying a $\Sigma_u^+ - \Sigma_g^+$ infrared vibrational transition in a $D_{\infty h}$ linear polyatomic molecule. For $C_{\infty v}$ the g and u subscripts and s and a labels should be dropped.

as for a diatomic molecule, and the resulting spectrum shows a P branch ($\Delta J = -1$) and an R branch ($\Delta J = +1$) with members of each separated by about $2B$, where B is an average rotational constant for the two vibrational states, and a spacing of about $4B$ between $R(0)$ and $P(1)$. Treatment of the wavenumbers of the P and R branches to give the rotational constants B_1 and B_0, for the upper and lower states respectively, follows exactly the same method of combination differences derived for diatomic molecules in equations (6.27) to (6.33).

Just as for diatomics, for a $D_{\infty h}$ polyatomic molecule rotational levels are symmetric (s) or antisymmetric (a) to nuclear exchange which, when nuclear spins are taken into account, may result in an intensity alternation with J. These labels are given in Figure 6.22.

Also given in Figure 6.22 are the parity labels, $+$ or $-$, and the alternative e or f labels for each rotational level. The general selection rules involving these properties are[†]

$$+\leftrightarrow-, \quad +\nleftrightarrow+, \quad -\nleftrightarrow- \tag{6.73}$$

or alternatively,

$$e\leftrightarrow f, e\nleftrightarrow e, f\nleftrightarrow f \text{ for } \Delta J = 0 \tag{6.74}$$
$$e\nleftrightarrow f, e\leftrightarrow e, f\leftrightarrow f \text{ for } \Delta J = \pm 1$$

but these are superfluous and can be ignored for a $\Sigma-\Sigma$ type of transition.

[†] $\leftrightarrow$ and $\nleftrightarrow$ indicate allowed and forbidden transitions, respectively, whichever is the upper state.

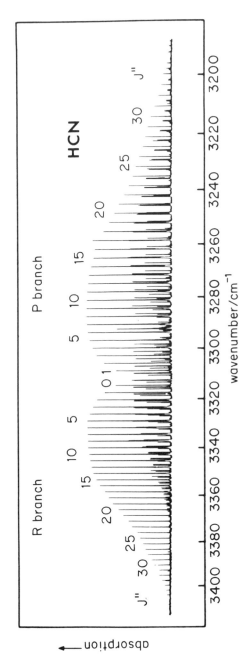

Figure 6.23 The 3^1_0, $\Sigma^+ - \Sigma^+$ infrared band of HCN and two weaker, overlapping bands. (Reproduced, with permission, from Cole, A. R. H., *Tables of Wavenumbers for the Calibration of Infrared Spectrometers*, 2nd ed., p. 28. Copyright (1977) Pergamon Press, Oxford.)

Figure 6.23 shows the 3_0^1 band of HCN in which v_3 is the C—H stretching vibration. The transition is of the Σ^+–Σ^+ type showing clear P- and R- branch structure like a diatomic molecule. However, unlike a diatomic molecule, there are two hot bands, involving lower states in which the low wavenumber bending vibration v_2 is excited, overlapping 3_0^1. One band shows a P, Q, and R branch with the band centre at about $3292\ \mathrm{cm}^{-1}$ and the other a P and R branch with the band centre at about $3290\ \mathrm{cm}^{-1}$.

Figure 6.24 shows the rotational levels associated with a Π_u and a Σ_g^+ vibrational level. The Π_u stack of levels differs from the Σ_g^+ stack in two respects: (a) there is no $J = 0$ level and (b) each of the levels is split, the splitting increasing with J. The reason for there being no $J = 0$ level is that there is one quantum of angular momentum in a Π vibrational state. Therefore J, which actually refers to the *total* angular momentum, cannot be less than 1. It is this angular momentum associated with a Π state which also accounts for the splitting of the rotational levels, an effect known as ℓ-type doubling and due to Coriolis forces, modifying the term value expression to

$$F_v(J) = B_v J(J + 1) - B_v \pm \frac{q_i}{2} J(J + 1) \qquad (6.75)$$

for the $v_i = 1$ level of a π vibration, where q_i is a parameter which determines the magnitude of the splitting of levels and centrifugal distortion has been neglected. The rotational selection rule is

$$\Delta J = 0,\ \pm 1 \qquad (6.76)$$

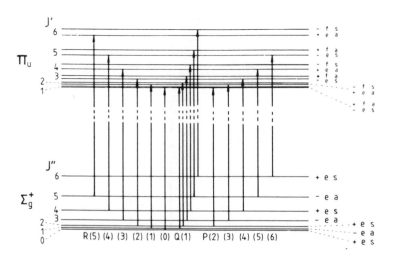

Figure 6.24 Rotational transitions accompanying a $\Pi_u - \Sigma_g^+$ infrared vibrational transition in a $D_{\infty h}$ linear polyatomic molecule. For $C_{\infty v}$ the g and u subscripts and s and a labels should be dropped.

giving a central Q branch ($\Delta J = 0$) as well as P and R branches as Figure 6.24 shows. We can see now that, for a Π–Σ type of transition, the additional selection rules of equation (6.73) or (6.74) are necessary in order to decide whether the P and R branches involve the lower components of the split levels in the Π state and the Q branch the upper components, as is the case in Figure 6.24, or vice versa.

Figure 6.25 shows the $1_0^1 5_0^1$ infrared combination band of acetylene, where v_1 is the symmetric CH stretching vibration and v_5 the *cis* bending vibration, as an example of a Π_u–Σ_g^+ band of a linear molecule. Note that the P branch starts with $P(2)$, rather than $P(1)$ as it would in a Σ–Σ type of transition, and that there is an intensity alternation of 1:3 for J'' even:odd because the two equivalent protons, each with a nuclear spin quantum number $I = \frac{1}{2}$, are exchanged on rotation of the molecule through π radians—exactly as for $^1H_2$ illustrated in Figure 5.18.

The separation of individual lines within the Q branch is small, causing the branch to stand out as more intense than the rest of the band. This appearance is typical of all Q branches in infrared spectra because of the similarity of the rotational constants in the upper and lower states of the transition.

The effective value of B_v, for the lower components of the doubled Π_u levels, can be obtained from the P and R branches by the same method of combination differences used for a Σ–Σ type of band and, for the upper components, from the Q branch. From these two quantities B_v and q_i may be calculated.

However, unless q_i is unusually large or the spectrometer is of fairly high resolution, the doubling of levels is relatively small and can be neglected.

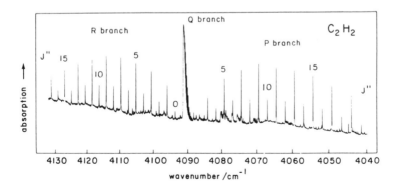

Figure 6.25 The $1_0^1 5_0^1$, $\Pi_u - \Sigma_g^+$ infrared band of acetylene. (Reproduced, with permission, from Cole, A. R. H., *Tables of Wavenumbers for the Calibration of Infrared Spectrometers*, 2nd ed., p. 12. Copyright (1977) Pergamon Press, Oxford.)

6.2.3.2 *Infrared spectra of symmetric rotors*

For a symmetric rotor molecule such as methyl fluoride, a prolate symmetric rotor belonging to the C_{3v} point group, in the zero-point level the vibrational selection rule in equation (6.56) and the character table (Table A.12 in the Appendix) show that only

$$A_1 - A_1 \quad \text{and} \quad E - A_1 \tag{6.77}$$

transitions are allowed. In an $A_1 - A_1$ transition the transition moment is along the top (C_3) axis of the molecule and gives rise to what is called a parallel band while, in an $E - A_1$ transition, the transition moment is perpendicular to the top axis and gives rise to a perpendicular band.

A parallel, $A_1 - A_1$, band involves rotational transitions between stacks of levels like those in Figure 5.6(a), associated with both A_1 states, and given by equation (5.32). The selection rules are

$$\Delta K = 0 \text{ and } \Delta J = \pm 1, \text{ for } K = 0 \tag{6.78}$$

$$\Delta K = 0 \text{ and } \Delta J = 0, \pm 1, \text{ for } K \neq 0$$

giving a P, Q, and R branch for each value of K except $K = 0$ which has no Q branch. As for all molecules, the rotational constants A_v and B_v change very little with v so the P, Q, and R branches for all values of K fall more or less on top of each other. The result is a band looking very much like a Π–Σ type of band of a linear molecule (see Figure 6.25). An example is the 1_0^1 A_1–A_1 infrared band of C^2H_3F, shown in Figure 6.26, where v_1 is the a_1 C—^2H stretching vibration. The separation of the central Q branches with different values of K is seen to be small due to the similarity of rotational constants in the upper and lower vibrational states.

In an E vibrational state there is some splitting of rotational levels, compared to those of Figure 5.6(a), due to Coriolis forces, rather like that found in a Π

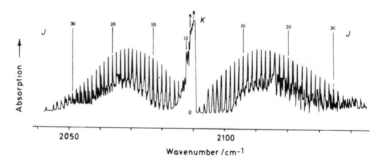

Figure 6.26 The 1_0^1, A_1–A_1 infrared parallel band of C^2H_3F. (Reproduced, with permission, from Jones, E. W., Popplewell, R. J., and Thompson, H. W., *Proc. R. Soc.*, **A290**, 490, 1966.)

vibrational state, but the main difference in an E–A_1 band from an A_1–A_1 band is due to the selection rules

$$\Delta K = \pm 1 \quad \text{and} \quad \Delta J = 0, \pm 1 \tag{6.79}$$

The effect of the $\Delta K = \pm 1$ selection rule, compared to $\Delta K = 0$ for an A_1–A_1 transition, is to spread out the sets of P, Q, and R branches with different values of K. Each Q branch consists, as usual, of closely spaced lines, so as to appear almost line-like, and the separation between adjacent Q branches is approximately $2(A'-B')$. Figure 6.27 shows such an example, the 6_0^1 E–A_1 band of the prolate symmetric rotor silyl fluoride (SiH_3F) where v_6 is the e rocking vibration of the SiH_3 group. The Q branches dominate this fairly low resolution spectrum, those with $\Delta K = \pm 1$ and -1 being on the high and low wavenumber sides respectively.

The selection rules are the same for oblate symmetric rotors and parallel bands appear similar to those of a prolate symmetric rotor. However, perpendicular bands of an oblate symmetric rotor show Q branches with $\Delta K = +1$ and -1 on the low and high wavenumber sides respectively, since the spacing, $2(C'-B')$, is negative.

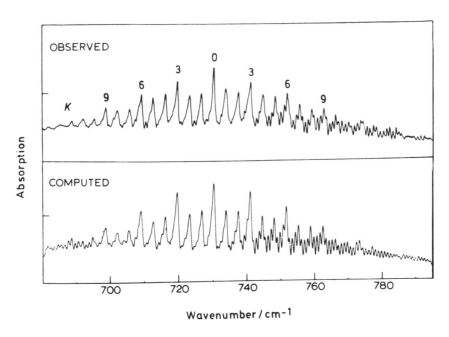

Figure 6.27 The 6_0^1, E–A_1 infrared perpendicular band of SiH_3F. (Reproduced, with permission, from Robiette, A. G., Cartwright, G. J., Hoy, A. R., and Mills, I. M., *Mol. Phys.*, **20**, 541, 1971.)

6.2.3.3 Infrared spectra of spherical rotors

As we proceed to molecules of higher symmetry the vibrational selection rules become more restrictive. A glance at the character table for the T_d point group (Table A.41 in the Appendix) together with equation (6.56) shows that, for regular tetrahedral molecules such as CH_4, the only type of allowed infrared vibrational transition is

$$T_2 - A_1 \tag{6.80}$$

where T_2 is a triply degenerate species.

Neglecting centrifugal distortion, the rotational term values for a spherical rotor in an A_1 vibrational state are

$$F_v(J) = B_v J(J + 1) \tag{6.81}$$

Although, as in linear and symmetric rotor molecules, the term values are slightly modified by Coriolis forces in a degenerate (T_2) state, the rotational selection rules

$$\Delta J = 0, \pm 1 \tag{6.82}$$

result in a T_2–A_1 band looking quite similar to a Π–Σ type of band in a linear molecule.

For a spherical rotor belonging to the octahedral O_h point group, Table A.43 in the Appendix, in conjunction with the vibrational selection rules of equation (6.56), shows that the only allowed transitions are

$$T_{1u} - A_{1g} \tag{6.83}$$

Such bands also obey the rotational selection rules in equation (6.82) and appear similar to a Π–Σ band of a linear molecule.

6.2.3.4 Infrared spectra of asymmetric rotors

As in Section 5.2.4 on rotational spectra of asymmetric rotors, we do not treat this important group of molecules in any detail, so far as their rotational motion is concerned, because of the great complexity of their rotational energy levels. Nevertheless, however complex the stack of rotational levels associated with the $v = 0$ level, there is a very similar stack associated with each excited vibrational level. The selection rules for transitions between the rotational stacks of the vibrational levels are also complex but include

$$\Delta J = 0, \pm 1 \tag{6.84}$$

giving what may be rather randomly distributed sets of P, Q, and R branches.

In a molecule like the asymmetric rotor formaldehyde, shown in Figure 5.1(f), the a, b, and c inertial axes, of lowest, medium, and highest moments of inertia respectively, are defined by symmetry, the a axis being the C_2 axis, the b axis being in the yz plane and the c axis being perpendicular to the yz plane. Vibrational transition moments are confined to the a, b or c axis and the rotational selection rules are characteristic. We call them

$$\text{type } A, \text{ type } B, \text{ or type } C \qquad (6.85)$$

selection rules corresponding to a transition moment along the a, b, or c axis respectively.

Whether the molecule is a prolate or oblate asymmetric rotor, type A, B, or C selection rules result in characteristic band shapes. These shapes, or contours, are particularly important in gas phase infrared spectra of large asymmetric rotors, whose rotational lines cannot be resolved, for assigning symmetry species to observed fundamentals.

Ethylene ($H_2C{=}CH_2$) is a prolate asymmetric rotor which is sufficiently small for some rotational structure to be resolved even when only a medium resolution spectrometer is used. Figures 6.28, 6,29, and 6.30 show the 11_0^1 (type A), 9_0^1 (type B), and 7_0^1 (type C) bands of ethylene respectively, and Figure 6.31 shows the form and symmetry species of the vibrations involved. Without looking at the detail we can see that the type A band is dominated by a strong central peak and fairly weak wings, the type B band by a central minimum and moderately strong wings, and the type C band by a strong central peak and moderately strong wings.

This general behaviour is characteristic of type A, B, and C bands and is further illustrated by the much less well-resolved bands, one type A, one type B, and three type C, of the prolate asymmetric rotor perdeuteronaphthalene ($C_{10}D_8$) shown in the far-infrared spectrum in Figure 6.32.

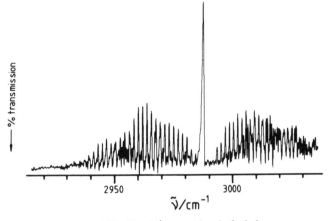

Figure 6.28 The 11_0^1 type A band of ethylene.

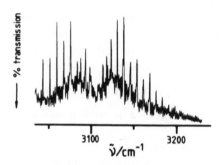

Figure 6.29 The 9_0^1 type B band of ethylene.

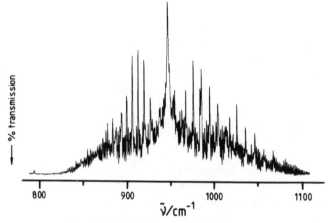

Figure 6.30 The 7_0^1 type C band of ethylene.

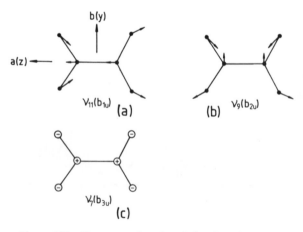

Figure 6.31 Three normal modes of vibration of ethylene.

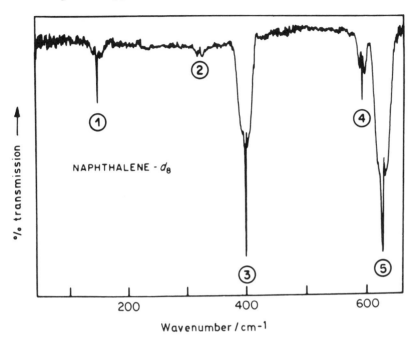

Figure 6.32 Part of the far-infrared spectrum of perdeuteronaphthalene. Band 4 is type *A*, band 2 type *B*, and bands 1, 3, and 5 type *C*. (Reproduced, with permission, from Duckett, J. A., Smithson, T. L., and Wieser, H., *J. Mol. Struct.*, **44**, 97, 1978.)

6.2.4 Anharmonicity

6.2.4.1 Potential energy surfaces

In a diatomic molecule one of the main effects of mechanical anharmonicity, the only type that concerns us in detail, is to cause the vibrational energy levels to close up smoothly with increasing v, as shown in Figure 6.4. The separation of the levels becomes zero at the limit of dissociation.

The potential energy curve in Figure 6.4 is a two-dimensional plot, one dimension for the potential energy V and a second for the vibrational co-ordinate r. For a polyatomic molecule, with $3N - 6$ (non-linear) or $3N - 5$ (linear) normal vibrations, it requires a $[(3N - 6) + 1]$- or $[(3N - 5) + 1]$-dimensional surface to illustrate the variation of V with all the normal coordinates. Such a surface is known as a hypersurface and clearly cannot be illustrated in diagrammatic form. What we can do is take a section of the surface in two dimensions, corresponding to V and each of the normal coordinates in turn, thereby producing a potential energy curve for each normal coordinate.

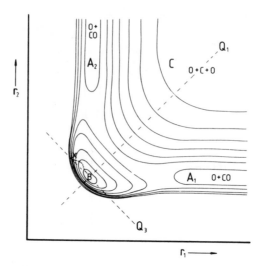

Figure 6.33 Contour of potential energy as a function of the two C–O bond lengths, r_1 and r_2, in CO_2.

If we use a contour map to represent a three-dimensional surface, with each contour line representing constant potential energy, two vibrational coordinates can be illustrated. Figure 6.33 shows such a map for the linear molecule CO_2. The coordinates used here are not normal coordinates but the two CO bond lengths r_1 and r_2 shown in Figure 6.34(a). It is assumed that the molecule does not bend.

In Figure 6.33 the region labelled B is a deep pocket, the bottom of which represents the CO_2 molecule at equilibrium with $r_1 = r_2 = r_e$. Regions A_1 and A_2 are valleys which are higher in energy than B since they correspond to the removal of one of the oxygen atoms, leaving behind a CO molecule. In the atom–molecule reaction

$$O + CO \rightarrow OCO \rightarrow OC + O \qquad (6.86)$$

the reaction coordinate, which is not a normal coordinate, represents the pathway of minimum energy for the reaction in going from A_1 to B to A_2. The variation of energy along this coordinate is shown in Figure 6.35(a).

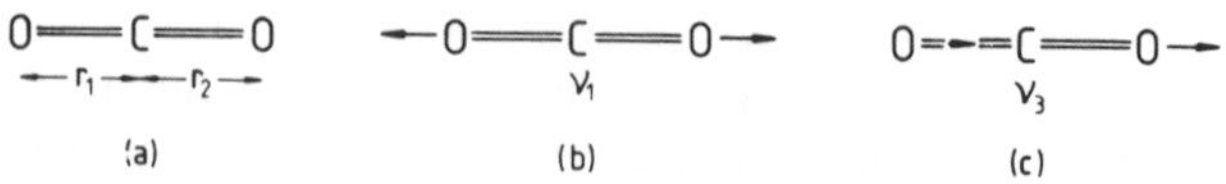

Figure 6.34 (a) The coordinates r_1 and r_2, (b) the symmetric stretching vibration v_1, and (c) the antisymmetric stretching vibration v_3 of CO_2.

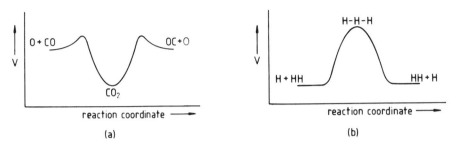

Figure 6.35 Potential energy variation along the reaction coordinate for the reactions between (a) O and CO, (b) H and H_2.

In a reaction such as

$$H + H_2 \rightarrow H \cdots H \cdots H \rightarrow H_2 + H \qquad (6.87)$$

in which the linear triatomic molecule is unstable, there is an energy *maximum* corresponding to the region B in Figure 6.33 and the variation of potential energy with reaction coordinate is like that in Figure 6.35(b).

In order to see how V varies with the two normal coordinates Q_1 and Q_3, corresponding to the symmetric and antisymmetric stretching vibrations v_1 and v_3 of CO_2 in Figure 6.34(b) and (c), we proceed along the dashed line labelled Q_1 in Figure 6.33, which corresponds to changing r_1 and r_2 identically, or along that labelled Q_3, which corresponds to increasing r_1 and decreasing r_2 (or vice versa) by equal amounts. The resulting potential energy curves are shown in Figure 6.36. The one for v_1 appears similar to that in Figure 6.4 for a diatomic molecule. The horizontal part of the curve corresponds to region C in Figure 6.33 in which dissociation to $O + C + O$ has occurred, a very high energy process involving the breaking of two double bonds.

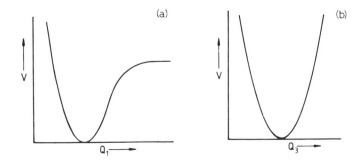

Figure 6.36 General shapes of the potential energy curves for the vibrations (a) v_1 and (b) v_3 of CO_2.

On the other hand the curve for v_3 in Figure 6.36(b) is symmetrical about the centre. It is approximately parabolic but shows steeper sides corresponding to the reluctance of an oxygen nucleus to approach the carbon nucleus at either extreme of the vibrational motion.

The vibrations v_1 and v_3 of CO_2 illustrate the important general point that, in polyatomic molecules, some vibrations are dissociative, such as v_1, and others are non-dissociative, such as v_3. The bending vibration v_2 is also non-dissociative and the corresponding potential energy curve is similar in shape to that in Figure 6.36(b). Both types of curve in Figure 6.36 support vibrational levels which are anharmonic, but in different ways. Those for v_1 converge to the dissociation limit while those for v_2 and v_3 are more equally spaced but may show some divergence.

Figure 6.33 illustrates how anharmonicity mixes the two vibrations v_1 and v_3 of CO_2. If the molecule starts from the point X and proceeds to B it will tend to follow the line of maximum slope shown. In doing so it deviates considerably from the dashed line representing Q_3 and so involves an admixture of Q_1 and Q_3.

6.2.4.2 Vibrational term values

The vibrational term values for a polyatomic anharmonic oscillator with only non-degenerate vibrations are modified from the harmonic oscillator values of equation (6.41) to

$$\sum_i G(v_i) = \sum_i \omega_i(v_i + \tfrac{1}{2}) + \sum_{i \leqslant j} x_{ij}(v_i + \tfrac{1}{2})(v_j + \tfrac{1}{2}) + \cdots \qquad (6.88)$$

where the x_{ij} are anharmonic constants. For $i = j$ the x_{ij} are analogous to $-\omega_e x_e$ in a diatomic molecule but the x_{ij}, when $i \neq j$, have no such analogues and illustrate the approximation involved in taking a two-dimensional section in a potential energy hypersurface and assuming that we can treat that normal coordinate independently of all others.

For an anharmonic oscillator with degenerate vibrations the term values are modified from those of equation (6.88) to

$$\sum_i G(v_i) = \sum_i \omega_i\left(v_i + \frac{d_i}{2}\right) + \sum_{i \leqslant j} x_{ij}\left(v_i + \frac{d_i}{2}\right)\left(v_j + \frac{d_j}{2}\right) + \sum_{i \leqslant j} g_{ij}\ell_i\ell_j + \cdots \qquad (6.89)$$

where d_i is the degree of degeneracy of vibration i and $g_{ij}\ell_i\ell_j$ is an additional anharmonic term.

6.2.4.3 Local mode treatment of vibrations

Normal modes of vibration, with their corresponding normal coordinates, are very satisfactory in describing the low-lying vibrational levels, usually those with $v = 1$ or 2, which can be investigated by traditional infrared absorption

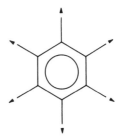

Figure 6.37 The totally symmetric CH-stretching vibration v_2 of benzene.

or Raman spectroscopy. For certain types of vibration, particularly stretching vibrations involving more than one symmetrically equivalent terminal atom, this description becomes less satisfactory as v increases.

Consider the CH, stretching vibrations of benzene, for example. Since there are six identical C—H bonds there are six CH-stretching vibrations. These belong to various symmetry species but only one, v_2[†] illustrated in Figure 6.37, is totally symmetric a_{1g}.

It might be supposed that, since the potential energy curve for v_2 is of a similar shape to that in Figure 6.36(a), if we excite the molecule with sufficiently high energy it will eventually dissociate losing six hydrogen atoms in the process:

$$C_6H_6 \rightarrow C_6 + 6H \qquad (6.90)$$

This seems reasonable when we think only in terms of normal vibrations but intuition suggests that, since the dissociation in equation (6.90) would require something like six times the C—H bond dissociation energy (ca 6×412 kJ mol^{-1}), the process

$$C_6H_6 \rightarrow C_6H_5 + H \qquad (6.91)$$

is surely more likely to occur. It is now known that this is what happens but the difficulty is that no normal vibration of benzene leads to CH stretching being localized in only one bond. A new type of description of vibrational behaviour in such cases has been devised which is to treat high overtone levels of CH stretching by a local mode, rather than a normal mode model. Each C—H bond stretching is taken to behave as in the Morse model (equation 6.23) which describes fairly satisfactorily the anharmonic vibration of a diatomic molecule. The extremely sensitive technique of photoacoustic spectroscopy has enabled transitions up to $v = 6$ in the CH-stretching vibration to be observed and the transition wavenumbers have been shown to fit those of a Morse oscillator quite well.

This local mode behaviour applies to vibrations of many other molecules with two or more equivalent terminal atoms, and CO_2 is such an example.

_____

[†] Using the Wilson numbering system which is frequently used for benzene (see the bibliography).

The probability of going, in Figure 6.33, from region B, where CO_2 is in the equilibrium configuration, to region C, in which it has lost both oxygen atoms, by exciting more and more quanta of v_1 and proceeding along the dashed line Q_1, seems energetically unlikely. The molecule will prefer to go from B to A_1 (or A_2), losing one oxygen atom only, by local mode behaviour. Then it will go from A_1 (or A_2) to C by dissociation of CO. Therefore the high overtone levels of CO stretching in CO_2 will be better described by a local mode model, involving anharmonic stretching of one $C{=}O$ bond, rather than a normal mode model.

It should be realized, though, that either model can be used for levels for *all* values of v for such stretching vibrations but the normal mode model is more practically useful at low v and the local mode model more useful at higher values of v.

Although any stretching vibration involving symmetrically equivalent atoms should show local mode behaviour at high v, the X—H stretching vibrations are more amenable to investigation. The reason for this is that their vibrational quanta are large and, therefore, a relatively low value of v is nearer to the dissociation energy than, say, for CF-stretching vibrations, and the transition probability decreases rapidly as v increases.

6.2.4.4 Vibrational potential functions with more than one minimum

The only types of anharmonic potential function we have encountered so far are the two illustrated in Figure 6.36, both of which show only a single minimum. There are, however, some vibrations whose potential functions do not resemble either of those but show more than one minimum and whose term values are neither harmonic, nor are they given by equation (6.88) or equation (6.89). Such vibrations can be separated into various types which will now be discussed individually.

6.2.4.4(a) Inversion vibrations   The inversion, or umbrella, vibration v_2 of ammonia is shown in Figure 4.16. The equilibrium configuration of NH_3 is pyramidal with $\angle HNH = 106.7°$ but, for large amplitude motion in v_2, the molecule may go through the planar configuration to an identical, but inverted, pyramidal configuration. The planar and the two equivalent pyramidal configurations are shown in Figure 6.38. The pyramidal configurations (i) and (iii) obviously correspond to two identical minima in the potential energy and the planar configuration (ii) to a maximum. In the resulting W-shaped potential curve, with an energy barrier between the two minima, b is the barrier height and is the energy which is required, classically, to go from the pyramidal to the planar configuration.

In a quantum mechanical system it may not be necessary to surmount the barrier in order to go from (i) to (iii). The phenomenon of quantum mechanical

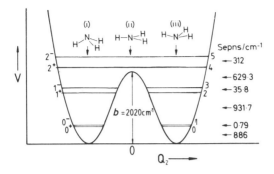

Figure 6.38 Potential energy curve for the inversion vibration ν_2 of NH_3.

tunnelling allows a degree of penetration of the barrier. If the barrier is sufficiently low or narrow (or both) the penetration may be so great that interaction occurs between the identical sets of vibrational levels in the two limbs of the potential curve. This interaction splits the levels into two components, the splitting being greater towards the top of the barrier where tunnelling is more effective. The splittings are shown for a general case in Figure 6.39(b) and for NH_3 in Figure 6.38: both show that even above the barrier, there is some staggering of the levels.

Figure 6.39(a) shows how the splittings are removed, and the levels become evenly spaced, when the barrier is very high. When the barrier is reduced to zero the equilibrium configuration is planar, as for BF_3, and the levels again become evenly spaced, as shown in Figure 6.39(c). The figure shows how the vibrational energy levels correlate as the barrier is decreased smoothly from being very high to being zero.

As Figure 6.39 shows, there are two alternative vibrational numbering schemes: 0, 1, 2, 3,..., which emphasizes the relation to the zero barrier case, and $0^+, 0^-, 1^+, 1^-,...$, which relates to the high barrier case.

The time, τ, taken to tunnel through the barrier, i.e. the time for the molecule to invert from one pyramidal form to the other, is given by

$$\tau = (2\Delta\nu)^{-1} \tag{6.92}$$

where $\Delta\nu$ is the splitting of the levels. For $^{14}N^1H_3$, $\Delta\nu = 23.786$ GHz for the $\nu_2 = 1$ to $\nu_2 = 0$ (or 0^- to 0^+) splitting, giving $\tau = 2.1 \times 10^{-11}$ s.

There have been several suggested forms of the potential function which reproduce the way in which the potential energy $V(Q)$ depends on the vibrational coordinate Q which relates to the inversion motion. Perhaps the most successful form for the potential function is

$$V(Q) = \tfrac{1}{2}aQ^2 + b\exp(-cQ^2) \tag{6.93}$$

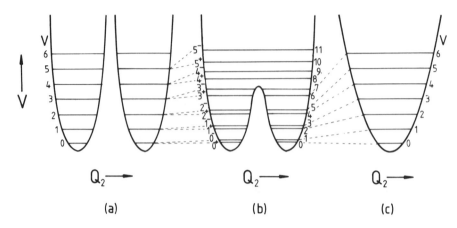

Figure 6.39 Potential energy curves and vibrational energy levels for an inversion vibration when the barrier to planarity is (a) infinite, (b) moderately low, and (c) zero.

The first term $\frac{1}{2}aQ^2$ would, by itself, give an ordinary harmonic oscillator potential, like the dashed curve in Figure 6.4. The second term introduces an energy barrier, of height b, at $Q = 0$ and the resulting $V(Q)$ resembles the W-shaped potential in Figure 6.39(b). The parameters a, b and c can be varied until the resulting vibrational energy levels match those that are observed. In the case of NH_3 the resulting barrier to planarity b is 2020 cm^{-1}.

A useful alternative W-shaped potential function to that in equation (6.93) is

$$V(Q) = AQ^2 + BQ^4 \qquad (6.94)$$

The BQ^4 term alone, with B positive, would give a potential resembling the harmonic oscillator potential in Figure 6.4 (dashed curve) but with steeper sides. The inclusion of the AQ^2 term, with A negative, adds an upside-down parabola at $Q = 0$ and the result is a W-shaped potential. The barrier height b is given by

$$b = A^2/4B \qquad (6.95)$$

In the case of NH_3, the potential in equation (6.93) is rather more successful in fitting the experimental data but that in equation (6.94) has been used for inversion vibrations in other molecules.

Molecules with an inversion vibration which is qualitatively similar to that of NH_3 are formamide (NH_2CHO) and aniline ($C_6H_5NH_2$) in which the vibration involves primarily the hydrogen atoms of the NH_2-group. Both molecules are non-planar, having a pyramidal configuration about the nitrogen atom, with barriers to planarity of 370 cm^{-1} (4.43 kJ mol^{-1}) and 547 cm^{-1} (6.55 kJ mol^{-1}), respectively.

6.2.4.4(b) Ring-puckering vibrations Cyclic molecules, which are at least partially saturated and contain such groups as $-CH_2-$, $-O-$, or $-S-$, have low wavenumber vibrations involving a bending motion of the group out of the plane of the ring (or what would be the plane if it were a planar molecule). Such a vibration is called a ring-puckering vibration. Two such examples are to be found in cyclobutane and cyclopentene, shown in Figure 6.40(a) and (b) respectively.

In cyclobutane the puckering vibration is an out-of-plane bending, butterfly-like motion of the molecule about a line joining two opposite carbon atoms as shown. The equilibrium configuration of the carbon atoms is non-planar with a so-called dihedral angle, shown in Figure 6.40(a), of 35°. The potential energy curve for the vibration shows, therefore, two identical minima corresponding to the ring being puckered 'upwards' or 'downwards'.

Cyclopentene, shown in Figure 6.40(b), behaves, in respect of the ring-puckering vibration, rather like cyclobutane and is referred to as a pseudo four-membered ring. This is because the $C=C$ bond is resistant to twisting compared to a $C-C$ bond so that, so far as ring puckering is concerned, the $HC=CH$ part of the ring behaves as a single rigid group. Again, puckering of the ring 'upwards' or 'downwards' corresponds to the same energy and the potential energy curve shows two identical minima.

The potential energy curves for ring-puckering vibrations in cyclobutane, cyclopentene, and similar cyclic molecules with at least four single bonds in the ring, are W-shaped rather like that for an inversion vibration (section 6.2.4.4a). Consequently similar forms for the potential function have been used although,

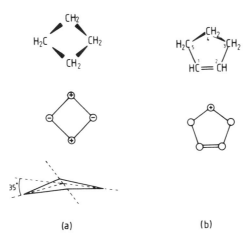

(a)   (b)

Figure 6.40 (a) Cyclobutane, its ring-puckering vibration and dihedral angle. (b) Cyclopentene and its ring-puckering vibration.

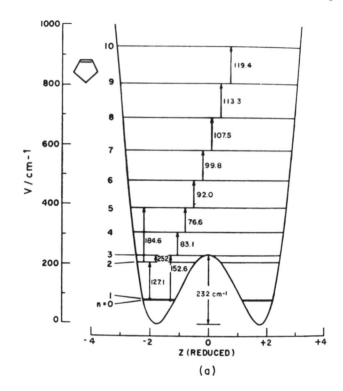

(a)

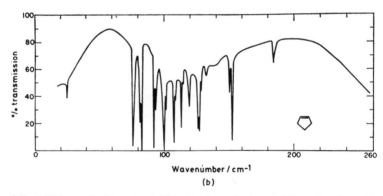

(b)

Figure 6.41 (a) Ring-puckering potential function for cyclopentene. The reduced coordinate z is proportional to the normal coordinate. (b) Far-infrared absorption spectrum of cyclopentene vapour. (Reproduced, with permission, from Laane, J., and Lord, R. C., *J. Chem. Phys.*, **47**, 4941, 1967.)

for ring puckering, the form for $V(Q)$ given in equation (6.94) is the more successful.

The ring-puckering potential for cyclopentene is shown in Figure 6.41(a) where the transitions marked, in cm^{-1}, have been observed in the far-infrared spectrum which is shown in Figure 6.41(b). Splitting of the levels due to tunnelling through the barrier, which has a height of only $232\ cm^{-1}$ ($2.78\ kJ\ mol^{-1}$), occurs as for an inversion vibration.

6.2.4.4(c)   Torsional vibrations   Figure 6.42 shows a few molecules in which there is a vibration involving motion about a bond, other than a terminal bond, in which one part of the molecule, known as the top, vibrates backwards and forwards in a twisting, or torsional, motion relative to the rest of the molecule, known as the frame.

In phenol (Figure 6.42b) the resistance to torsion about the C—O bond is provided by a degree of conjugation, giving some π-bonding character involving the $2p$ orbital on C perpendicular to the ring, and a lone pair orbital on O. Plotting potential energy $V(\phi)$ against torsional angle ϕ, taking ϕ to be zero for the planar configuration, gives a potential energy curve of the type shown in Figure 6.43. This is a repetitive curve extending from $\phi = 0$ to ∞ and repeats every π radians with an identical energy barrier at $\phi = \pi/2,\ 3\pi/2,\ \dots$ The physical reason for the barrier is that when the O—H plane is perpendicular to the ring, all the stabilizing conjugation in the C—O bond is lost. On the other hand, the destabilizing steric hindrance between the hydrogen atom of OH and that on carbon in the 2-position of the ring is minimized. However, because we

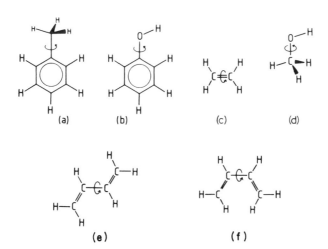

Figure 6.42 Torsional vibrations in (a) toluene, (b) phenol, (c) ethylene, (d) methyl alcohol, (e) *s-trans*-buta-1,3-diene, and (f) *s-cis*-buta-1,3-diene.

Table 6.5 Barrier heights V for some torsional vibrations.

Molecule	V/cm^{-1}	$V/kJ\ mol^{-1}$	Molecule	V/cm^{-1}	$V/kJ\ mol^{-1}$
C_6H_5OH	1207	14.44	$C_6H_5CH_3$	4.9	0.059
$CH_2{=}CH_2$	22 750[‡]	272.2[†]	CH_3NO_2	2.1	0.025
	or 14 000[†]	167.5[†]	$C_6H_5CH{=}CH_2$[‡]	1070	12.8
CH_3OH	375	4.49	$CH_2{=}CH{-}CH{=}CH_2$	2660	31.8
			(*s-trans* to *s-cis*)		
CH_3CH_3	960	11.5	$CH_2{=}CH{-}CH{=}CH_2$	1060	12.7
			(*s-cis* to *s-trans*)		

[†] Independent estimates.
[‡] Torsion about the (ring)C–(substituent)C bond.

know that the stable configuration of phenol is planar, the effects of steric hindrance are small compared to those of conjugation. The torsional barrier height is given in Table 6.5.

Torsional barriers are referred to as *n*-fold barriers, where the torsional potential function repeats every $2\pi/n$ radians. As in the case of inversion vibrations (Section 6.2.4.4a) quantum mechanical tunnelling through an *n*-fold torsional barrier may occur, splitting a vibrational level into *n* components. The splitting into two components near the top of a two-fold barrier is shown in Figure 6.43. When the barrier is surmounted free internal rotation takes place, the energy levels then resembling those for rotation rather than vibration.

Table 6.5 gives a few other examples of torsional barrier heights. That for ethylene is high, typical of a double bond, but its value is uncertain. The barriers for methyl alcohol and ethane are three-fold which can be confirmed using molecular models, while those of toluene and nitromethane are six-fold. The decrease in barrier height on going to a higher-fold barrier is typical. Rotation about the C—C bond in toluene and the C—N bond in nitromethane is very nearly free.

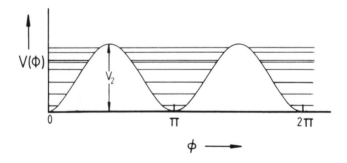

Figure 6.43 Torsional potential function showing a two-fold barrier.

Buta-1,3-diene is one of many examples of molecules in which torsional motion may convert a stable isomer into another, less stable, isomer. The more stable isomer in this case is the *s-trans* form, shown in Figure 6.42(e), and the less stable one is the *s-cis* form[†], shown in Figure 6.42(f). Both isomers are planar due to the stabilizing effect of conjugation across the central C—C bond. The *s-trans* isomer is stabilized further by minimizing steric hindrance between hydrogen atoms. The resulting potential energy curve is similar to that in Figure 6.43 except that, if $\phi = 0$ corresponds to the *s-trans* isomer, the minimum at $\phi = \pi$, corresponding to the *s-cis* isomer, is higher in energy than that for the *s-trans* form. One result of this is that the barrier, given in Table 6.5, for going from *s-trans* to *s-cis* is higher than that for the opposite process.

The most useful general form of the potential function $V(\phi)$ is

$$V(\phi) = \frac{1}{2} \sum_n V_n(1 - \cos n\phi) \qquad (6.96)$$

where n is an integer. Which terms in the summation are dominant depends on the molecule concerned. In the case of toluene (Figure 6.42a) it is V_6 because of the six-fold barrier. Similarly, in phenol (Figure 6.42b), ethylene (Figure 6.42c) and methanol (Figure 6.42d) it is V_2, V_2 and V_3, respectively. In buta-1,3-diene (Figure 6.42e and f) both V_1 and V_2 are most important because of the two possible isomers, one (*cis*) being much less stable than the other (*trans*).

Questions

1. From the following wavenumbers of the P and R branches of the 1–0 infrared vibrational band of $^2H^{35}Cl$ obtain values for the rotational constants B_0, B_1, and B_e, the band centre ω_0, the vibration–rotation interaction constant α, and the internuclear distance r_e. Given that the band centre of the 2–0 band is at 4128.6 cm^{-1} determine ω_e and, using this value, the force constant k.

J	$\tilde{v}[R(J)]/\text{cm}^{-1}$	$\tilde{v}[P(J)]/\text{cm}^{-1}$	J	$\tilde{v}[R(J)]/\text{cm}^{-1}$	$\tilde{v}[P(J)]/\text{cm}^{-1}$
0	2107.5	—	7	2174.0	2016.8
1	2117.8	2086.0	8	2183.2	2003.2
2	2127.3	2074.3	9	2191.5	1991.0
3	2137.5	2063.0	10	2199.5	1978.5
4	2147.4	2052.0	11	2207.5	1966.0
5	2156.9	2040.0	12	2214.9	1952.5
6	2166.2	2027.7	13	—	1938.8

[†] There is evidence that the second isomer may be a non-planar *gauche* form.

2. From the values of B_0, B_1, and ω_0 obtained in question 1 calculate the wavenumbers of the first two members of each of the O and S branches, in the Raman vibration–rotation spectrum, and their relative intensities.

3. From the following separations of vibrational levels in the ground electronic state of CO obtain values for ω_e and $\omega_e x_e$ and also for the dissociation energy D_e.

$v' - v''$	1–0	2–1	3–2	4–3	5–4	6–5
$[G(v + 1) - G(v)]/\text{cm}^{-1}$	2138	2115	2091	2063	2038	2011

4. What group vibrations would you hope to identify in the infrared and Raman

spectra of
$$\begin{array}{c} CH_3 \\ \diagdown \\ \diagup \\ F \end{array} C{=}CH{-}C{\equiv}C{-}OH?$$

5. Having assigned symmetry species to each of the six vibrations of formaldehyde shown in question 3 of Chapter 4 use the appropriate character table to show which are allowed in (a) the infrared spectrum and (b) the Raman spectrum. In each case state the direction of the transition moment for the infrared-active vibrations and which component of the polarizability is involved for the Raman-active vibrations.

6. For a molecule belonging to the D_{2h} point group deduce whether the following vibrational transitions, all from the zero-point level, are allowed in the infrared and/or Raman spectrum, stating the direction of the transition moment and/or the component of the polarizability involved:
   (a) to the $v = 2$ level of a b_{1g} vibration;
   (b) to the $v = 1$ level of an a_u or b_{2u} vibration;
   (c) to the combination level involving $v = 1$ of a b_{1u} and $v = 1$ of a b_{3g} vibration;
   (d) to the combination level involving $v = 2$ of an a_u vibration and $v = 1$ of a b_{2g} vibration.

7. Given that, for $^{12}C^{16}O_2$:

$$\omega_1 = 1354.07 \text{ cm}^{-1}, \quad \omega_2 = 672.95 \text{ cm}^{-1}, \quad \omega_3 = 2396.30 \text{ cm}^{-1},$$

$$x_{11} = -3.10 \text{ cm}^{-1}, \quad x_{22} = 1.59 \text{ cm}^{-1}, \quad x_{33} = -12.50 \text{ cm}^{-1},$$

$$x_{12} = -5.37 \text{ cm}^{-1}, \quad x_{13} = -19.27 \text{ cm}^{-1}, \quad x_{23} = -12.51 \text{ cm}^{-1},$$

$$g_{22} = -0.62 \text{ cm}^{-1}$$

calculate the wavenumbers of the $v_1 = 1$ level and the $v_2 = 2$ ($\ell_2 = 0$) level. The resulting levels both have Σ_g^+ symmetry and, because of this and the fact that they would otherwise be close together, they interact by a process called Fermi resonance. As a result they are pushed much further apart. (ω_1, ω_2, and ω_3 are *equilibrium* values corresponding to the vibrations v_1, v_2, and v_3).

8. Sketch the form of the potential energy as a function of torsional angle ϕ for the torsional vibration in (a) ethane, (b) CH_3NO_2, (c) 2-fluorophenol, (d) CH_2FOH, and (e) 1,2-dichloroethane.

Bibliography

Allen, H. C., Jr. and Cross, P. C. (1963). *Molecular Vib-Rotors*, Wiley, New York.
Bellamy, L. J. (1980). *The Infrared Spectra of Complex Molecules*, Vol. 2, *Advances in Infrared Group Frequencies*, Chapman and Hall, London.
Gans, P. (1971). *Vibrating Molecules*, Chapman and Hall, London.
Herzberg, G. (1945). *Infrared and Raman Spectra*, Van Nostrand, New York.
Herzberg, G. (1950). *Spectra of Diatomic Molecules*, Van Nostrand, New York.
Long, D. A. (1977). *Raman Spectroscopy*, McGraw-Hill, London.
Wilson, E. B. (1934). *Phys. Rev.*, **45**, 706.
Wilson, E. B., Decius, J. C., and Cross, P. C. (1955). *Molecular Vibrations*, McGraw-Hill, New York.
Woodward, L. A. (1972). *Introduction to the Theory of Molecular Vibrations and Vibrational Spectroscopy*, Oxford University Press, Oxford.

CHAPTER 7 ——————————————————————

Electronic Spectroscopy

7.1 Atomic Spectroscopy

Electronic spectroscopy is the study of transitions, in absorption or emission, between electronic states of an atom or molecule. Atoms are unique in this respect as they have only electronic degrees of freedom, apart from translation and nuclear spin, whereas molecules have, in addition, vibrational and rotational degrees of freedom. One result is that electronic spectra of atoms are very much simpler than those of molecules.

7.1.1 The Periodic Table

For the hydrogen atom, and for hydrogen-like atoms such as He^+, Li^{2+}, ..., with a single electron in the field of a nucleus with charge $+Ze$, the hamiltonian (the quantum mechanical form of the energy) is given by

$$H = -\frac{\hbar^2}{2\mu}\nabla^2 - \frac{Ze^2}{4\pi\varepsilon_0 r} \tag{7.1}$$

analogous to equation (1.30) for the hydrogen atom where the terms are explained.

For a polyelectronic atom the hamiltonian becomes

$$H = -\frac{\hbar^2}{2m_e}\sum_i \nabla_i^2 - \sum_i \frac{Ze^2}{4\pi\varepsilon_0 r_i} + \sum_{i<j} \frac{e^2}{4\pi\varepsilon_0 r_{ij}} \tag{7.2}$$

where the summation is over all electrons i. The first two terms are simply sums of terms like those for one electron in equation (7.1). The third term is new and is due to coulombic repulsions between all possible pairs of electrons a distance r_{ij} apart. It contrasts with the second term which represents the coulombic attraction between each electron and the nucleus at a distance r_i.

Because of the electron–electron repulsion term in equation (7.2) the

184

hamiltonian cannot be broken down into a sum of contributions from each electron and the Schrödinger equation (equation 1.28) can no longer be solved exactly. Various approximate methods of solution have been devised and that of Hartree who rewrote the hamiltonian in the form

$$H \simeq -\frac{\hbar^2}{2m_e} \sum_i \nabla_i^2 - \sum_i \frac{Ze^2}{4\pi\varepsilon_0 r_i} + \sum_i V(r_i) \tag{7.3}$$

is one of the most useful. He approximated the contributions to the potential energy due to electron repulsions as a sum of contributions from individual electrons. The Schrödinger equation is then soluble. The method is known as the self-consistent field (SCF) method.

An important effect of electron repulsions is to remove the degeneracy of those orbitals, such as $2s$, $2p$ and $3s$, $3p$, $3d$ which are degenerate in the hydrogen atom (see Figure 1.1), to give a set of orbitals with relative energies similar to those in Figure 7.1. The orbital energies E_i vary not only with the principal quantum number n, as in the hydrogen atom, but also with the orbital angular

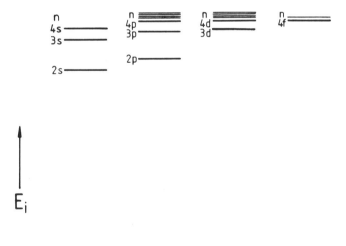

Figure 7.1 Orbital energies, E_i, typical of a polyelectronic atom.

momentum quantum number $\ell(= 0, 1, 2, \ldots$ for $s, p, d, \ldots$ orbitals). The value of E_i for a particular orbital increases with the nuclear charge of the atom. This is illustrated by the fact that the energy required to remove an electron from the 1s orbital, i.e. the ionization energy, is 13.6 eV for H and 870.4 eV for Ne.

Electrons in the atom concerned may be fed into the orbitals in Figure 7.1 in order of increasing energy until the electrons are used up, to give what is referred to as the ground configuration of the atom. This feeding in of the electrons in order of increasing orbital energy follows the *aufbau* or building-up principle, but must also obey the Pauli exclusion principle. This states that no two electrons can have the same set of quantum numbers n, ℓ, m_ℓ, m_s. Since m_ℓ can take $(2\ell + 1)$ values (see equation 1.45) and $m_s = \pm\frac{1}{2}$ (see equation 1.47), each orbital characterized by particular values of n and ℓ can accommodate $2(2\ell + 1)$ electrons. It follows that an ns orbital with $\ell = 0$ can accommodate two electrons, an np orbital with $\ell = 1$ can take six, an nd orbital with $\ell = 2$ can take ten, and so on.

Whereas an orbital refers to a particular set of values of n and ℓ, a shell refers to all orbitals having the same value of n. For $n = 1, 2, 3, 4, \ldots$ the shells are labelled $K, L, M, N, \ldots$.

The distinction between a configuration and a state is an important one. A configuration describes the way in which the electrons are distributed among various orbitals but a configuration may give rise to more than one state. For example, as we shall see in Section 7.1.2.3(b), the ground configuration $1s^2 2s^2 2p^2$ of the carbon atom gives rise to *three* electronic states of different energies.

Table 7.1 gives the ground configurations and other useful data for all the elements. This table illustrates features which are common to the electron configurations of elements which are known to be chemically similar. The alkali metals, Li, Na, K, Rb, and Cs, all have an outer ns^1 configuration consistent with their being monovalent, and the alkaline earth metals, Be, Mg, Ca, Sr, and Ba, all have an outer ns^2 configuration and are divalent. The noble (inert) gases, Ne, Ar, Kr, Xe, and Rn, all have an outer np^6 configuration, the filled orbital (or sub-shell, as it is sometimes called) conferring chemical inertness, and the filled K shell in He has a similar effect.

The first transition series, Sc, Ti, V, Cr, Mn, Fe, Co, Ni, Cu, and Zn, is characterized by the filling up of the 3d orbital. Figure 7.1 shows that the 3d and 4s orbitals are very similar in energy but their separation changes along the series. The result is that, although there is a preference by most of these elements for having two electrons in the 4s orbital, Cu has a $\ldots 3d^{10}4s^1$ ground configuration because of the innate stability associated with a filled 3d orbital. There is also some stability associated with a half-filled orbital ($2p^3, 3d^5$, etc.) resulting in the $\ldots 3d^5 \, 4s^1$ ground configuration of Cr. There is no simple reason for the stability of the half-filled orbital but, when this occurs, all electrons have parallel spins, i.e. all have $m_s = +\frac{1}{2}$ or $m_s = -\frac{1}{2}$ and this situation modifies the electron–nucleus interactions resulting in a lowering of the energy.

Table 7.1 Ground configurations and ground states of atoms.

Atom	Atomic number (Z)	Ground configuration	First ionization energy/eV[†]	Ground state
H	1	$1s^1$	13.598	$^2S_{1/2}$
He	2	$1s^2$	24.587	$^1S_0$
Li	3	$K2s^1$	5.392	$^2S_{1/2}$
Be	4	$K2s^2$	9.322	$^1S_0$
B	5	$K2s^22p^1$	8.298	$^2P^o_{1/2}$
C	6	$K2s^22p^2$	11.260	$^3P_0$
N	7	$K2s^22p^3$	14.534	$^4S^o_{3/2}$
O	8	$K2s^22p^4$	13.618	$^3P_2$
F	9	$K2s^22p^5$	17.422	$^2P^o_{3/2}$
Ne	10	$K2s^22p^6$	21.564	$^1S_0$
Na	11	$KL3s^1$	5.139	$^2S_{1/2}$
Mg	12	$KL3s^2$	7.646	$^1S_0$
Al	13	$KL3s^23p^1$	5.986	$^2P^o_{1/2}$
Si	14	$KL3s^23p^2$	8.151	$^3P_0$
P	15	$KL3s^23p^3$	10.486	$^4S^o_{3/2}$
S	16	$KL3s^23p^4$	10.360	$^3P_2$
Cl	17	$KL3s^23p^5$	12.967	$^2P^o_{3/2}$
Ar	18	$KL3s^23p^6$	15.759	$^1S_0$
K	19	$KL3s^23p^64s^1$	4.341	$^2S_{1/2}$
Ca	20	$KL3s^23p^64s^2$	6.113	$^1S_0$
Sc	21	$KL3s^23p^63d^14s^2$	6.54	$^2D_{3/2}$
Ti	22	$KL3s^23p^63d^24s^2$	6.82	$^3F_2$
V	23	$KL3s^23p^63d^34s^2$	6.74	$^4F_{3/2}$
Cr	24	$KL3s^23p^63d^54s^1$	6.766	$^7S_3$
Mn	25	$KL3s^23p^63d^54s^2$	7.435	$^6S_{5/2}$
Fe	26	$KL3s^23p^63d^64s^2$	7.870	$^5D_4$
Co	27	$KL3s^23p^63d^74s^2$	7.86	$^4F_{9/2}$
Ni	28	$KL3s^23p^63d^84s^2$	7.635	$^3F_4$
Cu	29	$KLM4s^1$	7.726	$^2S_{1/2}$
Zn	30	$KLM4s^2$	9.394	$^1S_0$
Ga	31	$KLM4s^24p^1$	5.999	$^2P^o_{1/2}$
Ge	32	$KLM4s^24p^2$	7.899	$^3P_0$
As	33	$KLM4s^24p^3$	9.81	$^4S^o_{3/2}$
Se	34	$KLM4s^24p^4$	9.752	$^3P_2$
Br	35	$KLM4s^24p^5$	11.814	$^2P^o_{3/2}$
Kr	36	$KLM4s^24p^6$	13.999	$^1S_0$
Rb	37	$KLM4s^24p^65s^1$	4.177	$^2S_{1/2}$
Sr	38	$KLM4s^24p^65s^2$	5.695	$^1S_0$
Y	39	$KLM4s^24p^64d^15s^2$	6.38	$^3D_{3/2}$
Zr	40	$KLM4s^24p^64d^25s^2$	6.84	$^3F_2$
Nb	41	$KLM4s^24p^64d^45s^1$	6.88	$^6D_{1/2}$
Mo	42	$KLM4s^24p^64d^55s^1$	7.099	$^7S_3$
Tc	43	$KLM4s^24p^64d^55s^2$	7.28	$^6S_{5/2}$
Ru	44	$KLM4s^24p^64d^75s^1$	7.37	$^5F_5$
Rh	45	$KLM4s^24p^64d^85s^1$	7.46	$^4F_{9/2}$
Pd	46	$KLM4s^24p^64d^{10}$	8.34	$^1S_0$
Ag	47	$KLM4s^24p^64d^{10}5s^1$	7.576	$^2S_{1/2}$
Cd	48	$KLM4s^24p^64d^{10}5s^2$	8.993	$^1S_0$
In	49	$KLM4s^24p^64d^{10}5s^25p^1$	5.786	$^2P^o_{1/2}$
Sn	50	$KLM4s^24p^64d^{10}5s^25p^2$	7.344	$^3P_0$

(continued)

Table 7.1 (*continued*)

Atom	Atomic number (Z)	Ground configuration	First ionization energy/eV[†]	Ground state
Sb	51	$KLM4s^24p^64d^{10}5s^25p^3$	8.641	$^4S^o_{3/2}$
Te	52	$KLM4s^24p^64d^{10}5s^25p^4$	9.009	$^3P_2$
I	53	$KLM4s^24p^64d^{10}5s^25p^5$	10.451	$^2P^o_{3/2}$
Xe	54	$KLM4s^24p^64d^{10}5s^25p^6$	12.130	$^1S_0$
Cs	55	$KLM4s^24p^64d^{10}5s^25p^66s^1$	3.894	$^2S_{1/2}$
Ba	56	$KLM4s^24p^64d^{10}5s^25p^66s^2$	5.212	$^1S_0$
La	57	$KLM4s^24p^64d^{10}5s^25p^65d^16s^2$	5.577	$^2D_{3/2}$
Ce	58	$KLM4s^24p^64d^{10}4f^15s^25p^65d^16s^2$	5.47	$^1G^o_4$
Pr	59	$KLM4s^24p^64d^{10}4f^35s^25p^66s^2$	5.42	$^4I^o_{9/2}$
Nd	60	$KLM4s^24p^64d^{10}4f^45s^25p^66s^2$	5.49	$^5I_4$
Pm	61	$KLM4s^24p^64d^{10}4f^55s^25p^66s^2$	5.55	$^6H^o_{5/2}$
Sm	62	$KLM4s^24p^64d^{10}4f^65s^25p^66s^2$	5.63	$^7F_0$
Eu	63	$KLM4s^24p^64d^{10}4f^75s^25p^66s^2$	5.67	$^8S^o_{7/2}$
Gd	64	$KLM4s^24p^64d^{10}4f^75s^25p^65d^16s^2$	6.14	$^9D^o_2$
Tb	65	$KLM4s^24p^64d^{10}4f^95s^25p^66s^2$	5.85	$^6H^o_{15/2}$
Dy	66	$KLM4s^24p^64d^{10}4f^{10}5s^25p^66s^2$	5.93	$^5I_8$
Ho	67	$KLM4s^24p^64d^{10}4f^{11}5s^25p^66s^2$	6.02	$^4I^o_{15/2}$
Er	68	$KLM4s^24p^64d^{10}4f^{12}5s^25p^66s^2$	6.10	$^3H_6$
Tm	69	$KLM4s^24p^64d^{10}4f^{13}5s^25p^66s^2$	6.18	$^2F^o_{7/2}$
Yb	70	$KLMN5s^25p^66s^2$	6.254	$^1S_0$
Lu	71	$KLMN5s^25p^65d^16s^2$	5.426	$^2D_{3/2}$
Hf	72	$KLMN5s^25p^65d^26s^2$	7.0	$^3F_2$
Ta	73	$KLMN5s^25p^65d^36s^2$	7.89	$^4F_{3/2}$
W	74	$KLMN5s^25p^65d^46s^2$	7.98	$^5D_0$
Re	75	$KLMN5s^25p^65d^56s^2$	7.88	$^6S_{5/2}$
Os	76	$KLMN5s^25p^65d^66s^2$	8.7	$^5D_4$
Ir	77	$KLMN5s^25p^65d^76s^2$	9.1	$^4F_{9/2}$
Pt	78	$KLMN5s^25p^65d^96s^1$	9.0	$^3D_3$
Au	79	$KLMN5s^25p^65d^{10}6s^1$	9.225	$^2S_{1/2}$
Hg	80	$KLMN5s^25p^65d^{10}6s^2$	10.437	$^1S_0$
Tl	81	$KLMN5s^25p^65d^{10}6s^26p^1$	6.108	$^2P^o_{1/2}$
Pb	82	$KLMN5s^25p^65d^{10}6s^26p^2$	7.416	$^3P_0$
Bi	83	$KLMN5s^25p^65d^{10}6s^26p^3$	7.289	$^4S^o_{3/2}$
Po	84	$KLMN5s^25p^65d^{10}6s^26p^4$	8.42	$^3P_2$
At	85	$KLMN5s^25p^65d^{10}6s^26p^5$	—	$^2P^o_{3/2}$
Rn	86	$KLMN5s^25p^65d^{10}6s^26p^6$	10.748	$^1S_0$
Fr	87	$KLMN5s^25p^65d^{10}6s^26p^67s^1$	—	$^2S_{1/2}$
Ra	88	$KLMN5s^25p^65d^{10}6s^26p^67s^2$	5.279	$^1S_0$
Ac	89	$KLMN5s^25p^65d^{10}6s^26p^66d^17s^2$	6.9	$^2D_{3/2}$
Th	90	$KLMN5s^25p^65d^{10}6s^26p^66d^27s^2$	—	$^3F_2$
Pa	91	$KLMN5s^25p^65d^{10}5f^26s^26p^66d^17s^2$	—	$^4K_{11/2}$
U	92	$KLMN5s^25p^65d^{10}5f^36s^26p^66d^17s^2$	—	$^5L^o_6$
Np	93	$KLMN5s^25p^65d^{10}5f^46s^26p^66d^17s^2$	—	$^6L_{11/2}$
Pu	94	$KLMN5s^25p^65d^{10}5f^66s^26p^67s^2$	5.8	$^7F_0$
Am	95	$KLMN5s^25p^65d^{10}5f^76s^26p^67s^2$	6.0	$^8S^o_{7/2}$
Cm	96	$KLMN5s^25p^65d^{10}5f^76s^26p^66d^17s^2$	—	$^9D^o_2$
Bk	97	$KLMN5s^25p^65d^{10}5f^96s^26p^67s^2$	—	$^6H^o_{15/2}$
Cf	98	$KLMN5s^25p^65d^{10}5f^{10}6s^26p^67s^2$	—	$^5I_8$
Es	99	$KLMN5s^25p^65d^{10}5f^{11}6s^26p^67s^2$	—	$^4I^o_{15/2}$
Fm	100	$KLMN5s^25p^65d^{10}5f^{12}6s^26p^67s^2$	—	$^3H_6$

(*continued*)

Table 7.1 (*continued*)

Atom	Atomic number (Z)	Ground configuration	First ionization energy/eV[†]	Ground state
Md	101	$KLMN5s^25p^65d^{10}5f^{13}6s^26p^67s^2$	—	$^2F^{\circ}_{3/2}$
No	102	$KLMNO6s^26p^67s^2$	—	$^1S_0$
Lr	103	$KLMNO6s^26p^66d^17s^2$	—	$^2D_{3/2}$
—	104	$KLMNO6s^26p^66d^27s^2$	—	$^3F_2$

[†] For the process $A \rightarrow A^+ + e$, where A^+ is in its ground state.

Filling up the $4f$ orbital is a feature of the lanthanides. The $4f$ and $5d$ orbitals are of similar energy so that occasionally, as in La, Ce, and Gd, one electron goes into $5d$ rather than $4f$. Similarly in the actinides, Ac to No, the $5f$ subshell is filled in competition with $6d$.

7.1.2 Vector Representation of Momenta and Vector Coupling Approximations

7.1.2.1 Angular momenta and magnetic moments

We have seen in Section 1.3.2 and in Figure 1.5 how the orbital angular momentum of an electron can be represented by a vector, the direction of which is determined by the right-hand screw rule.

Each electron in an atom has two possible kinds of angular momenta, one due to its orbital motion and the other to its spin motion. The magnitude of the orbital angular momentum vector for a single electron is given, as in equation (1.44), by

$$[\ell(\ell + 1)]^{1/2}\hbar = \ell^*\hbar \tag{7.4}$$

where $\ell = 0, 1, 2, \ldots, (n - 1)$. (We shall come across the quantity $[Q(Q + 1)]^{1/2}$, where Q is a quantum number, so often that it is convenient to abbreviate it to Q^*.) Similarly, the magnitude of the spin angular momentum vector for a single electron is, as in equation (1.46),

$$[s(s + 1)]^{1/2}\hbar = s^*\hbar \tag{7.5}$$

where $s = \frac{1}{2}$ only.

For an electron having orbital and spin angular momentum there is a quantum number j associated with the total (orbital + spin) angular momentum which is a vector quantity whose magnitude is given by

$$[j(j + 1)]^{1/2}\hbar = j^*\hbar \tag{7.6}$$

where j can take the values

$$j = \ell + s, \ell + s - 1, \ldots, |\ell - s| \tag{7.7}$$

Since $s = \frac{1}{2}$, only, j is not a very useful quantum number for one-electron atoms, unless we are concerned with the fine detail of their spectra, but the analogous quantum number J, in polyelectronic atoms, is very important.

A charge of $-e$ circulating in an orbit is equivalent to a current flowing in a wire and therefore causes a magnetic moment. That due to orbital motion, μ_ℓ, is a vector which is opposed to the corresponding orbital angular momentum vector l, as shown in Figure 7.2(a). The classical picture of an electron spinning on its own axis indicates that there is a magnetic moment μ_s associated with this angular momentum also. Figure 7.2(a) shows that this vector is opposed to s. The magnetic moments μ_ℓ and μ_s can be regarded as acting rather like tiny bar magnets. For each electron they may be parallel, as in Figure 7.2(a), or opposed, as in Figure 7.2(b).

If the nucleus has a non-zero spin quantum number I (see Table 1.3) there is an additional, nuclear spin, angular momentum given by equation (1.48) but, because of the large mass of the nucleus compared to that of the electron, this angular momentum is typically very small. We shall neglect this angular momentum here together with the small effects, in the form of so-called hyperfine splitting, that it may have on an observed atomic spectrum.

7.1.2.2 Coupling of angular momenta

It follows from the fact that the magnetic moments due to orbital and spin angular momenta of each electron can be regarded as very small bar magnets that they interact with each other, just as a set of bar magnets would. We refer to this interaction as coupling of the angular momenta and the greater are the magnetic moments, the stronger is the coupling. However, some couplings are so weak that they may be neglected.

Coupling between two vectors a and b produces a resultant vector c, as shown in Figure 7.3(a). If the vectors represent angular momenta, a and b precess around c, as in Figure 7.3(b), the rate of precession increasing with the strength of coupling. In practice c precesses about an arbitrary direction in space and, when an electric or magnetic field (Stark or Zeeman effect) is introduced, the effects of space quantization (Section 1.3.2) may be observed.

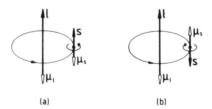

(a)   (b)

Figure 7.2 Vectors l and s and magnetic moments μ_ℓ and μ_s associated with orbital and spin angular momenta when the motions are (a) in the same direction and (b) in opposite directions.

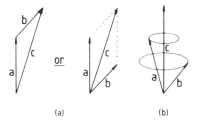

Figure 7.3 (a) Addition of two vectors *a* and *b* to give *c*. (b) Precession of *a* and *b* around *c*.

The strength of coupling between the spin and orbital motions of the electrons, referred to as spin–orbit coupling, depends on the atom concerned.

The spin of one electron can interact with (a) the spins of the other electrons, (b) its own orbital motion, and (c) the orbital motions of the other electrons. This last is called spin-other-orbit interaction and is normally too small to be taken into account. Interactions (a) and (b) are more important and the methods of treating them involve two types of approximation representing two extremes of coupling.

One approximation assumes that coupling between spin momenta is sufficiently small to be neglected, as also is the coupling between orbital momenta. On the other hand, coupling between the spin of an electron and its own angular momentum, to give a resultant total angular momentum j, is assumed to be strong and the coupling between the j's for all the electrons to be less strong but appreciable. This coupling treatment is known as the jj-coupling approximation but its usefulness is limited mainly to a few states of heavy atoms.

A second approximation neglects coupling between the spin of an electron and its orbital momentum but assumes that coupling between orbital momenta is strong and that between spin momenta relatively weak but appreciable. This represents the opposite extreme to the jj-coupling approximation. It is known as the Russell–Saunders coupling approximation and serves as a useful basis for describing most states of most atoms and is the only one we shall consider in detail.

7.1.2.3 Russell–Saunders coupling approximation

7.1.2.3(a) Non-equivalent electrons Non-equivalent electrons are those which have different values of *either n or ℓ* so that, for example, those in a $3p^1 3d^1$ or $3p^1 4p^1$ configuration are non-equivalent while those in a $2p^2$ configuration are equivalent. Coupling of angular momenta of non-equivalent electrons is rather more straightforward than for equivalent electrons.

First we consider, for non-equivalent electrons, the strong coupling between orbital angular momenta, referred to as $\ell\ell$ coupling, using a particular example.

Consider just two non-equivalent electrons in an atom, e.g. the helium atom in the highly excited configuration $2p^1 3d^1$. We shall label the $2p$ and $3d$ electrons '1' and '2' respectively, so that $\ell_1 = 1$ and $\ell_2 = 2$, the ℓ_1 and ℓ_2 vectors representing these orbital angular momenta are of magnitudes $2^{1/2}\hbar$ and $6^{1/2}\hbar$ respectively (see equation 7.4). These vectors couple, as in Figure 7.3(a), to give a resultant L of magnitude

$$[L(L + 1)]^{1/2}\hbar = L^*\hbar \tag{7.8}$$

However, the values of the total orbital angular momentum quantum number, L, are limited; or, in other words, the relative orientations of ℓ_1 and ℓ_2 are limited. The orientations which they can take up are governed by the values that the quantum number L can take. L is associated with the total orbital angular momentum for the two electrons and is restricted to the values

$$L = \ell_1 + \ell_2, \ell_1 + \ell_2 - 1, \dots, |\ell_1 - \ell_2| \tag{7.9}$$

In the present case $L = 3$, 2, or 1 and the magnitude of L is $12^{1/2}\hbar$, $6^{1/2}\hbar$, or $2^{1/2}\hbar$. These are illustrated by the vector diagrams of Figure 7.4(a).

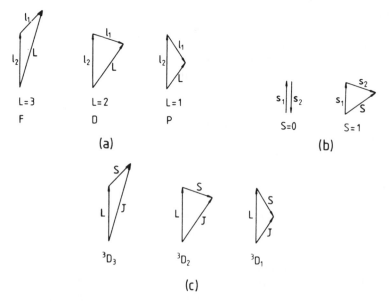

Figure 7.4 Russell–Saunders coupling of (a) orbital angular momenta ℓ_1 and ℓ_2, (b) spin angular momenta s_1 and s_2, and (c) total orbital and total spin angular momenta, L and S, of a p and a d electron.

The terms of the atom are labelled $S, P, D, F, G, \ldots$ corresponding to $L = 0, 1, 2, 3, 4, \ldots$, analogous to the labelling of one-electron orbitals $s, p, d, f, g, \ldots$ according to the value of ℓ. It follows that the $2p^1 3d^1$ configuration gives rise to P, D, and F terms.

In a similar way the coupling of a third vector to any of the L in Figure 7.4(a) will give the terms arising from three non-equivalent electrons, and so on.

It can be shown quite easily that, for a filled sub-shell such as $2p^6$ or $3d^{10}$, $L = 0$. Space quantization of the total orbital angular momentum produces $2L + 1$ components with $M_L = L, L - 1, \ldots, -L$, analogous to space quantization of ℓ. In a filled sub-shell $\sum_i (m_\ell)_i = 0$, where the sum is over all electrons in the sub-shell. Since $M_L = \sum_i (m_\ell)_i$, it follows that $L = 0$. Therefore the excited configurations

$$\begin{array}{ll} \text{C} & 1s^2 2s^2 2p^1 3d^1 \\ \text{Si} & 1s^2 2s^2 2p^6 3s^2 3p^1 3d^1 \end{array} \tag{7.10}$$

of C and Si both give P, D, and F terms.

The coupling between the spin momenta is referred to as ss coupling. The results of coupling of the s vectors can be obtained in a similar way to $\ell\ell$ coupling with the difference that, since s is always $\frac{1}{2}$, the vector for each electron is *always* of magnitude $3^{1/2}\hbar/2$ according to equation (7.5). The two s vectors can only take up orientations relative to each other such that the resultant S is of magnitude

$$[S(S + 1)]^{1/2}\hbar = S^*\hbar \tag{7.11}$$

where S, the total spin quantum number[†], is restricted to the values:

$$S = s_1 + s_2, s_1 + s_2 - 1, \ldots, |s_1 - s_2| \tag{7.12}$$

In the case of two electrons this means that $S = 0$ or 1 only. The vector sums, giving resultant S vectors of magnitude 0 and $2^{1/2}\hbar$, are illustrated in Figure 7.4(b).

The labels for the terms indicate the value of S by having $2S + 1$ as a pre-superscript to the $S, P, D, \ldots$ label. The value of $2S + 1$ is known as the multiplicity and is the number of values that M_S can take: these are

$$M_S = S, S - 1, \ldots, -S \tag{7.13}$$

Since, for two electrons, $S = 0$ or 1 the value of $2S + 1$ is 1 or 3 and the resulting terms are called singlet or triplet respectively. Just as $L = 0$ for a filled orbital, $S = 0$ also since $M_S = \sum_i (m_s)_i = 0$.

[†] This use of S should not be confused with its use as a term symbol to imply that $L = 0$.

It follows from this that the excited configurations of C and Si in equation (7.10) give ^1P, ^3P, ^1D, ^3D, ^1F, and ^3F terms. It follows also that the noble gases, in which all occupied orbitals are filled, have only ^1S terms arising from their ground configurations.

Table 7.2 lists the terms which arise from various combinations of two non-equivalent electrons.

There is appreciable coupling between the resultant orbital and resultant spin momenta. This is referred to as LS coupling and is due to spin–orbit interaction. This interaction is caused by the positive charge Ze on the nucleus and is proportional to Z^4. The coupling between L and S gives the total angular momentum vector J.

The phrase 'total angular momentum' is commonly used to refer to a number of different quantities. Here it implies 'orbital plus electron spin' but it is also used to imply 'orbital plus electron spin plus nuclear spin' when the symbol F is used.

The resultant vector J has the magnitude

$$[J(J + 1)]^{1/2}\hbar = J^*\hbar \tag{7.14}$$

where J is restricted to the values

$$J = L + S, L + S - 1, \ldots, |L - S| \tag{7.15}$$

from which it follows that, if $L > S$, then J can take $2S + 1$ values but, if $L < S$, J can take $2L + 1$ values.

As an example we consider LS coupling in a ^3D term. Since $S = 1$ and $L = 2$, then $J = 3, 2$, or 1 and Figure 7.4(c) illustrates the three ways of coupling L and S. The value of J is attached to the term symbol as a post-subscript so that the three components of ^3D are $^3D_3$, $^3D_2$, and $^3D_1$.

Table 7.2 Terms arising from some configurations of non-equivalent and equivalent electrons.

Non-equivalent electrons		Equivalent electrons	
Configuration	Terms	Configuration	Terms[†]
s^1s^1	$^{1,3}S$	p^2	$^1S, ^3P, ^1D$
s^1p^1	$^{1,3}P$	p^3	$^4S, ^2P, ^2D$
s^1d^1	$^{1,3}D$	d^2	$^1S, ^3P, ^1D, ^3F, ^1G$
s^1f^1	$^{1,3}F$	d^3	$^2P, ^4P, ^2D(2), ^2F,$
p^1p^1	$^{1,3}S, ^{1,3}P, ^{1,3}D$		$^4F, ^2G, ^2H$
p^1d^1	$^{1,3}P, ^{1,3}D, ^{1,3}F$	d^4	$^1S(2), ^3P(2), ^1D(2),$
p^1f^1	$^{1,3}D, ^{1,3}F, ^{1,3}G$		$^3D, ^5D, ^1F, ^3F(2),$
d^1d^1	$^{1,3}S, ^{1,3}P, ^{1,3}D, ^{1,3}F, ^{1,3}G$		$^1G(2), ^3G, ^3H, ^1I$
d^1f^1	$^{1,3}P, ^{1,3}D, ^{1,3}F, ^{1,3}G, ^{1,3}H$	d^5	$^2S, ^6S, ^2P, ^4P, ^2D(3),$
f^1f^1	$^{1,3}S, ^{1,3}P, ^{1,3}D, ^{1,3}F, ^{1,3}G,$		$^4D, ^2F(2), ^4F, ^2G(2),$
	$^{1,3}H, ^{1,3}I$		$^4G, ^2H, ^2I$

[†] The numbers in brackets indicate that a particular term occurs more than once.

The total number of states arising from the C or Si configuration of equation (7.10) is now seen to comprise

$$^1P_1, {}^3P_0, {}^3P_1, {}^3P_2, {}^1D_2, {}^3D_1, {}^3D_2, {}^3D_3, {}^1F_3, {}^3F_2, {}^3F_3, {}^3F_4$$

At this stage it is appropriate to digress for a moment on the subject of 'configurations', 'terms', and 'states'.

It is important that electron configurations, such as those in Table 7.1 and equation (7.10), should never be confused with terms or states. An electron configuration represents a gross but useful approximation in which the electrons have been fed into orbitals whose energies have been calculated neglecting the last term in equation (7.2). Nearly all configurations (all those with at least one unfilled orbital) give rise to more than one term or state and so it is quite wrong to speak of, for example, the $1s^1 2s^1$ state of helium: this is properly called a configuration and it gives rise to two states, $^1S_0$ and $^3S_1$.

The use of the words 'term' and 'state' is not so clear cut. The word 'term' was originally used in the early days of spectroscopy in the sense of equation (1.4), where the frequency of a line in an atomic spectrum is expressed as the difference between two terms which are simply terms in an equation.

Nowadays there is a tendency to use the word 'term' to describe what arises from an approximate treatment of an electron configuration, whereas the word 'state' is used to describe something which is observable experimentally. For example, we can say that the $1s^2 2s^2 2p^1 3d^1$ configuration of C gives rise to a ^3P term which, when spin–orbit coupling is taken into account, splits into $^3P_1$, $^3P_2$, and $^3P_3$ states. Since spin–orbit coupling can be excluded only in theory but never in practice there can be no experimental observation associated with the ^3P term[†].

If the nucleus possesses a spin angular momentum, these states are further split and therefore, perhaps, should not have been called states in the first place! However the splitting due to nuclear spin is small and it is normal to refer to nuclear spin *components* of states.

7.1.2.3(b) Equivalent electrons The Russell–Saunders coupling scheme for two, or more, equivalent electrons, i.e. with the same n and ℓ, is rather more lengthy to apply. We shall use the example of two equivalent p electrons, as in the ground configuration of carbon:

$$\text{C} \quad 1s^2 2s^2 2p^2 \tag{7.16}$$

Again, for the filled orbitals $L = 0$ and $S = 0$, so we have to consider only the $2p$ electrons. Since $n = 2$ and $\ell = 1$ for both electrons the Pauli exclusion

---

[†] It is unfortunately the case, however, that ^3P, for example, is sometimes referred to as a state.

principle is in danger of being violated unless the two electrons have different values of either m_ℓ or m_s. For non-equivalent electrons we do not have to consider the values of these two quantum numbers because, as either n or ℓ is different for the electrons, there is no danger of violation.

For the $2p$ electron, which we shall label 1, we have $\ell_1 = 1$ and $(m_\ell)_1 = +1$, 0 or -1 and, in addition, $s_1 = \frac{1}{2}$ and $(m_s)_1 = +\frac{1}{2}$ or $-\frac{1}{2}$; similarly for electron 2. The Pauli exclusion principle requires that the pair of quantum numbers $(m_\ell)_1$ and $(m_s)_1$ cannot simultaneously have the same values as $(m_\ell)_2$ and $(m_s)_2$. The result is that there are fifteen allowed combinations of values and they are all given in Table 7.3.

It should be noted that the indistinguishability of the electrons has been taken into account in the table so that, for example, the combination $(m_\ell)_1 = (m_\ell)_2 = 1$, $(m_s)_1 = -\frac{1}{2}$, $(m_s)_2 = \frac{1}{2}$ cannot be included in addition to $(m_\ell)_1 = (m_\ell)_2 = 1$, $(m_s)_1 = \frac{1}{2}$, $(m_s)_2 = -\frac{1}{2}$, which is obtained from the first by electron exchange.

The values of M_L $(= \Sigma_i (m_\ell)_i)$ and M_S $(= \Sigma_i (m_s)_i)$ are given in Table 7.3. The highest value of M_L is 2 and this indicates that this is also the highest value of L and that there is a D term. Since $M_L = 2$ is associated only with $M_S = 0$ it must be a ^1D term. This term accounts for five of the combinations, as shown at the bottom of the table. In the remaining combinations, the highest value of L is 1 and, since this is associated with $M_S = 1, 0$, and -1, there is a ^3P term. This accounts for a further nine combinations, leaving only $M_L = 0$, $M_S = 0$, which implies a ^1S term.

It is interesting to note that of the ^1S, ^3S, ^1P, ^3P, ^1D, and ^3D terms which arise from two non-equivalent p electrons, as in the $1s^2 2s^2 2p^1 3p^1$ configuration of the carbon atom, only ^1S, ^3P, and ^1D are allowed for two equivalent p electrons: the Pauli exclusion principle forbids the other three.

Terms arising from three equivalent p electrons and also from various equivalent d electrons can be derived using the same methods, but this can be a very lengthy operation. The results are given in Table 7.2.

In deriving the terms arising from non-equivalent or equivalent electrons there is a very useful rule that, in this respect, a vacancy in a sub-shell behaves like an electron. For example, the ground configurations of C and O:

$$\left. \begin{array}{ll} \text{C} & 1s^2 2s^2 2p^2 \\ \text{O} & 1s^2 2s^2 2p^4 \end{array} \right\} {}^1S,\ {}^3P,\ {}^1D \qquad (7.17)$$

give rise to the same terms, as do the excited configurations of C and Ne:

$$\left. \begin{array}{ll} \text{C} & 1s^2 2s^2 2p^1 3d^1 \\ \text{Ne} & 1s^2 2s^2 2p^5 3d^1 \end{array} \right\} {}^{1,3}P,\ {}^{1,3}D,\ {}^{1,3}F \qquad (7.18)$$

In 1927 Hund formulated two empirical rules which enable us to determine which of the terms arising from equivalent electrons lies lowest in

Table 7.3 Derivation of terms arising from two equivalent p electrons.

Quantum number	Values														
$(m_l)_1$	1	1	1	1	1	1	1	1	0	1	0	0	0	0	-1
$(m_l)_2$	1	0	0	0	0	-1	-1	-1	0	-1	-1	-1	-1	-1	-1
$(m_s)_1$	$\frac{1}{2}$	$\frac{1}{2}$	$\frac{1}{2}$	$-\frac{1}{2}$	$-\frac{1}{2}$	$\frac{1}{2}$	$\frac{1}{2}$	$-\frac{1}{2}$	$\frac{1}{2}$	$-\frac{1}{2}$	$\frac{1}{2}$	$\frac{1}{2}$	$-\frac{1}{2}$	$-\frac{1}{2}$	$\frac{1}{2}$
$(m_s)_2$	$-\frac{1}{2}$	$\frac{1}{2}$	$-\frac{1}{2}$	$\frac{1}{2}$	$-\frac{1}{2}$	$\frac{1}{2}$	$-\frac{1}{2}$	$\frac{1}{2}$	$-\frac{1}{2}$	$-\frac{1}{2}$	$\frac{1}{2}$	$-\frac{1}{2}$	$\frac{1}{2}$	$-\frac{1}{2}$	$-\frac{1}{2}$
$M_L = \sum_i (m_l)_i$	2	1	1	1	1	0	0	0	0	0	-1	-1	-1	-1	-2
$M_s = \sum_i (m_s)_i$	0	1	0	0	-1	1	0	0	0	-1	1	0	0	-1	0

Pairs of values of M_L and M_S can be rearranged as follows:

M_L	2	1	0	-1	-2	1	0	-1	1	0	-1	1	0	-1	0
M_S	0	0	0	0	0	1	1	1	0	0	0	-1	-1	-1	0
			^1D							^3P					^1S

energy. This means that for a ground configuration with only equivalent electrons in partly filled orbitals we can determine the lowest energy, or ground, term. These rules are:

1. Of the terms arising from equivalent electrons those with the highest multiplicity lie lowest in energy.
2. Of these, the lowest is that with the highest value of L.

Using these rules it follows that, for the ground configurations of both C and O in equation (7.17), the ^3P term is the lowest in energy.

The ground configuration of Ti is

$$\text{Ti} \quad KL3s^2 3p^6 3d^2 4s^2 \tag{7.19}$$

Of the terms arising from the d^2 configuration given in Table 7.2 the rules indicate that ^3F is the lowest in energy.

The splitting of a term by spin–orbit interaction is proportional to J:

$$E_J - E_{J-1} = AJ \tag{7.20}$$

where E_J is the energy corresponding to J, and a multiplet results. If A is positive, the component with the smallest value of J lies lowest in energy and the multiplet is said to be normal whereas, if A is negative, the multiplet is inverted.

There are two further rules for ground terms which tell us whether a multiplet arising from equivalent electrons is normal or inverted.

3. Normal multiplets arise from equivalent electrons when a partially filled orbital is *less* than half full.
4. Inverted multiplets arise from equivalent electrons when a partially filled orbital is *more* than half full.

It follows that the lowest energy term of Ti, ^3F, is split by spin–orbit coupling into a normal multiplet and therefore the ground state is $^3F_2$. Similarly the lowest energy term of C, ^3P, splits into a normal multiplet resulting in a $^3P_0$ ground state, whereas that of O, with an inverted multiplet, is $^3P_2$.

Atoms with a ground configuration in which an orbital is exactly half-filled, as for example in N($2p^3$), Mn($3d^5$), and Eu($4f^7$), always have an S ground state. Since such states have only one component the problem of a normal or inverted multiplet does not arise.

Table 7.1 gives the ground states of all atoms in the periodic table.

For excited terms split by spin–orbit interaction there are no general rules regarding normal or inverted multiplets. For example, in He, excited states form mostly inverted multiplets while in the alkaline earth metals, Be, Mg, Ca, ..., they are mostly normal.

There is one further addition to the state symbolism which we have not mentioned so far. This is the superscript 'o' as in the ground state of boron, $^2P^o_{1/2}$. The symbol implies that the arithmetic sum $\Sigma_i \ell_i$ for all the electrons in the atom is an odd number, 1 in this case. When there is no such superscript this implies that the sum is an even number; for example it is 4 in the case of oxygen.

7.1.3 Spectra of Alkali Metal Atoms

The hydrogen atom and one-electron ions are the simplest systems in the sense that, having only one electron, there are no inter-electron repulsions. However, this unique property leads to degeneracies, or near-degeneracies, which are absent in all other atoms and ions. The result is that the spectrum of the hydrogen atom, although very simple in its coarse structure (Figure 1.1) is more unusual in its fine structure than those of polyelectronic atoms. For this reason we shall defer a discussion of its spectrum to the next section.

The alkali metal atoms all have one valence electron in an outer ns orbital, where $n = 2, 3, 4, 5, 6$ for Li, Na, K, Rb, and Cs. If we consider only orbital changes involving this electron the behaviour is expected to resemble that of the hydrogen atom. The reason for this is that the core, consisting of the nucleus of charge $+Ze$ and filled orbitals containing $(Z - 1)$ electrons, has a net charge of $+e$ and therefore has an effect on the valence electron which is similar to that of the nucleus of the hydrogen atom on its electron.

Whereas the emission spectrum of the hydrogen atom shows only one series, the Balmer series (see Figure 1.1), in the visible region the alkali metals show at least three. The spectra can be excited in a discharge lamp containing a sample of the appropriate metal. One series was called the principal series because it could also be observed in absorption through a column of the vapour. The other two were called sharp and diffuse because of their general appearance. A part of a fourth series, called the fundamental series, can sometimes be observed.

Figure 7.5 shows these series schematically for Li. All such series converge smoothly towards high energy (low wavelength) in a way that resembles the series in the hydrogen spectrum.

Figure 7.6 shows an energy level diagram, a so-called Grotrian diagram, for Li for which the ground configuration is $1s^22s^1$. The lowest energy level corresponds to this configuration. The higher energy levels are labelled according to the orbital to which the valence electron has been promoted: e.g. the level labelled $4p$ corresponds to the configuration $1s^24p^1$.

The relatively large energy separation between configurations in which the valence electron is in orbitals differing only in the value of ℓ, e.g. $1s^23s^1$, $1s^23p^1$, $1s^23d^1$, is characteristic of all atoms except hydrogen (and one-electron ions).

The selection rules governing the promotion of the electron to an excited

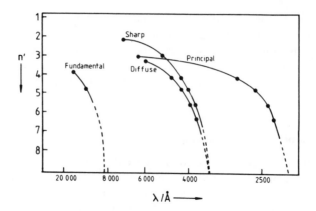

Figure 7.5 Four series in the emission spectrum of lithium.

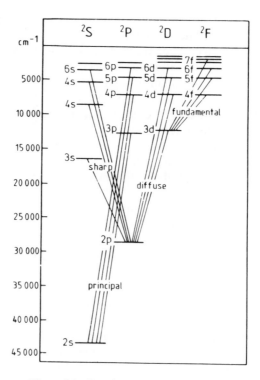

Figure 7.6 Grotrian diagram for lithium.

orbital, and also its falling back from an excited orbital, are

$$\text{(a)} \quad \Delta n \text{ is unrestricted}$$
$$\text{(b)} \quad \Delta \ell = \pm 1 \tag{7.21}$$

These selection rules lead to the sharp, principal, diffuse, and fundamental series, shown in Figures 7.5 and 7.6, in which the promoted electron is in an s, p, d, and f orbital respectively. Indeed these rather curious orbital symbols originate from the first letters of the corresponding series observed in the spectrum.

Some excited configurations of the lithium atom, involving promotion of only the valence electron, are given in Table 7.4, which also lists the states arising from these configurations. Similar states can easily be derived for other alkali metals.

Spin–orbit coupling splits apart the two components of the ^2P, ^2D, ^2F, ... terms. The splitting decreases with L and n and increases with atomic number but is not large enough, for any terms of the lithium atom, to show on the Grotrian diagram of Figure 7.6. The resulting fine structure is difficult to resolve in the lithium spectrum but is more easily observed in other alkali metal spectra.

In the sodium atom pairs of $^2P_{1/2}$, $^2P_{3/2}$ states result from the promotion of the $3s$ valence electron to any np orbital with $n > 2$. It is convenient to label the states with this value of n, as $n^2P_{1/2}$ and $n^2P_{3/2}$, the n label being helpful for states which arise when only one electron is promoted and the unpromoted electrons are either in filled orbitals or in an s orbital. The n label can be used, therefore, for hydrogen, the alkali metals, helium, and the alkaline earths. In other atoms it is usual to precede the state symbols by the configuration of the electrons in unfilled orbitals, as in the $2p3p\ ^1S_0$ state of carbon.

The splitting of the $3^2P_{1/2}$, $3^2P_{3/2}$ states of sodium is $17.2\ \text{cm}^{-1}$ and this reduces to 5.6, 2.5, and $1.3\ \text{cm}^{-1}$ for $n = 4, 5$, and 6 respectively. The splitting decreases rapidly with L as exemplified by the splitting of only $0.1\ \text{cm}^{-1}$ for the $3^2D_{3/2}$, $3^2D_{5/2}$ states. All these ^2P and ^2D multiplets are normal, the state with lowest J lying lowest in energy.

The fine structure selection rule is

$$\Delta J = 0, \pm 1 \text{ except, } J = 0 \nleftrightarrow J = 0 \tag{7.22}$$

Table 7.4 Configurations and states of the lithium atom.

Configuration	States	Configuration	States
$1s^2 2s^1$	$^2S_{1/2}$	$1s^2 nd^1\ (n = 3, 4, \ldots)$	$^2D_{3/2}, ^2D_{5/2}$
$1s^2 ns^1\ (n = 3, 4, \ldots)$	$^2S_{1/2}$	$1s^2 nf^1\ (n = 3, 4, \ldots)$	$^2F_{5/2}, ^2F_{7/2}$
$1s^2 np^1\ (n = 2, 3, \ldots)$	$^2P_{1/2}, ^2P_{3/2}$		

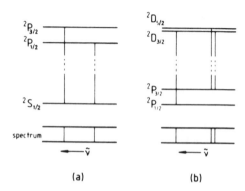

Figure 7.7 (a) A simple doublet and (b) a compound doublet in the spectrum of, for example, the sodium atom

The result is that the principal series consists of pairs of $^2P_{1/2} - {}^2S_{1/2}$, $^2P_{3/2} - {}^2S_{1/2}$ transitions[†], as illustrated in Figure 7.7(a). The pairs are known as simple doublets. The first member of this series in sodium, called the sodium D line, appears in the yellow region of the spectrum with components at 589.592 and 588.995 nm.

The $3^2P_{1/2}$, $3^2P_{3/2}$ excited states involved in the sodium D line are the lowest energy excited states of the atom. Consequently, in a discharge in the vapour at a pressure which is sufficiently high for collisional deactivation of excited states to occur readily, a majority of atoms find themselves in these states before emission of radiation has taken place. Therefore the D line is prominent in emission, which explains the predominantly yellow colour of sodium discharge lamps.

The sharp series members are all simple doublets which all show the same splitting, namely that of the $3^3P_{1/2}$ and $3^2P_{3/2}$ states.

All members of a diffuse series consist of compound doublets, as illustrated in Figure 7.7(b), but the splitting of the $^2D_{3/2}$, $^2D_{5/2}$ states may be too small for the close pair of transitions to be resolved. It is for this reason that the set of three transitions has become known as a compound doublet rather than a triplet.

7.1.4 Spectrum of the Hydrogen Atom

The hydrogen atom presented a unique opportunity in the development of quantum mechanics. The single electron moves in a coulombic field, free from

---

[†] The convention of indicating a transition involving an *upper* electronic state N and a *lower* electronic state M by N–M is analogous to that used in rotational and vibrational spectroscopy.

the effects of interelectron repulsions. This has two important consequences which do not apply to any atom with two or more electrons:

1. The Schrödinger equation (equation 1.28) is exactly soluble with the hamiltonian of equation (1.30).
2. The orbital energies, at this level of approximation, are independent of the quantum number ℓ, as Figure 1.1 shows.

When the members of the Balmer and Paschen series (see Figure 1.1) are observed at high resolution they show closely-spaced fine structure and it was an important test of quantum mechanical methods to explain this.

Figure 7.8 shows that Dirac, by including the effects of relativity in the quantum mechanical treatment, predicted the splitting of the $n = 2$ level into two components 0.365 cm^{-1} apart. For $n = 2$, ℓ can be 0 or 1. Equation (7.7) shows that, since $s = \frac{1}{2}$, j can be $\frac{3}{2}$ or $\frac{1}{2}$, for $\ell = 1$, and $\frac{1}{2}$, for $\ell = 0$. One of the components of the $n = 2$ level has $j = \frac{3}{2}$, $\ell = 1$ and the other is doubly degenerate with $j = \frac{1}{2}$, $\ell = 0, 1$.

In 1947 Lamb and Retherford observed the $2^2P_{3/2} - 2^2S_{1/2}$ transition using microwave techniques and found it to have a wavenumber 0.0354 cm^{-1} less than predicted by Dirac. The corresponding shift of the energy level, known as the Lamb shift, is shown in Figure 7.8: the $2^2P_{1/2}$ level is not shifted. Later, Lamb and Retherford observed the $2^2S_{1/2} - 2^2P_{1/2}$ transition directly with a wavenumber of 0.0354^{-1}. Quantum electrodynamics is the name given to the modified Dirac theory which accounts for the Lamb shift.

Figure 7.8 shows that the $1^2S_{1/2}$ state is shifted, but not split, when quantum

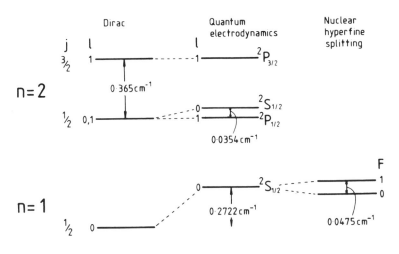

Figure 7.8 The $n = 1$ and $n = 2$ levels of the hydrogen atom

electrodynamics is applied. It is, however, split into two components, 0.0457 cm^{-1} apart, by the effects of nuclear spin ($I = \frac{1}{2}$ for ^{1}H).

The hydrogen atom and its spectrum are of enormous importance in astrophysics because of the large abundance of hydrogen atoms both in stars, including the sun, and in the interstellar medium.

Hydrogen is easily the most abundant element in stars and may be detected by its absorption spectrum. The temperature of the interior of a star is of the order of 10^6 K but that of the exterior, the photosphere, is only about 10^3 K. The absorption spectrum observed from a star involves the interior acting as a continuum source and the photosphere as the absorber. Whereas in an earthbound absorption experiment only the Lyman series would be observed, because all the atoms are in the $n = 1$ level, the absorption spectrum of a star shows the Balmer series also in absorption. Although the ratio of the population of the $n = 2$ to the $n = 1$ level, obtained from equation (2.11), is only 2.9×10^{-5} at a temperature of 10^3 K, the high concentration of hydrogen atoms, together with the long absorption pathlength in the photosphere, combine to make the observation possible.

Although Figure 1.1 shows the first five series in the spectrum of atomic hydrogen there are an infinite number of series. As the value of n'' for the lower level common to a series increases, the levels become closer together so that series with high values of n'' appear in the radiofrequency region (see Figure 3.1). Scanning of the interstellar medium (see Section 5.2.6) with radiotelescopes has resulted in many members of such series being observed in emission from the abundant hydrogen atoms. For example, the first member of each of the series with $n'' = 90, 104, 109, 126, 156, 158, 159,$ and 166 have been observed.

For monitoring the quantity of hydrogen atoms in stars or the interstellar medium the $F = 1$–0 nuclear hyperfine transition, shown in Figure 7.8, is commonly observed. This occurs in the radiofrequency region at a wavelength of 21 cm and may be seen in emission or absorption, the background radiation acting as a continuum source. Results for the interstellar medium show highest concentrations of hydrogen atoms to be in regions, such as the Milky Way in our own galaxy, where there are large concentrations of stars.

7.1.5 Spectra of Helium and the Alkaline Earth Metal Atoms

The emission spectrum of a discharge in helium gas in the visible and near-ultraviolet regions appears rather like the spectrum of two alkali metals. It shows series of lines converging smoothly to high energy (low wavelength) which can be divided into two groups. One group consists of single lines and the other, at low resolution, double lines.

Derivation of an energy level diagram shows that it consists of two sets of energy levels, one corresponding to the single lines and the other to the double lines, and that no transitions between the two sets of levels are observed. For this reason it was suggested that helium exists in two separate forms. In 1925 it became clear that, when account is taken of electron spin, the two forms are really singlet helium and triplet helium.

For hydrogen and the alkali metal atoms in their ground configurations, or excited configurations involving promotion of the valence electron, there is only one electron with an unpaired spin. For this electron $m_s = +\frac{1}{2}$ or $-\frac{1}{2}$ and the corresponding electron spin part of the total wave function is conventionally given the symbol α or β respectively. All states arising are doublet states.

In helium we need to look more closely at the consequences of electron spin since this is the prototype of all atoms and molecules having easily accessible states with two different multiplicities.

If the spin–orbit coupling is small, as it is in helium, the total electronic wave function ψ_e can be factorized into an orbital part ψ_e^o and a spin part ψ_e^s:

$$\psi_e = \psi_e^o \psi_e^s \tag{7.23}$$

The spin part ψ_e^s can be derived by labelling the electrons 1 and 2 and remembering that, in general, each can have an α or β spin wave function giving four possible combinations: $\alpha(1)\beta(2)$, $\beta(1)\alpha(2)$, $\alpha(1)\alpha(2)$, and $\beta(1)\beta(2)$. Because the first two are neither symmetric nor antisymmetric to the exchange of electrons, which is equivalent to the exchange of the labels 1 and 2, they must be replaced by linear combinations giving

$$\psi_e^s = 2^{-1/2}[\alpha(1)\beta(2) - \beta(1)\alpha(2)] \tag{7.24}$$

and

$$\begin{aligned}\psi_e^s = \quad &\alpha(1)\alpha(2) \\ \text{or} \quad &\beta(1)\beta(2) \\ \text{or} \quad &2^{-1/2}\,[\alpha(1)\beta(2) + \beta(1)\alpha(2)]\end{aligned} \tag{7.25}$$

where the factors $2^{-1/2}$ are normalization constants. The wave function ψ_e^s in equation (7.24) is antisymmetric to electron exchange and is that of a singlet state, while the three in equation (7.25) are symmetric and are those of a triplet state.

Turning to the orbital part of ψ_e we consider the electrons in two different atomic orbitals χ_a and χ_b as, for example, in the $1s^1 2p^1$ configuration of helium. There are two ways of placing electrons 1 and 2 in these orbitals giving wave functions $\chi_a(1)\chi_b(2)$ and $\chi_a(2)\chi_b(1)$ but, once again, we have to use, instead, the linear combinations

$$\psi_e^o = 2^{-1/2}[\chi_a(1)\chi_b(2) + \chi_a(2)\chi_b(1)] \tag{7.26}$$

$$\psi_e^o = 2^{-1/2}[\chi_a(1)\chi_b(2) - \chi_a(2)\chi_b(1)] \tag{7.27}$$

The wave functions in equations (7.26) and (7.27) are symmetric and antisymmetric respectively to electron exchange.

The most general statement of the Pauli principle for electrons and other fermions is that the total wave function must be antisymmetric to electron (or fermion) exchange. For bosons it must be symmetric to exchange.

For helium, therefore, the singlet spin wave function of equation (7.24) can combine only with the orbital wave function of equation (7.26) giving, for singlet states,

$$\psi_e = 2^{-1}[\chi_a(1)\chi_b(2) + \chi_a(2)\chi_b(1)][\alpha(1)\beta(2) - \beta(1)\alpha(2)] \tag{7.28}$$

Similarly, for triplet states,

$$\psi_e = 2^{-1/2}[\chi_a(1)\chi_b(2) - \chi_a(2)\chi_b(1)]\alpha(1)\alpha(2)$$
$$\text{or } 2^{-1/2}[\chi_a(1)\chi_b(2) - \chi_a(2)\chi_b(1)]\beta(1)\beta(2)$$
$$\text{or } 2^{-1}[\chi_a(1)\chi_b(2) - \chi_a(2)\chi_b(1)][\alpha(1)\beta(2) + \beta(1)\alpha(2)] \tag{7.29}$$

For the ground configuration, $1s^2$, the orbital wave function is given by

$$\psi_e^{\circ} = \chi_a(1)\chi_b(2) \tag{7.30}$$

which is symmetric to electron exchange. This configuration leads, therefore, to only a singlet term, whereas each excited configuration arising from the promotion of one electron gives rise to a singlet and a triplet. Each triplet term lies lower in energy than the corresponding singlet.

The Grotrian diagram in Figure 7.9 gives the energy levels for all the terms arising from the promotion of one electron in helium to an excited orbital.

The selection rules are

$$\Delta\ell = \pm 1, \text{ for the promoted electron}$$
$$\Delta S = 0 \tag{7.31}$$

The latter rule is rigidly obeyed in the observed spectrum of helium. From the accurately known energy levels it is known precisely where to look for transitions between singlet and triplet states but none has been found.

Because of this spin selection rule, atoms which get into the lowest triplet state, 2^3S_1, do not easily revert to the ground 1^1S_0 state: the transition is forbidden by both the orbital and spin selection rules. The lowest triplet state is therefore metastable. In a typical discharge it has a lifetime of the order of 1 ms.

The first excited singlet state, 2^1S_0, is also metastable in the sense that a transition to the ground state is forbidden by the $\Delta\ell$ selection rule but, because the transition is not spin forbidden, this state is not so long-lived as the 2^3S_1 metastable state.

Owing to the effects of spin–orbit coupling all the triplet terms, except ^3S, are split into three components. For example, in the case of a ^3P term, with $L = 1$ and $S = 1$, J can take the values 2, 1, 0 (equation 7.15).

The splitting of triplet terms of helium is unusual in two respects. Firstly,

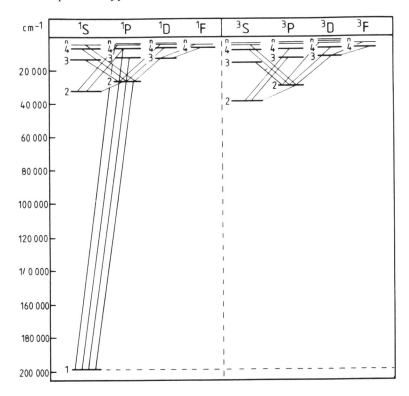

Figure 7.9 Grotrian diagram for helium. The scale is too small to show splittings due to spin–orbit coupling.

multiplets may be inverted and, secondly, the splittings of the multiplet components do not obey equation (7.20). For this reason we shall discuss fine structure due to spin–orbit coupling in the context of the alkaline earth atomic spectra where multiplets are usually normal and also obey equation (7.20). The alkaline earth metal spectra resemble that of helium quite closely. All the atoms have an ns^2 valence orbital configuration and, when one of these electrons is promoted, give a series of singlet and triplet states.

The fine structure of a $^3P - {}^3S$ transition of an alkaline earth metal is illustrated in Figure 7.10(a). The ΔJ selection rule (equation 7.22) results in a simple triplet. (The very small separation of $2{}^3P_1$ and $2{}^3P_2$ in helium accounts for the early description of the low resolution spectrum of triplet helium as consisting of 'doublets'.)

A $^3D - {}^3P$ transition, shown in Figure 7.10(b), has six components. As with doublet states the multiplet splitting decreases rapidly with L so the resulting six lines in the spectrum appear, at medium resolution, as a triplet. For this reason the fine structure is often called a compound triplet.

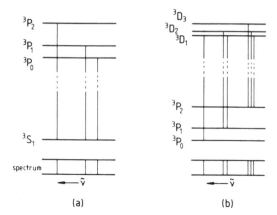

Figure 7.10 (a) A simple triplet and (b) a compound triplet in the spectrum of an alkaline earth metal atom.

7.1.6 Spectra of Other Polyelectronic Atoms

So far we have considered only hydrogen, helium, the alkali metals, and the alkaline earth metals but the selection rules and general principles encountered can be extended quite straightforwardly to any other atom.

An obvious difference between the emission spectra of most atoms and those we have considered so far is their complexity, the spectra showing very many lines and no obvious series. An extreme example is the spectrum of iron which is so rich in lines that it is commonly used as a calibration spectrum throughout the visible and ultraviolet regions.

However complex the atom, we can use the Russell–Saunders coupling approximation (or jj coupling, if necessary) to derive the states which arise from any configuration. The general selection rules which apply to transitions between these states are:

1. $\Delta L = 0, \pm 1$ except $L = 0 \nleftrightarrow L = 0$ (7.32)

Previously we have considered the promotion of only one electron, for which $\Delta \ell = \pm 1$ applies, but the general rule given here involves the total orbital angular momentum quantum number L and applies to the promotion of any number of electrons.

2. Even $\nleftrightarrow$ even, odd $\nleftrightarrow$ odd, even $\leftrightarrow$ odd   (7.33)

Here 'even' and 'odd' refer to the arithmetic sum $\Sigma_i \ell_i$ over all the electrons and this selection rule is called the Laporte rule. An important result of this is that transitions are forbidden between states arising from the same configuration. For example, of the terms given in equation (7.18) arising from the $1s^2 2s^2 2p^1 3d^1$ configuration of the carbon atom, a $^1P - {}^1D$ transition would be allowed if we considered only the ΔS (see rule 4 below) and ΔL selection rules, but the Laporte rule forbids it. Similarly any transitions between states arising from the $1s^2 2s^2 2p^1 3d^1$ configuration, with $\Sigma_i \ell_i = 3$, and those arising from the $1s^2 2s^2 3d^1 4f^1$ configuration, with $\Sigma_i \ell_i = 5$, are also forbidden[†].

The Laporte rule is consistent with $\Delta \ell = \pm 1$ when only one electron is promoted from the ground configuration.

3. $\Delta J = 0, \pm 1$ except $J = 0 \not\leftrightarrow J = 0$ \hfill (7.34)

This rule is the same for all atoms.

4. $\Delta S = 0$ \hfill (7.35)

applies only to atoms with a small nuclear charge. In atoms with a large nuclear charge it breaks down so that the factorization of ψ_e in equation (7.23) no longer applies due to spin–orbit interaction and states can no longer be described accurately as singlet, doublet, etc. The mercury atom provides a good example of the breakdown of the $\Delta S = 0$ selection rule. Having the ground configuration $KLMN5s^2 5p^6 5d^{10} 6s^2$ it is rather like an alkaline earth metal. Promotion of an electron from the $6s$ to the $6p$ orbital produces $6^1P_1, 6^3P_0, 6^3P_1$, and 6^3P_2 states, the three components of 6^3P being widely split by spin–orbit interaction. This interaction also breaks down the spin selection rule to such an extent that the $6^3P_1 - 6^1S_0$ transition at 253.652 nm is one of the strongest in the mercury emission spectrum.

7.2 Electronic Spectroscopy of Diatomic Molecules

7.2.1 Molecular Orbitals

7.2.1.1 Homonuclear diatomic molecules

The molecular orbital (MO) approach to the electronic structure of diatomic, and also polyatomic, molecules is not the only one which is used but it lends itself to a fairly qualitative description which we require here.

The approach adopted in the MO method is to consider the two nuclei, without their electrons, a distance apart equal to the equilibrium internuclear distance and to construct MOs around them—rather as we construct atomic orbitals (AOs) around a single bare nucleus. Electrons are then fed into the MOs in pairs (with the electron spin quantum number, $m_s = \pm \frac{1}{2}$) in order of

---

[†] The $2^2P_{3/2} - 2^2S_{1/2}$ and $2^2S_{1/2} - 2^2P_{1/2}$ transitions observed in the hydrogen atom violate the Laporte rule because they are magnetic dipole transitions and the rule applies only to electric dipole transitions.

increasing energy using the *aufbau* principle, just as for atoms (Section 7.1.1), to give the ground configuration of the molecule.

The basis of constructing the MOs is the linear combination of atomic orbitals (LCAO) method. This takes account of the fact that, in the region close to a nucleus, the MO wave function resembles an AO wave function for the atom of which the nucleus is a part. It is reasonable, then, to express an MO wave function ψ as a linear combination of AO wave functions χ_i on both nuclei:

$$\psi = \Sigma_i c_i \chi_i \qquad (7.36)$$

where c_i is the coefficient of the wave function χ_i. However, not all linear combinations are effective in the sense of producing an MO which is appreciably different from the AOs from which it is formed. For effective linear combinations:

1. The energies of the AOs must be comparable.
2. The AOs should overlap as much as possible.
3. The AOs must have the same symmetry properties with respect to certain symmetry elements of the molecule.

For a homonuclear diatomic molecule with nuclei labelled 1 and 2 the LCAO method gives the MO wave function

$$\psi = c_1 \chi_1 + c_2 \chi_2 \qquad (7.37)$$

Using the N_2 molecule as an example we can see that, for instance, the nitrogen $1s$ AOs satisfy condition (1), since their energies are identical, but not condition (2) because the high nuclear charge causes the $1s$ AOs to be close to the nuclei, resulting in little overlap. On the other hand, the $2s$ AOs satisfy both conditions and, since they are spherically symmetrical, condition (3) as well. Examples of AOs which satisfy conditions 1 and 2 but not 3 are the $2s$ and $2p_x$ orbitals in Figure 7.11. The $2s$ AO is symmetric to reflection across a plane containing the internuclear z axis and is perpendicular to the figure, while the $2p_x$ AO is antisymmetric to this reflection. In fact we can easily see that any overlap between $2s$ and the positive lobe of $2p_x$ is exactly cancelled by that involving the negative lobe.

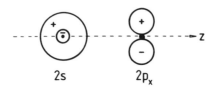

Figure 7.11 Illustration of zero overlap between a $2s$ and a $2p_x$ (or $2p_y$) AO.

Important properties of the MO are the energy E associated with it and the values of c_1 and c_2 in equation (7.37). They are obtained from the Schrödinger equation

$$H\psi = E\psi \tag{7.38}$$

Multiplying both sides by ψ^*, the complex conjugate of ψ obtained from it by replacing all $i(=\sqrt{-1})$ by $-i$, and integrating over all space gives

$$E = \frac{\int \psi^* H\psi \, d\tau}{\int \psi^* \psi \, d\tau} \tag{7.39}$$

E can be calculated only if ψ is known, so what is done is to make an intelligent guess at the MO wave function, say ψ_n, and calculate the corresponding value of the energy $\bar{E}_n$ from equation (7.39). A second guess at ψ, say ψ_m, gives a corresponding energy $\bar{E}_m$. The variation principle states that, if $\bar{E}_m < \bar{E}_n$, then ψ_m is closer than ψ_n to the true MO wave function. This applies to the ground state only. In this way the true ground state wave function can be approached as closely as we choose but this is usually done by varying parameters in the chosen wave function until they have their optimum values.

Combining equation (7.37) and (7.39) and assuming that χ_1 and χ_2 are not complex gives

$$\bar{E} = \frac{\int (c_1^2\chi_1 H\chi_1 + c_1c_2\chi_1 H\chi_2 + c_1c_2\chi_2 H\chi_1 + c_2^2\chi_2 H\chi_2)d\tau}{\int (c_1^2\chi_1^2 + 2c_1c_2\chi_1\chi_2 + c_2^2\chi_2^2)d\tau} \tag{7.40}$$

If the AO wave functions are normalized

$$\int \chi_1^2 \, d\tau = \int \chi_2^2 \, d\tau = 1 \tag{7.41}$$

and, as H is a hermitian operator,

$$\int \chi_1 H\chi_2 \, d\tau = \int \chi_2 H\chi_1 \, d\tau = H_{12} \text{ (say)} \tag{7.42}$$

The quantity

$$\int \chi_1\chi_2 \, d\tau = S \tag{7.43}$$

and is called the overlap integral as it is a measure of the degree to which χ_1 and χ_2 overlap each other. In addition, integrals such as $\int \chi_1 H\chi_1 d\tau$ are abbreviated to H_{11}. All these simplifications and abbreviations reduce equation (7.40) to

$$\bar{E} = \frac{c_1^2 H_{11} + 2c_1c_2 H_{12} + c_2^2 H_{22}}{c_1^2 + 2c_1c_2 S + c_2^2} \tag{7.44}$$

Using the variation principle to optimize c_1 and c_2 we obtain $\partial \bar{E}/\partial c_1$ and $\partial \bar{E}/\partial c_2$ from equation (7.44) and put them equal to zero, giving

$$c_1(H_{11} - E) + c_2(H_{12} - ES) = 0$$
$$c_1(H_{12} - ES) + c_2(H_{22} - E) = 0 \tag{7.45}$$

where we have replaced $\bar{E}$ by E since, although it is probably not the true energy, it is the nearest approach to it with the wave function of equation (7.37). The equations (7.45) are the secular equations and the two values of E which satisfy them are obtained by solution of the two simultaneous equations or, more simply, from the secular determinant

$$\begin{vmatrix} H_{11} - E & H_{12} - ES \\ H_{12} - ES & H_{22} - E \end{vmatrix} = 0 \tag{7.46}$$

H_{12} is the resonance integral, usually symbolized by β. In a homonuclear diatomic molecule $H_{11} = H_{22} = \alpha$, which is known as the Coulomb integral, and the secular determinant becomes

$$\begin{vmatrix} \alpha - E & \beta - ES \\ \beta - ES & \alpha - E \end{vmatrix} = 0 \tag{7.47}$$

which gives

$$(\alpha - E)^2 - (\beta - ES)^2 = 0 \tag{7.48}$$

from which we get two values of E, namely E_+ and E_-, where

$$E_\pm = (\alpha \pm \beta)/(1 \pm S) \tag{7.49}$$

If we are interested only in very approximate MO wave functions and energies we can assume that $S = 0$ (a typical value is about 0.2) and that the hamiltonian H is the same as in the atom giving $\alpha = E_A$, the AO energy. These assumptions result in

$$E_\pm \simeq E_A \pm \beta \tag{7.50}$$

At this level of approximation the two MOs are symmetrically displaced from E_A with a separation of 2β. Since β is a negative quantity the orbital with energy $(E_A + \beta)$ lies lowest, as shown in Figure 7.12.

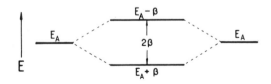

Figure 7.12   Formation of two MOs from two identical AOs.

With the approximation that $S = 0$, the secular equations of equation (7.45) become

$$c_1(\alpha - E) + c_2\beta = 0 \qquad (7.51)$$

$$c_1\beta + c_2(\alpha - E) = 0$$

Putting $E = E_+$ or E_- we get $c_1/c_2 = 1$ or -1 respectively and therefore the corresponding wave functions ψ_+ and ψ_- are given by

$$\psi_+ = N_+(\chi_1 + \chi_2) \qquad (7.52)$$

$$\psi_- = N_-(\chi_1 - \chi_2)$$

N_+ and N_- are normalization constants obtained from the conditions

$$\int \psi_+^2 \, d\tau = \int \psi_-^2 \, d\tau = 1 \qquad (7.53)$$

Neglecting $\int \chi_1\chi_2 \, d\tau$, the overlap integral, gives $N_+ = N_- = 2^{-1/2}$ and

$$\psi_\pm = 2^{-1/2}(\chi_1 \pm \chi_2) \qquad (7.54)$$

In this way every linear combination of two identical AOs gives two MOs, one higher and the other lower in energy than the AOs. Figure 7.13 illustrates the MOs from $1s$, $2s$, and $2p$ AOs showing the approximate forms of the MO wave functions.

The designation $\sigma_g 1s$, $\sigma_u^* 1s$, etc., includes the AO from which the MO was derived, $1s$ in these examples, and also a symmetry species label, here σ_g or σ_u from the $D_{\infty h}$ point group, for the MO. However, the use of symmetry species labels may be more or less avoided by using the fact that only σ- and π-type MOs are usually encountered, and these can be easily distinguished by the σ orbitals being cylindrically symmetrical about the internuclear axis while the π orbitals are not: this can be confirmed from the examples in orbitals Figure 7.13. The 'g' or 'u' subscripts imply symmetry or antisymmetry respectively to inversion through the centre of the molecule (see Section 4.1.3) but are often dropped in favour of the asterisk as in, say, σ^*2s or π^*2p. The asterisk implies antibonding character due to the nodal plane, perpendicular to the internuclear axis, of such orbitals: those without an asterisk are bonding orbitals.

The MOs from $1s$, $2s$, and $2p$ AOs are arranged in order of increasing energy in Figure 7.14 which is applicable to all first-row diatomic molecules except O_2 and F_2. Because of the symmetrical arrangement, illustrated in Figure 7.12, of the two MOs with respect to the AOs from which they are formed, and because the resonance integral for the MO formed from the $2p_z$ AOs is larger than for those formed from $2p_x$ and $2p_y$ MOs, the expected order of MO energies is

$$\sigma_g 1s < \sigma_u^* 1s < \sigma_g 2s < \sigma_u^* 2s < \sigma_g 2p < \pi_u 2p < \pi_g^* 2p < \sigma_u^* 2p \qquad (7.55)$$

In fact this order is maintained only for O_2 and F_2. For all the other

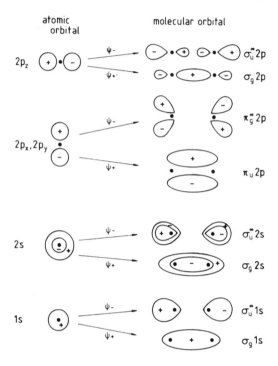

Figure 7.13 Formation of MOs from $1s$, $2s$, and $2p$ AOs.

first-row diatomic molecules $\sigma_g 2s$ and $\sigma_g 2p$ interact (because they are of the same symmetry) and push each other apart to such an extent that $\sigma_g 2p$ is now above $\pi_u 2p$ giving the order

$$\sigma_g 1s < \sigma_u^* 1s < \sigma_g 2s < \sigma_u^* 2s < \pi_u 2p < \sigma_g 2p < \pi_g^* 2p < \sigma_u^* 2p \qquad (7.56)$$

This is the order shown in Figure 7.14.

The electronic structure of any first-row diatomic molecule can be obtained by feeding the available electrons into the MOs in order of increasing energy and taking account of the fact that π orbitals are doubly degenerate and can accommodate four electrons each. For example, the ground configuration of the fourteen-electron nitrogen molecule is

$$(\sigma_g 1s)^2 (\sigma_u^* 1s)^2 (\sigma_g 2s)^2 (\sigma_u^* 2s)^2 (\pi_u 2p)^4 (\sigma_g 2p)^2 \qquad (7.57)$$

There is a general rule that the bonding character of an electron in a bonding orbital is approximately cancelled by the antibonding character of an electron in an antibonding orbital. In nitrogen, therefore, the bonding of the two electrons in $\sigma_g 1s$ is cancelled by the antibonding of two electrons in $\sigma_u^* 1s$ and similarly with $\sigma_g 2s$ and $\sigma_u^* 2s$. So there remain six electrons in the bonding

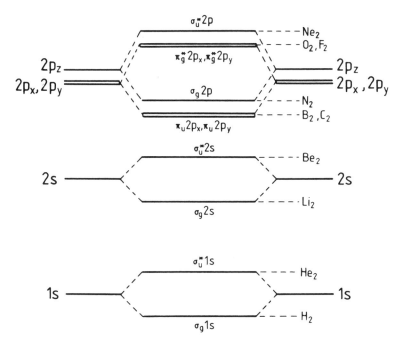

Figure 7.14 MO energy level diagram for first-row homonuclear diatomic molecules. The $2p_x$, $2p_y$, $2p_z$ AOs are degenerate in an atom and have been separated for convenience. (In O_2 and F_2 the order of $\sigma_g 2p$ and $\pi_u 2p$ is reversed.)

orbitals $\pi_u 2p$ and $\sigma_g 2p$ and since

$$\text{Bond order} = \tfrac{1}{2} \times \text{net number of bonding electrons} \qquad (7.58)$$

the bond order is three, consistent with the triple bond which we associate with nitrogen.

The ground configuration of oxygen

$$(\sigma_g 1s)^2 (\sigma_u^* 1s)^2 (\sigma_g 2s)^2 (\sigma_u^* 2s)^2 (\sigma_g 2p)^2 (\pi_u 2p)^4 (\pi_g^* 2p)^2 \qquad (7.59)$$

is consistent with a double bond. Just as for the ground configuration of an atom the first of Hund's rules (page 198) applies and, if states of more than one multiplicity arise from a ground configuration, the ground state has the higher multiplicity. In oxygen, the two electrons in the unfilled $\pi_g^* 2p$ orbital may have their spins parallel, giving $S = 1$ and a multiplicity of three (equation 7.13), or antiparallel, giving $S = 0$ and a multiplicity of one. It follows that the ground state of oxygen is a triplet state. One of the resulting paramagnetic properties may be demonstrated by the deviation observed of a stream of liquid oxygen flowing between the poles of a magnet.

Singlet states with the same configuration as in equation (7.59) with antiparallel spins of the electrons in the $\pi_g^*\,2p$ orbital are higher in energy and are low-lying excited states.

Fluorine has the ground configuration

$$(\sigma_g 1s)^2(\sigma_u^*1s)^2(\sigma_g 2s)^2(\sigma_u^*2s)^2(\sigma_g 2p)^2(\pi_u 2p)^4(\pi_g^*2p)^4 \qquad (7.60)$$

consistent with a single bond.

Just as for atoms, excited configurations of molecules are likely to give rise to more than one state. For example the excited configuration

$$(\sigma_g 1s)^2(\sigma_u^*1s)^2(\sigma_g 2s)^2(\sigma_u^*2s)^2(\pi_u 2p)^1(\sigma_g 2p)^1 \qquad (7.61)$$

of the short-lived molecule C_2 results in a singlet and a triplet state because the two electrons in partially filled orbitals may have parallel or antiparallel spins.

7.2.1.2 Heteronuclear diatomic molecules

Some heteronuclear diatomic molecules, such as nitric oxide (NO), carbon monoxide (CO), and the short-lived CN molecule, contain atoms which are sufficiently similar that the MOs resemble quite closely those of homonuclear diatomics. In nitric oxide the fifteen electrons can be fed into MOs, in the order relevant to O_2 and F_2, to give the ground configuration

$$(\sigma 1s)^2(\sigma^*1s)^2(\sigma 2s)^2(\sigma^*2s)^2(\sigma 2p)^2(\pi 2p)^4(\pi^*2p)^1 \qquad (7.62)$$

(Note that, for heteronuclear diatomics, the 'g' and 'u' subscripts in Figure 7.14 must be dropped as there is no centre of inversion, but the asterisk still retains its significance.) The net number of bonding electrons is five giving a bond order (equation 7.58) of two and a half. This configuration gives rise to a doublet state, because of the unpaired electron in the π^*2p orbital, and paramagnetic properties.

Even molecules such as the short-lived SO and PO molecules can be treated, at the present level of approximation, rather like homonuclear diatomics. The reason is that the outer shell MOs can be constructed from $2s$ and $2p$ AOs on the oxygen atom and $3s$ and $3p$ AOs on the sulphur or phosphorus atom and appear similar to those in Figure 7.13. These linear combinations, such as $2p$ on oxygen and $3p$ on sulphur, obey all the conditions given on p. 210—notably the first one, which states that the energies of the AOs should be comparable.

There is, in principle, no reason why linear combinations should not be made between AOs which have the correct symmetry but very different energies, such as the $1s$ orbital on the oxygen atom and the $1s$ orbital on the phosphorus atom. The result would be that the resonance integral β (see Figure 7.12) would be extremely small so that the MOs would be virtually unchanged from the AOs and the linear combination would be ineffective.

In a molecule such as hydrogen chloride (HCl) the MOs bear no resemblance

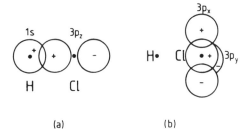

Figure 7.15 In HCl (a) the single bond MO is formed by a linear combination of $1s$ on H and $3p_z$ on Cl and (b) electrons in the $3p_x$ and $3p_y$ AOs on Cl remain as lone pairs.

to those in Figure 7.13 but the rules for making effective linear combinations still hold good.

The electron configuration of the chorine atom is $KL3s^2 3p^5$ and it is only the $3p$ electrons (ionization energy 12.967 eV) which are comparable in energy with the hydrogen $1s$ electron (ionization energy 13.598 eV). Of the $3p$ orbitals it is only $3p_z$ which has the correct symmetry for a linear combination to be formed with the hydrogen $1s$ orbital, as Figure 7.15(a) shows. The MO wave function is of the form given in equation (7.37) but c_1/c_2 is no longer ± 1. Because of the higher electronegativity of Cl compared to H there is considerable concentration near to Cl of the electrons which go into the resulting σ bonding orbital. The two electrons in this orbital form the single bond.

The $3p_x$ and $3p_y$ AOs of Cl cannot overlap with the $1s$ AO of H and the electrons in them remain as lone pairs in orbitals which are very little changed in the molecule, as shown in Figure 7.15(b).

7.2.2 Classification of Electronic States

We have seen in Chapters 1 and 5 that there is an angular momentum associated with end-over-end rotation of a diatomic molecule. In this section we consider ony the non-rotating molecule so that we are concerned only with angular momenta due to orbital and spin motions of the electrons. As in polyelectronic atoms the orbital and spin motions of each electron create magnetic moments which behave like small bar magnets. The way in which the magnets interact with each other represents the coupling between the motions of the electrons.

For all diatomic molecules the coupling approximation which best describes electronic states is that which is analogous to the Russell–Saunders approximation in atoms discussed in Section 7.1.2.3. The orbital angular momenta of all the electrons in the molecule are coupled to give a resultant L and all the electron spin momenta to give a resultant S. However, if there is no highly charged nucleus in the molecule, the spin–orbit coupling between L and S is sufficiently weak that, instead of being coupled to each other, they couple instead to the electrostatic field produced by the two nuclear charges. This situation is shown in Figure 7.16(a) and is referred to as Hund's case (a).

The vector L is so strongly coupled to the electrostatic field and the consequent frequency of precession about the internuclear axis is so high that the magnitude of L is not defined: in other words L is not a good quantum number. Only the component $\Lambda\hbar$ of the orbital angular momentum along the internuclear axis is defined, where the quantum number Λ can take the values

$$\Lambda = 0, 1, 2, 3, \ldots \tag{7.63}$$

All electronic states with $\Lambda > 0$ are doubly degenerate. Classically this degeneracy can be thought of as being due to the electrons orbiting clockwise or anticlockwise around the internuclear axis, the energy being the same in both cases. If $\Lambda = 0$ there is no orbiting motion and no degeneracy.

The value of Λ, like that of L in an atom, is indicated by the main part of the symbol for an electronic state which is designated $\Sigma, \Pi, \Delta, \Phi, \Gamma, \ldots$ corresponding to $\Lambda = 0, 1, 2, 3, 4, \ldots$. The letters $\Sigma, \Pi, \Delta, \Phi, \Gamma, \ldots$ are the Greek equivalents of $S, P, D, F, G, \ldots$ used for atoms.

The coupling of S to the internuclear axis is caused not by the electrostatic field, which has no effect on it, but by the magnetic field along the axis due to the orbital motion of the electrons. Figure 7.16(a) shows that the component of S along the internuclear axis is $\Sigma\hbar$. The quantum number Σ is analogous to M_S in an atom and can take the values

$$\Sigma = S, S - 1, \ldots, -S \tag{7.64}$$

S remains a good quantum number and, for states with $\Lambda > 0$, there are $2S + 1$ components corresponding to the number of values that Σ can take. The multiplicity of the state is the value of $2S + 1$ and is indicated, as in atoms, by a pre-superscript as, for example, in $^3\Pi$.

The component of the total (orbital plus electron spin) angular momentum along the internuclear axis is $\Omega\hbar$, shown in Figure 7.16(a), where the quantum number Ω is given by

$$\Omega = |\Lambda + \Sigma| \tag{7.65}$$

Since $\Lambda = 1$ and $\Sigma = 1, 0, -1$ the three components of $^3\Pi$ have $\Omega = 2, 1, 0$ and are symbolized by $^3\Pi_2$, $^3\Pi_1$, and $^3\Pi_0{}^\dagger$.

Spin–orbit interaction splits the components so that the energy level after interaction is shifted by

$$\Delta E = A\Lambda\Sigma \tag{7.66}$$

where A is the spin–orbit coupling constant. The splitting produces what is

---

† It is not important in this example but, in general, the value of the post-subscript is $\Lambda + \Sigma$ *not* $|\Lambda + \Sigma|$.

called a normal multiplet if the component with the lowest Ω has the lowest energy (that is A positive) and an inverted multiplet if the component with the lowest Ω has the highest energy (that is A negative).

For Σ states there is no orbital angular momentum and therefore no resulting magnetic field to couple S to the internuclear axis. The result is that a Σ state has only one component, whatever the multiplicity.

Hund's case (a), in Figure 7.16(a), is the most commonly encountered case but, like all assumed coupling of angular momenta, it is an approximation. However, when there is at least one highly charged nucleus in the molecule, spin–orbit coupling may be sufficiently large that L and S are not uncoupled by the electrostatic field of the nuclei. Instead, as shown in Figure 7.16(b), L and S couple to give J, as in an atom, and J couples to the internuclear axis along which the component is $\Omega\hbar$. This coupling approximation is known as Hund's case (c), in which Λ is no longer a good quantum number. The main label for a state is now the value of Ω. This approximation is by no means as useful as Hund's case (a) and, even when it is used, states are often labelled with the value that Λ would have if it were a good quantum number.

For atoms, electronic states may be classified and selection rules specified entirely by use of the quantum numbers L, S, and J. In diatomic molecules the quantum numbers Λ, S, and Ω are not quite sufficient. We must also use one (for heteronuclear) or two (for homonuclear) symmetry properties of the electronic wave function ψ_e.

The first is the 'g' or 'u' symmetry property which indicates that ψ_e is symmetric or antisymmetric respectively to inversion through the centre of the molecule (see Section 4.1.3). Since the molecule must have a centre of inversion for this property to apply, states are labelled 'g' or 'u' for homonuclear diatomics only. The property is indicated by a post-subscript, as in $^4\Pi_g$.

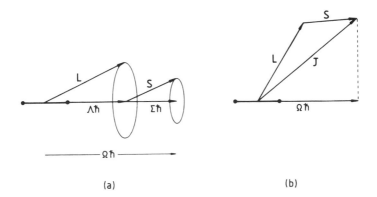

(a)   (b)

Figure 7.16 (a) Hund's case (a) and (b) Hund's case (c) coupling of orbital and electron spin angular momenta in a diatomic molecule.

The second symmetry property applies to all diatomics and concerns the symmetry of ψ_e with respect to reflection across any (σ_v) plane containing the internuclear axis. If ψ_e is symmetric to, i.e. unchanged by, this reflection the state is labelled ' $+$ ' and if it is antisymmetric to, i.e. changed in sign by, this reflection the state is labelled ' $-$ ', as in $^3\Sigma_g^+$ or $^2\Sigma_g^-$. This symbolism is normally used only for Σ states. Although Π, Δ, Φ, ... states do not usually have such symbols attached, one component of such a doubly degenerate state is ' $+$ ' and the other is ' $-$ '; however the symbolism $\Pi^\pm$, $\Delta^\pm$, ... is not often used.

7.2.3   Electronic Selection Rules

As is the case for vibrational transitions, electronic transitions are mostly of the electric dipole type for which the selection rules are:

1. $\Delta\Lambda = 0, \pm 1$ (7.67)

   For example, $\Sigma - \Sigma$, $\Pi - \Sigma$, $\Delta - \Pi$ transitions[†] are allowed but not $\Delta - \Sigma$ or $\Phi - \Pi$.

2. $\Delta S = 0$ (7.68)

   As in atoms, this selection rule breaks down as the nuclear charge increases. For example, triplet–singlet transitions are strictly forbidden in H_2 but, in CO, the $a^3\Pi-X^1\Sigma^+$ transition[‡] is observed weakly.

3. $\Delta\Sigma = 0; \qquad \Delta\Omega = 0, \pm 1$ (7.69)

   for transitions between multiplet components.

4. $+ \leftrightarrow\!\!\!/ -; \qquad + \leftrightarrow +; \qquad - \leftrightarrow -$ (7.70)

   This is relevant only for $\Sigma - \Sigma$ transitions so that only $\Sigma^+ - \Sigma^+$ and $\Sigma^- - \Sigma^-$ transitions are allowed. (Note that this selection rule is the opposite of the vibration–rotation selection rule of equation (6.73).)

5. $g \leftrightarrow u, \; g \leftrightarrow\!\!\!/ g, \; u \leftrightarrow\!\!\!/ u$ (7.71)

   For example, $\Sigma_g^+ - \Sigma_g^+$ and $\Pi_u - \Sigma_u^-$ transitions are forbidden but $\Sigma_u^+ - \Sigma_g^+$ and $\Pi_u - \Sigma_g^+$ transitions are allowed.

For Hund's case (c) coupling (Figure 7.16b) the selection rules are slightly

---

[†] As for all other types of transitions, the upper state is given first and the lower state second.

[‡] There is a convention, which is commonly but not always used, for labelling electronic states. The ground state is labelled X and higher states of the same multiplicity are labelled $A, B, C, \ldots$ in order of increasing energy. States with multiplicity different from that of the ground state are labelled $a, b, c, \ldots$ in order of increasing energy.

different. Those in equations (7.68) and (7.71) still apply but, because Λ and Σ are not good quantum numbers, that in equation (7.67) and $\Delta\Sigma = 0$ in equation (7.69) do not. In the case of the rule in equation (7.70) the ' $+$ ' or ' $-$ ' label now refers to the symmetry of ψ_e to reflection in a plane containing the internuclear axis only when $\Omega = 0$ (not $\Lambda = 0$, as in Hund's case (a)) and the selection rule is now

$$0^+ \leftrightarrow 0^+, \; 0^- \leftrightarrow 0^-, \; 0^+ \nleftrightarrow 0^- \tag{7.72}$$

where '0' refers to the value of Ω.

The lowest $^3\Pi_u$ term of I_2 approximates more closely to Hund's case (c) than case (a). Since, in a case (a) description, Λ and S would be 1 and Σ would be 1, 0, or -1 the value of Ω can be 2, 1, or 0. The $\Omega = 0$ and 1 components are split by 3881 cm^{-1}, a large value indicating appreciable spin–orbit interaction and a case (c) approximation. The transition[†] $B^3\Pi_{0_u^+} - X^1\Sigma_g^+$ involves the $\Omega = 0$ component of the B state for which ψ_e is symmetric to reflection across any plane containing the internuclear axis, i.e. the 0^+ component. The transition is allowed by the $0^+ \leftrightarrow 0^+$ selection rule. Note also that the $\Delta S = 0$ selection rule breaks down in a molecule with such high nuclear charges. Indeed, the transition is quite intense, occurring in the visible region and giving rise to the violet colour of I_2 vapour.

7.2.4 Derivation of States Arising from Configurations

In the case of atoms, deriving states from configurations, in the Russell–Saunders approximation (Section 7.1.2.3), simply involved juggling with the available quantum numbers. In diatomic molecules we have seen already that some symmetry properties must be included, in addition to the available quantum numbers, in a discussion of selection rules. In deriving states from orbital configurations, symmetry arguments are even more essential. However, those readers who do not require to be able to do this may proceed to Section 7.2.5.

We have seen how the ground configuration $\ldots (\pi_u 2p)^4 (\pi_g^* 2p)^2$ of O_2 in equation (7.59) gives rise to a triplet ground state but, in fact, it gives rise to three states, $^3\Sigma_g^-$, $^1\Delta_g$, and $^1\Sigma_g^+$, the $^1\Delta_g$ and $^1\Sigma_g^+$ being low-lying excited states 7918 and 13 195 cm^{-1} respectively above the ground state $^3\Sigma_g^-$. How can we derive these states from this configuration or, indeed, electronic states from MO configurations in general?

The symmetry species $\Gamma(\psi_e^o)$ of the orbital part of the electronic wave function corresponding to a particular configuration is given by

$$\Gamma(\psi_e^o) = \Pi_i \Gamma(\psi_i) \tag{7.73}$$

[†] The labelling of the $B^3\Pi_{0_u^+}$ state follows general usage rather than the convention which would label it $b^3\Pi_{0_u^+}$.

where $\Pi_i \Gamma(\psi_i)$ stands for the product, over all electrons i, of the symmetry species of all occupied MOs ψ_i. The product for filled orbitals gives the totally symmetric species and, since all electron spins are paired, $S = 0$ for these orbitals. The ground configuration of nitrogen is given in equation (7.57). Since all occupied orbitals are filled the ground state is $^1\Sigma_g^+$ as Σ_g^+ is the totally symmetric species[†] of the $D_{\infty h}$ point group (Table A.37 in the Appendix). For the excited configuration

$$\ldots (\pi_u 2p)^4 (\sigma_g 2p)^1 (\pi_g^* 2p)^1 \tag{7.74}$$

the orbital symmetry species is given by[‡]

$$\Gamma(\psi_e^0) = \sigma_g^+ \times \pi_g = \Pi_g \tag{7.75}$$

since filled orbitals need not be considered as they only have the effect of multiplying the result by the totally symmetric species. The multiplication is carried out as in Section 4.3.3. The two electrons in partly filled orbitals may have parallel or antiparallel spins so that two states arise from the configuration, $^3\Pi_g$ and $^1\Pi_g$.

For the excited configuration

$$\ldots (\pi_u 2p)^3 (\sigma_g 2p)^2 (\pi_g^* 2p)^1 \tag{7.76}$$

we recall that, as in atoms, a vacancy in an orbital can be treated like a single electron and we have

$$\Gamma(\psi_e^0) = \pi_u \times \pi_g = \Sigma_u^+ + \Sigma_u^- + \Delta_u \tag{7.77}$$

a multiplication similar to that in equation (4.33). Since the two electrons (or one electron and one vacancy) may have parallel or antiparallel spins, six states arise from the configuration in equation (7.76), namely $^{1,3}\Sigma_u^+$, $^{1,3}\Sigma_u^-$, $^{1,3}\Delta_u$.

The ground configuration of oxygen is

$$\ldots (\sigma_g 2p)^2 (\pi_u 2p)^4 (\pi_g^* 2p)^2 \tag{7.78}$$

and, in deriving the states arising, we need consider only the $(\pi_g^* 2p)^2$ electrons. For two or more electrons in a degenerate orbital we start off in the usual way to obtain

$$\Gamma(\psi_e^0) = \pi_g \times \pi_g = \Sigma_g^+ + \Sigma_g^- + \Delta_g \tag{7.79}$$

which is the same result as if the electrons were in two different π_g orbitals. The difference lies in the treatment of the two electron spins. When the electrons are in the same degenerate orbital the Pauli principle forbids some orbital and

---

[†] Upper case greek letters are used for states, lower case letters for orbitals.
[‡] Note that a σ MO is not usually labelled σ^+, which it should really be, as there are no σ^- MOs.

Table 7.5 States from ground configurations in diatomic molecules.

Point group	Configuration	States	Point group	Configuration	States
$C_{\infty v}$	$(\pi)^2$	$^3\Sigma^- + {}^1\Sigma^+ + {}^1\Delta$	$D_{\infty h}$	$(\pi_g)^2$ and $(\pi_u)^2$	$^3\Sigma_g^- + {}^1\Sigma_g^+ + {}^1\Delta_g$

spin combinations. This is similar to the problem encountered in equation (4.30) in determining $\Gamma(\psi_v)$ when a molecule is vibrating with two quanta of the same degenerate vibration: the Pauli principle forbids the antisymmetric part of the direct product.

The analogy is even closer when the situation in oxygen is compared to that in excited configurations of the helium atom summarized in equations (7.28) and (7.29). According to the Pauli principle for electrons the total wave function must be antisymmetric to electron exchange.

It has been explained in Section 4.3.2 that the direct product of two identical degenerate symmetry species contains a symmetric part and an antisymmetric part. The antisymmetric part is an A (or Σ) species and, where possible, *not* the totally symmetric species. Therefore, in the product in equation (7.79), Σ_g^- is the antisymmetric and $\Sigma_g^+ + \Delta_g$ the symmetric part.

Equation (7.23) expresses the total electronic wave function as the product of the orbital and spin parts. Since ψ_e must be antisymmetric to electron exchange the Σ_g^+ and Δ_g orbital wave functions of oxygen combine only with the antisymmetric (singlet) spin wave function which is the same as that in equation (7.24) for helium. Similarly the Σ_g^- orbital wave function combines only with the three symmetric (triplet) spin wave functions which are the same as those in equation (7.25) for helium.

So the states that arise from the ground configuration of oxygen are $^3\Sigma_g^-$, $^1\Sigma_g^+$, and $^1\Delta_g$. One of Hund's rules (rule 1 on page 198) tells us that $X^3\Sigma_g^-$ is the ground state. The Pauli principle forbids the $^1\Sigma_g^-$, $^3\Sigma_g^+$, and $^3\Delta_g$ states.

Table 7.5 lists the states arising from a few electron configurations in $D_{\infty h}$ and $C_{\infty v}$ diatomic molecules in which there are two electrons in the same degenerate orbital.

7.2.5 Vibrational Coarse Structure

7.2.5.1 *Potential energy curves in excited electronic states*

In Section 6.1.3.2 we have discussed the form of the potential energy curve in the ground electronic state of a diatomic molecule. Figure 6.4 shows a typical curve with a potential energy minimum at r_e, the equilibrium internuclear distance. Dissociation occurs at high energy and the dissociation energy is D_e, relative to the minimum in the curve, or D_0, relative to the zero-point level.

The vibrational term values, $G(v)$, are given in equation (6.3) for a harmonic oscillator and in equation (6.16) for an anharmonic oscillator.

For each excited electronic state of a diatomic molecule there is a potential energy curve and, for most states, the curve appears qualitatively similar to that in Figure 6.4.

As an example of such excited state potential energy curves Figure 7.17 shows curves for several excited states and also for the ground state of the short-lived C_2 molecule. The ground electron configuration is

$$(\sigma_g 1s)^2(\sigma_u^* 1s)^2(\sigma_g 2s)^2(\sigma_u^* 2s)^2(\pi_u 2p)^4 \tag{7.80}$$

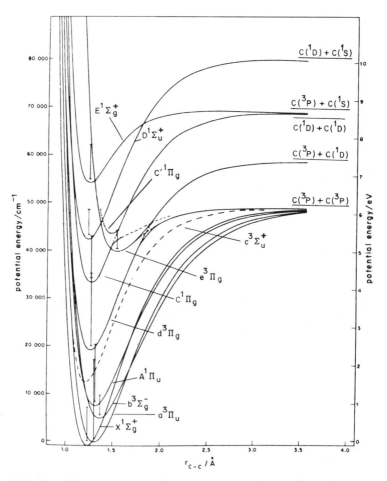

Figure 7.17 Potential energy curves for the ground and several excited states of C_2. (Reproduced, with permission, from Ballik, E. A., and Ramsay, D. A., *Astrophys. J.*, **137**, 84, 1963 published by the University of Chicago Press; Copyright (1963) The American Astronomical Society.)

giving the $X^1\Sigma_g^+$ ground state. The low-lying excited electronic states in Figure 7.17 arise from configurations in which an electron is promoted from the $\pi_u 2p$ or $\sigma_u^* 2s$ orbital to the $\sigma_g 2p$ orbital. The information contained in Figure 7.17 has been obtained using various experimental techniques to observe absorption or emission spectra. Table 7.6 lists the transitions which have been observed, the names of those associated with their discovery, the region of the spectrum where they were observed, and the nature of the source. The mixture of techniques for observing the spectra, including a high temperature furnace, flames, arcs, discharges, and, for a short-lived molecule, flash photolysis is typical. The table shows that several electronic systems of C_2 are observed in absorption with $a^3\Pi_u$ as the lower state. It is unusual for an absorption system to have any other than the ground state as its lower state. However, in C_2, the $a^3\Pi_u$ state is only 716 cm^{-1} above the ground state and so is appreciably populated at only moderate temperatures. All the transitions in Table 7.6 are allowed by the selection rules in Section 7.2.3 for Hund's case (a).

In addition to these laboratory-based experiments it is interesting to note that the Swan bands of C_2 are important in astrophysics. They have been observed in the emission spectra of comets and also in the absorption spectra of stellar atmospheres, including that of the sun, in which the interior of the star acts as the continuum source.

When C_2 dissociates it gives two carbon atoms which may be in their ground or excited states. We saw, for example, in equation (7.17) that the ground configuration $1s^2 2s^2 2p^2$ of the carbon atom gives rise to three terms: ^3P, the ground term, and ^1D and ^1S, which are successively higher excited terms. Figure 7.17 shows that six states of C_2, including the ground state, dissociate to give two ^3P carbon atoms. Other states give dissociation products involving one or both carbon atoms with ^1D or ^1S terms.

Table 7.6 Electronic transitions observed in C_2.

Transition	Names associated	Spectral region/nm	Source of spectrum
$b^3\Sigma_g^- \rightarrow a^3\Pi_u$	Ballik–Ramsay	2700–1100	King furnace
$A^1\Pi_g \leftrightharpoons X^1\Sigma_g^+$	Phillips	1549– 672	Discharges
$d^3\Pi_g \leftrightharpoons a^3\Pi_u$	Swan	785– 340	Numerous, including carbon arc
$C^1\Pi_g \rightarrow A^1\Pi_u$	Deslandres–d'Azambuja	411– 339	Discharges, flames
$e^3\Pi_g \rightarrow a^3\Pi_u$	Fox–Herzberg	329– 237	Discharges
$D^1\Sigma_u^+ \leftrightharpoons X^1\Sigma_g^+$	Mulliken	242– 231	Discharges, flames
$E^1\Sigma_g^+ \rightarrow A^1\Pi_u$	Freymark	222– 207	Discharge in acetylene
$f^3\Sigma_g^- \leftarrow a^3\Pi_u$	—	143– 137 ⎫	Flash photolysis of
$g^3\Delta_g \leftarrow a^3\Pi_u$	—	140– 137 ⎬	mixture of a
$F^1\Pi_u \leftarrow X^1\Sigma_g^+$	—	135– 131 ⎭	hydrocarbon and an inert gas

Just as in the ground electronic state a molecule may vibrate and rotate in excited electronic states. The total term value S for a molecule with an electronic term value T, corresponding to an electronic transition between equilibrium configurations and with vibrational and rotational term values $G(v)$ and $F(J)$, is given by

$$S = T + G(v) + F(J) \tag{7.81}$$

The vibrational term values for any electronic state, ground or excited, can be expressed, as in equation (6.16), by

$$G(v) = \omega_e(v + \tfrac{1}{2}) - \omega_e x_e(v + \tfrac{1}{2})^2 + \omega_e y_e(v + \tfrac{1}{2})^3 + \cdots \tag{7.82}$$

where the vibration wavenumber ω_e and the anharmonic constants $\omega_e x_e$, $\omega_e y_e, \ldots$ vary from one electronic state to another. Figure 7.17 shows that the equilibrium internuclear distance r_e is also different for each electronic state.

7.2.5.2 Progressions and sequences

Figure 7.18 shows sets of vibrational energy levels associated with two electronic states between which we shall assume an electronic transition is allowed. The vibrational levels of the upper and lower states are labelled by the quantum numbers v' and v'' respectively. We shall be discussing absorption as well as emission processes and it will be assumed, unless otherwise stated, that the lower state is the ground state.

In electronic spectra there is no restriction on the values that Δv can take

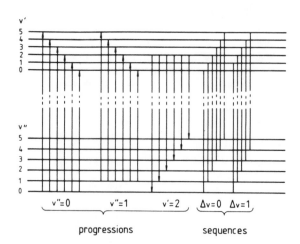

Figure 7.18 Vibrational progressions and sequences in the electronic spectrum of a diatomic molecule.

but, as we shall see in Section 7.2.5.3, the Franck–Condon principle imposes limitations on the intensities of the transitions.

Vibrational transitions accompanying an electronic transition are referred to as vibronic transitions. These vibronic transitions, with their accompanying rotational or, strictly, rovibronic transitions, give rise to bands in the spectrum and the set of bands associated with a single electronic transition is called an electronic band system. This terminology is usually adhered to in high resolution electronic spectroscopy but, in low resolution work, particularly in the liquid phase, vibrational structure may not be resolved and the whole band system is often referred to as an electronic band.

Vibronic transitions may be divided conveniently into progressions and sequences. A progression, as Figure 7.18 shows, involves a series of vibronic transitions with a common lower or upper level. For example, the $v'' = 0$ progression members all have the $v'' = 0$ level in common.

Quite apart from the necessity for Franck–Condon intensities of vibronic transitions to be appreciable, it is essential for the initial state of a transition to be sufficiently highly populated for a transition to be observed. Under equilibrium conditions the population $N_{v''}$ of any v'' level is related to that of the $v'' = 0$ level by

$$\frac{N_{v''}}{N_0} = \exp - \left\{ [G(v'') - G(0)] \frac{hc}{kT} \right\} \tag{7.83}$$

which follows from the Boltzmann equation (equation 2.11).

Because of the relatively high population of the $v'' = 0$ level the $v'' = 0$ progression is likely to be prominent in the absorption spectrum. In emission the relative populations of the v' levels depend on the method of excitation. In a low pressure discharge, in which there are not many collisions to provide a channel for vibrational deactivation, the populations may be somewhat random. On the other hand, higher pressure may result in most of the molecules being in the $v' = 0$ state and the $v' = 0$ progression being prominent.

The progression with $v'' = 1$ may also be observed in absorption but only in a molecule with a vibration wavenumber low enough for the $v'' = 1$ level to be sufficiently populated. This is the case in, for example, iodine for which $\omega_e'' = 214.50$ cm^{-1}. As a result the $B^3\Pi_{0_u^+} - X^1\Sigma_g^+$ visible system shows, in absorption at room temperature, not only a $v'' = 0$ but also a $v'' = 1$ and a $v'' = 2$ progression, as shown in Figure 7.19.

A progression with $v' = 2$, illustrated in Figure 7.18, could be observed only in emission. Its observation could result from a random population of v' levels or it could be observed on its own under rather special conditions involving monochromatic excitation from $v'' = 0$ to $v' = 2$ with no collisions occurring before emission. This kind of excitation could be achieved with a tunable laser.

If the emission is between states of the same multiplicity it is called fluorescence and if it is from only one vibrational level of the upper electronic state

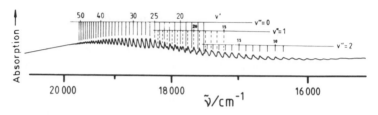

Figure 7.19 Progressions with $v'' = 0$, 1, and 2 in the $B^3\Pi_{0_u^+} - X^1\Sigma_g^+$ system of I_2.

it is single vibronic level fluorescence. Emission between states of different multiplicity is called phosphorescence.

A group of transitions with the same value of Δv is referred to as a sequence. Because of the population requirements long sequences are observed mostly in emission. For example, sequences of five or six members are observed in the $C^3\Pi_u - B^3\Pi_g$ band system of N_2 in emission in the visible and near-ultraviolet from a low pressure discharge in nitrogen gas. The vibration wavenumber ω_e is high (2047.18 cm^{-1}) in the C state and equilibrium population of the vibrational levels is not achieved before emission.

It is clear from Figure 7.18 that progressions and sequences are not mutually exclusive. Each member of a sequence is also a member of two progressions. However the distinction is useful because of the nature of typical patterns of bands found in a band system. Progression members are generally widely spaced with approximate separations of ω_e' in absorption and ω_e'' in emission. On the other hand, sequence members are more closely spaced with approximate separations of $\omega_e' - \omega_e''$.

The general symbolism for indicating a vibronic transition between an upper and lower level with vibrational quantum numbers v' and v'' respectively is $v' - v''$, consistent with the general spectroscopic convention. Thus the electronic transition is labelled $0-0$.

7.2.5.3 The Franck–Condon principle

In 1925, before the development of the Schrödinger equation, Franck put forward qualitative arguments to explain the various types of intensity distributions found in vibronic transitions. His conclusions were based on an appreciation of the fact that an electronic transition in a molecule takes place so much more rapidly than a vibrational transition that, in a vibronic transition, the nuclei have very nearly the same position and velocity before and after the transition.

Possible consequences of this are illustrated in Figure 7.20(a) which shows potential curves for the lower state, which is the ground state if we are considering an absorption process, and the upper state. The curves have been

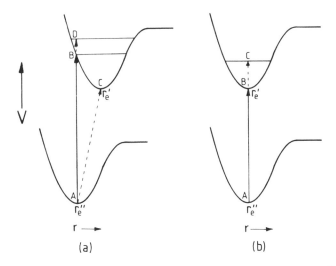

Figure 7.20 Illustration of the Franck principle for (a) $r'_e > r''_e$ and (b) r'_e, r''_e. The vibronic transition B–A is the most probable in both cases.

drawn so that $r'_e > r''_e$. When the lower state is the ground state this is very often the case since the electron promotion involved is often from a bonding orbital to an orbital which is less bonding, or even antibonding. For example, in nitrogen, promotion of an electron from the $\sigma_g 2p$ to the $\pi_g^* 2p$ orbital (Figure 7.14) gives two states[†], $a^1\Pi_g$ and $B^3\Pi_g$, in which r_e is 1.2203 and 1.2126 Å respectively, considerably increased from 1.0977 Å in the $X^1\Sigma_g^+$ ground state.

In absorption, from point A of the ground state in Figure 7.20(a) (zero-point energy can be neglected in considering Franck's semi-classical arguments) the transition will be to point B of the upper state. The requirement that the nuclei have the same position before and after the transition means that the transition is between points which lie on a vertical line in the figure: this means that r remains constant and such a transition is often referred to as a vertical transition. The second requirement, that the nuclei have the same velocity before and after the transition, means that a transition from A, where the nuclei are stationary, must go to B, as this is the classical turning point of a vibration, where the nuclei are also stationary. A transition from A to C is highly improbable because, although the nuclei are stationary at A and C, there is a large change of r. An A to D transition is also unlikely because, although r is unchanged, the nuclei are in motion at the point D.

Figure 7.20(b) illustrates the case where $r'_e \simeq r''_e$. An example of such a

[†] Unfortunately, in N_2 as in I_2, the conventional labelling of states in the footnote on page 220 is not adhered to.

transition is the $D^1\Sigma_u^+ - X^1\Sigma_g^+$ Mulliken band system of C_2 (see Table 7.6 and Figure 7.17). The value of r_e is 1.2380 Å in the D state and 1.2425 Å in the X state. Here the most probable transition is from A to B with no vibrational energy in the upper state. The transition from A to C maintains the value of r but the nuclear velocities are increased due to their having kinetic energy equivalent to the distance BC.

In 1928, Condon treated the intensities of vibronic transitions quantum mechanically.

The intensity of a vibronic transition is proportional to the square of the transition moment R_{ev}, which is given by (see equation 2.13)

$$R_{ev} = \int \psi_{ev}'^* \mu \psi_{ev}'' \, d\tau_{ev} \qquad (7.84)$$

where μ is the electric dipole moment operator and ψ_{ev}' and ψ_{ev}'' are the vibronic wave functions of the upper and lower states respectively. The integration is over electronic and vibrational coordinates. Assuming that the Born–Oppenheimer approximation (see Section 1.3.4) holds, ψ_{ev} can be factorized into $\psi_e \psi_v$. Then equation (7.84) becomes

$$R_{ev} = \iint \psi_e'^* \psi_v'^* \mu \psi_e'' \psi_v'' \, d\tau_e \, dr \qquad (7.85)$$

First we integrate over electron coordinates τ_e, giving

$$R_{ev} = \int \psi_v'^* R_e \psi_v'' \, dr \qquad (7.86)$$

where r is the internuclear distance and R_e is the electronic transition moment given by

$$R_e = \int \psi_e'^* \mu \psi_e'' \, d\tau_e \qquad (7.87)$$

Our ability to do the integration to give equation (7.86) is a consequence of the Born–Oppenheimer approximation which assumes that the nuclei can be regarded as stationary in relation to the much more fast-moving electrons. This approximation also allows us to take R_e outside the integral in equation (7.86), regarding it as a constant, independent of r, which is good enough for our purposes here. Thus we have

$$R_{ev} = R_e \int \psi_v'^* \psi_v'' \, dr \qquad (7.88)$$

The quantity $\int \psi_v'^* \psi_v'' \, dr$ is called the vibrational overlap integral, as it is a measure of the degree to which the two vibrational wave functions overlap: its square is known as the Franck–Condon factor. In carrying out the integration the requirement that r remains constant during the transition is necessarily taken into account.

The classical turning point of a vibration, where nuclear velocities are zero, is replaced in quantum mechanics by a maximum, or minimum, in ψ_v near to this turning point. As is illustrated in Figure 1.13 the larger is v the closer is the maximum, or minimum, in ψ_v to the classical turning point.

Figure 7.21 illustrates a particular case where the maximum of the $v' = 4$ wave function near to the classical turning point is vertically above that of the $v'' = 0$ wave function. The maximum contribution to the vibrational overlap integral is indicated by the solid line but appreciable contributions extend to values of r within the dashed lines. Clearly overlap integrals for v' close to four are also appreciable and give an intensity distribution in the $v'' = 0$ progression like that in Figure 7.22(b).

If $r'_e \gg r''_e$ there may be appreciable intensity involving the continuum of vibrational levels above the dissociation limit. This results in a $v'' = 0$ progression like that in Figure 7.22(c) where the intensity maximum is at a high value of v; or it may be in the continuum. An example of this is the $B^3\Pi_{0^+_u} - X^1\Sigma_g^+$ transition of iodine. In the B and X states r_e is 3.025 and 2.666 Å respectively, leading to the broad intensity maximum close to the continuum, as observed in Figure 7.19.

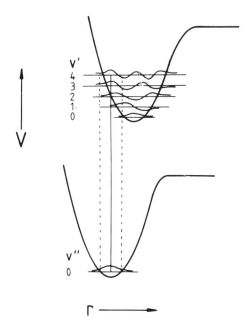

Figure 7.21 Franck–Condon principle applied to a case in which $r'_e > r''_e$ and the 4–0 transition is the most probable.

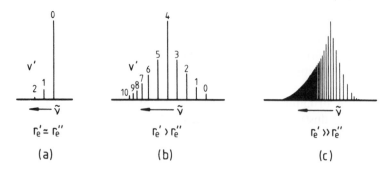

Figure 7.22 Typical vibrational progression intensity distributions.

Figure 7.22(a) shows the intensity maximum at $v' = 0$ for the case when $r'_e \simeq r''_e$. The intensity usually falls off rapidly in such a case.

Occasionally we encounter a case where $r'_e < r''_e$. When the lower state is the ground state this is unusual but it can happen when the electron promotion is from an antibonding to a non-bonding or bonding orbital. The situation is more likely to arise in a transition between two excited states. Qualitatively the situation is similar to that in Figure 7.21 except that the upper potential curve is displaced to low r so that the right hand maximum of, for example, $v' = 4$ is above the $v'' = 0$ maximum. The result is, again, an intensity distribution like that in Figure 7.22(b) so that an observation of a long $v'' = 0$ progression with an intensity maximum at $v' > 0$ indicates qualitatively an appreciable change in r_e from the lower to the upper state but does not indicate the sign of the change. This would be true, even quantitatively, if the molecule behaved as a harmonic oscillator but, due to anharmonicity, the intensity distribution along the progression is slightly different for $r'_e > r''_e$ than for $r'_e < r''_e$.

In the case where $r'_e > r''_e$ there is, when anharmonicity is taken into account, a relatively steep part of the excited state potential curve above $v'' = 0$, giving a relatively broad maximum in the progression intensity. On the other hand, for $r'_e < r''_e$, there is a shallower part of the excited state potential curve above $v'' = 0$ and a sharper intensity maximum results.

Accurate intensity measurements have been made in many cases and calculations of $r'_e - r''_e$ made, including the effects of anharmonicity and even allowing for breakdown of the Born–Oppenheimer approximation.

7.2.5.4 Deslandres tables

The illustration of various types of vibronic transitions in Figure 7.18 suggests that we can use the method of combination differences to obtain the separations of vibrational levels from observed transition wavenumbers. This method was introduced in Section 6.2.3.1 and was applied to obtaining rotational constants for two combining vibrational states. The method works on the simple principle that, if two transitions have an upper level in common, their wavenumber difference is a function of lower state parameters only, and vice versa if they have a lower level in common.

In using the combination difference method to obtain vibrational parameters, ω_e, $\omega_e x_e$, etc., for two electronic states between which vibronic transitions are observed, the first step is to organize all the vibronic transition wavenumbers into a Deslandres table. An example is shown in Table 7.7 for the $A^1\Pi - X^1\Sigma^+$ system of carbon monoxide. The electronic transition results from an electron promotion which, because of the similarity of the two nuclei, can be described approximately in terms of the MO diagram for homonuclear diatomic molecules in Figure 7.14. Carbon monoxide is isoelectronic with the nitrogen molecule so the lowest energy electron promotion is from $\sigma 2p$ to $\pi^* 2p$ (the 'g' and 'u' subscripts do not apply to heteronuclear diatomic molecules). The promotion gives two states, $A^1\Pi$ and $a^3\Pi$. The $A^1\Pi - X^1\Sigma^+$ band system lies in the far-ultraviolet region of the spectrum, with the 0–0 band at 154.4 nm.

In Table 7.7, all the transition wavenumbers have been arranged in rows and columns so that the differences between wavenumbers in adjacent columns correspond to vibrational level separations in the lower (ground) electronic state and the differences between adjacent rows to separations in the upper electronic state. These differences are shown in parentheses. The variations of the differences, e.g. between the first two columns, are a result of uncertainties in the experimental measurements.

From the table a series of averaged values of vibrational term value differences, $G(v + 1) - G(v)$, can be obtained for both electronic states and, from equation (6.18), values of ω_e, $\omega_e x_e$, etc., for both states. For example, in the $A^1\Pi$ and $X^1\Sigma^+$ states of CO, ω_e is 1518.2 and 2169.8 cm^{-1} respectively. The large decrease in ω_e in the A state is a consequence of promoting an electron from a bonding to an antibonding orbital, greatly reducing the force constant.

7.2.5.5 Dissociation energies

If a sufficient number of vibrational term values are known in any electronic state the dissociation energy D_0 can be obtained from a Birge–Sponer extrapolation, as discussed in Section 6.1.3.2 and illustrated in Figure 6.5. The possible inaccuracies of the method were made clear and it was stressed that these are reduced by obtaining term values near to the dissociation limit. Whether this can be done depends very much on the relative dispositions of the various potential curves in a particular molecule and whether electronic transitions between them are allowed. How many ground state vibrational term values can be obtained from an emission spectrum is determined by the Franck–Condon principle. If $r'_e \simeq r''_e$ then progressions in emission are very short and few term values result, but if r'_e is very different from r''_e, as in the $A^1\Pi - X^1\Sigma^+$ system

Table 7.7 Deslandres table for the $A^1\Pi - X^1\Sigma^+$ system of carbon monoxide[†].

v' \ v''	0		1		2		3		4		5		6
0	64 758	(2145)	62 613	(2117)	60 496	(2092)	58 404	(2063)	56 341	(2037)	54 304		—
	(1 476)		(1 485)				(1 487)		(1 486)		(1 487)		
1	66 234	(2136)	64 098		—		59 891	(2064)	57 827	(2036)	55 791	(2010)	53 781
	(1 448)		(1 441)				(1 444)				(1 443)		(1 443)
2	67 682	(2143)	65 539	(2115)	63 424	(2089)	61 335		—		57 234	(2010)	55 224
	(1 407)		(1 413)		(1 414)						(1 410)		
3	69 089	(2137)	66 952	(2114)	64 838		—		60 683	(2039)	58 644		—
	(1 378)		(1 382)		(1 370)				(1 379)				
4	70 467	(2133)	68 334	(2126)	66 208	(2085)	64 123	(2061)	62 062		—		58 011
	(1 341)		(1 338)		(1 350)		(1 343)						(1 340)
5	71 808	(2136)	69 672	(2114)	67 558	(2092)	65 466		—		61 365	(2014)	59 351
	(1 307)		(1 305)		(1 303)		(1 299)				(1 307)		
6	73 115	(2138)	70 977	(2116)	68 861	(2096)	66 765	(2053)	64 712	(2040)	62 672		—

† Units are cm^{-1} throughout. Measurements are of band heads, formed by the rotational structure, not band origins.

of carbon monoxide discussed in Section 7.2.5.4, then long progressions are observed in emission and a more accurate value of D_0'' can be obtained.

To obtain an accurate value of D_0'' for the ground electronic state is virtually impossible by vibrational spectroscopy because of the problems of a rapidly decreasing population with increasing v. In fact most determinations are made from electronic emission spectra from one, or more, excited electronic states to the ground state.

Obtaining D_0' for an excited electronic state depends on observing progressions either in absorption, usually from the ground state, or in emission from higher excited states. Again, the length of a progression limits the accuracy of the dissociation energy.

If the values of r_e in the combining states are very different the dissociation limit of a progression may be observed directly as an onset of diffuseness. However, the onset is not always particularly sharp: this is the case in the $B^3\Pi_{0_u^+} - X^1\Sigma_g^+$ absorption system of iodine shown in Figure 7.19, where the wavenumber $\tilde{v}_{\text{limit}}$ illustrated in Figure 7.23 is obtained more accurately by extrapolation than by direct observation.

Figure 7.23 shows that

$$\tilde{v}_{\text{limit}} = D_0' + \tilde{v}_0 = D_0'' + \Delta\tilde{v}_{\text{atomic}} \tag{7.89}$$

Hence, D_0' can be obtained from $\tilde{v}_{\text{limit}}$ if $\tilde{v}_0$, the wavenumber of the 0–0 band, is known. Figure 7.19 shows that extrapolation may be required to obtain $\tilde{v}_0$, limiting the accuracy of D_0'.

Equation (7.89) also shows that D_0'' may be obtained from $\tilde{v}_{\text{limit}}$ since $\Delta\tilde{v}_{\text{atomic}}$ is the wavenumber difference between two atomic states, the ground state $^2P_{3/2}$

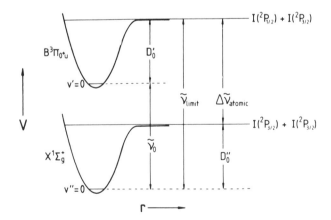

Figure 7.23 Dissociation energies D_0' and D_0'' may be obtained from $\tilde{v}_{\text{limit}}$, the wavenumber of the onset of a continuum in a progression in I_2.

and the first excited state $^2P_{1/2}$ of the iodine atom, known accurately from the atomic spectrum. Thus the accuracy of D_0'' is limited only by that of $\tilde{\nu}_{\text{limit}}$.

D_e' and D_e'', the dissociation energies relative to the minima in the potential curves, are obtained from D_0' and D_0'' by

$$D_e = D_0 + G(0) \tag{7.90}$$

where $G(0)$ is the zero-point term value given by equation (6.21).

7.2.5.6 Repulsive states and continuous spectra

The ground configuration of the He_2 molecule is, according to Figure 7.14, $(\sigma_g 1s)^2(\sigma_u^* 1s)^2$ and is expected to be unstable because of the cancelling of the bonding character of a $\sigma_g 1s$ orbital by the antibonding character of a $\sigma_u^* 1s$ orbital. The potential energy curve for the resulting $X^1\Sigma_g^+$ state shows no minimum but the potential energy decreases smoothly as r increases, as shown in Figure 7.24(a). Such a state is known as a repulsive state since the atoms repel each other. In this type of state, either there are no discrete vibrational levels or there may be a few in a very shallow minimum (see Section 9.2.7). All, or most, of the vibrational states form a continuum of levels.

Promotion of an electron in He_2 from $\sigma_u^* 1s$ to a bonding orbital produces bound states of the molecule of which several have been characterized in emission spectroscopy. For example, the configuration $(\sigma_g 1s)^2(\sigma_u^* 1s)^1(\sigma_g 2s)^1$ gives rise to the $A^1\Sigma_u^+$ and $a^3\Sigma_u^+$ bound states. Figure 7.24(a) shows the form of the potential curve for the $A^1\Sigma_u^+$ state. The A–X transition is allowed and gives rise to an intense continuum in emission between 60 and 100 nm. This is used as a far-ultraviolet continuum source (see Section 3.4.5) as are the corresponding continua from other noble gas diatomic molecules.

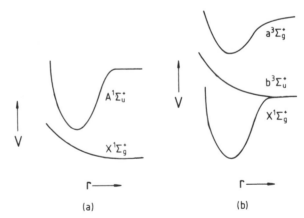

Figure 7.24 (a) The repulsive ground state and a bound excited state of He_2. (b) Two bound states and one repulsive state of H_2.

Another example of a continuous emission spectrum is that from a discharge in molecular hydrogen. It covers the range from 160 to 500 nm and is used as a visible and near-ultraviolet continuum source (see Section 3.4.4). The transition involved is from the bound $a^3\Sigma_g^+$ to the repulsive $b^3\Sigma_u^+$ state, shown in Figure 7.24(b).

The $b^3\Sigma_u^+$ state arises from the excited configuration $(\sigma_g 1s)^1 (\sigma_u^* 1s)^1$ and is a repulsive state. The dissociation products are two ground state $(1^2S_{1/2})$ hydrogen atoms, the same as for the $X^1\Sigma_g^+$ ground state. The $a^3\Sigma_g^+$ state arises from the configuration $(\sigma_g 1s)^1 (\sigma_g 2s)^1$.

7.2.6 Rotational Fine Structure

For electronic or vibronic transitions there is a set of accompanying rotational transitions between the stacks of rotational levels associated with the upper and lower electronic or vibronic states, in a rather similar way to infrared vibrational transitions (Section 6.2.3.1). The main differences are caused by there being a wider range of electronic or vibronic transitions: they are not confined to $\Sigma - \Sigma$ types and the upper and lower states may not be singlet states nor need their multiplicities to be the same. These possibilities result in a variety of types of rotational fine structure, but we shall confine ourselves to $^1\Sigma - ^1\Sigma$ and $^1\Pi - ^1\Sigma$ types of transitions only.

7.2.6.1 $^1\Sigma - ^1\Sigma$ Electronic and vibronic transitions

In Figure 7.25 are shown stacks of rotational levels associated with two $^1\Sigma^+$ electronic states between which a transition is allowed by the $+ \leftrightarrow +$ and, if it is a homonuclear diatomic, $g \leftrightarrow u$ selection rules of equations (7.70) and (7.71). The sets of levels would be similar if both were Σ^- states or if the upper state were 'g' and the lower state 'u'. The rotational term values for any $^1\Sigma$ state are given by the expression encountered first in equation (5.23), namely

$$F_v(J) = B_v J(J + 1) - D_v J^2(J + 1)^2 \qquad (7.91)$$

where B_v is the rotational constant (equations 5.11 and 5.12), D_v is a centrifugal distortion constant, and the subscripts 'v' indicate the vibrational dependence of these. Only the v-dependence of B_v is important here and it is given by

$$B_v = B_e - \alpha(v + \tfrac{1}{2}) \qquad (7.92)$$

as for the ground electronic state (equation 5.25). The constants B_e, α, and D_v are characteristic of a particular electronic state, be it the ground state or an excited state. The quantum number $J = 0, 1, 2, \ldots$ applies, as always, to the total angular momentum, excluding nuclear spin: in $^1\Sigma$ states it applies to rotation since this is the only type of angular momentum the molecule has when $\Lambda = 0$ and $S = 0$.

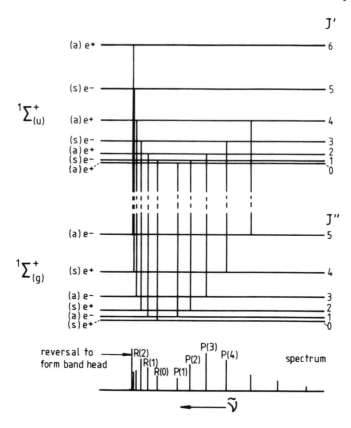

Figure 7.25 Rotational fine structure of a $^1\Sigma^+ - {}^1\Sigma^+$ electronic or vibronic transition in a diatomic molecule for which $r'_e > r''_e$. The g and u subscripts and the s and a labels apply only to a homonuclear molecule: the $+$, $-$, e, and f labels can be ignored.

The $+$, $-$, e, and f labels attached to the levels in Figure 7.25 have the same meaning as those in Figure 6.22 showing rotational levels associated with Σ_u^+ and Σ_g^+ *vibrational* levels of a linear polyatomic molecule. However, just as in that case, they can be ignored for a $^1\Sigma - {}^1\Sigma$ type of electronic transition.

The rotational section rule is

$$\Delta J = \pm 1 \qquad\qquad (7.93)$$

just as for a vibrational transition in a diatomic or linear polyatomic molecule. The result is that the rotational fine structure forms a P branch ($\Delta J = -1$) and an R branch ($\Delta J = +1$). Each branch member is labelled $P(J'')$ or $R(J'')$ as in Figure 7.25. An example of such a band, the $A^1\Sigma^+ - X^1\Sigma^+$ electronic transition of the short-lived molecule CuH, is shown in absorption in Figure 7.26. The

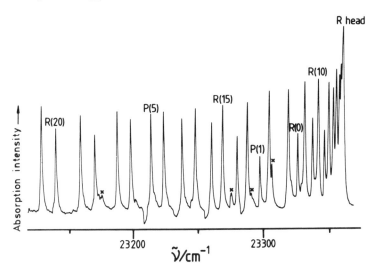

Figure 7.26 The $A^1\Sigma^+ - X^1\Sigma^+$ electronic transition of CuH in absorption. Lines marked with a cross are not due to CuH.

molecule is produced by heating metallic copper in hydrogen gas in a high temperature furnace. The absorption is at about 428 nm.

In principle this band is extremely similar to the $v = 1$–0 infrared band of HCl in Figure 6.8 and the 3_0^1 infrared vibrational band of HCN in Figure 6.23 but, in practice, we see that the CuH electronic band is very unsymmetrical about the band centre, between $R(0)$ and $P(1)$, which is where the forbidden $J' = 0 - J'' = 0$ transition would be. The reason for the asymmetry is that the rotational constants B' and B'' are typically very different in different electronic states, contrasting with their similarity in different vibrational states within the same electronic state. For reasons discussed in Section 7.2.5.3 it is likely that, if the lower state is the ground electronic state, $r' > r''$ and, therefore, $B' < B''$. This means that the rotational levels diverge more slowly in the upper than in the lower state. Figure 7.25 has been drawn for such a case which applies also to the CuH band in Figure 7.26. The result is seen to be a considerable asymmetry. There is a strong convergence to form a band head, due to reversal of the R branch, and a corresponding divergence in the P branch. The band is said to be degraded (or shaded) to low wavenumber or to the red; if $B' > B''$ the P branch forms a head and the band is degraded to the blue.

In the case of the CuH band r_e increases from 1.463 Å in the $X^1\Sigma^+$ state to 1.572 Å in the $A^1\Sigma^+$ state, resulting in the strong degradation to the red which is apparent in Figure 7.26.

The intensity distribution among the rotational transitions is governed by the population distribution among the rotational levels of the *initial* electronic or vibronic state of the transition. For absorption, the relative populations at a temperature T are given by the Boltzmann distribution law (equation 5.15) and intensities show a characteristic rise and fall, along each branch, as J increases.

If the spectrum is observed in emission it is the rotational populations in the upper state which determine relative intensities. They may or may not be equilibrium Boltzmann populations, depending on the conditions under which the molecule got into the upper state.

Obtaining the rotational constants B'' and B', or, more accurately, B'', D'' and B', D', for a $^1\Sigma - ^1\Sigma$ transition proceeds exactly as for an infrared vibration–rotation band of a diatomic molecule, as described in Section 6.2.3.1 and summarized in equations (6.29 to 6.33). If only B'' or B' is required, $\Delta_2''F(J)$ or $\Delta_2'F(J)$ is plotted against $(J + \frac{1}{2})$ and the slope of the straight line is $4B''$ or $4B'$ (see equation 6.29 and 6.30). To obtain B and D for either state the corresponding $\Delta_2F(J)/(J + \frac{1}{2})$ is plotted against $(J + \frac{1}{2})^2$ to give a straight line of slope $8D$ and intercept $(4B - 6D)$, as in equations (6.32) and (6.33).

In a homonuclear diatomic molecule there may be an intensity alternation with J for the same reasons that were discussed in Section 5.3.4 and illustrated in Figure 5.18.

By obtaining values for B_v in various vibrational states within the ground electronic state (usually from an emission spectrum) or an excited electronic state (usually from an absorption spectrum) the vibration–rotation interaction constant α and, more importantly, B_e may be obtained, from equation (7.92), for that electronic state. From B_e the value of r_e for that state easily follows.

It is important to realize that electronic spectroscopy provides the fifth method, for heteronuclear diatomic molecules, of obtaining the internuclear distance in the ground electronic state. The other four arise through the techniques of rotational spectroscopy (microwave, millimetre wave or far-infrared, and Raman) and vibration–rotation spectroscopy (infrared and Raman). In homonuclear diatomics, only the Raman techniques may be used. However if the molecule is short-lived, as is the case, for example, with CuH and C_2, electronic spectroscopy, because of its high sensitivity, is often the *only* means of determining the ground state internuclear distance.

7.2.6.2 $^1\Pi - ^1\Sigma$ Electronic and vibronic transitions

In a $^1\Pi$ state the rotational levels are different from those in a $^1\Sigma$ state because there are now two angular momenta. There is an angular momentum $\mathbf{R}$ due to end-over-end rotation of the molecule, the vector being directed along the axis of rotation as shown in Figure 7.27. In addition there is the component, $\Lambda\hbar$, of the orbital angular momentum along the internuclear axis, where $\Lambda = 1$ for a Π state. Figure 7.27 also shows $\mathbf{J}$, the resultant total angular momentum whose magnitude $B_v J(J + 1)$ depends on the value of the quantum

Figure 7.27 Resultant J of the rotational angular momentum R and the component, $\Lambda\hbar$, of the orbital angular momentum.

number J. Since $J \geqslant K$ it cannot be less than 1 so that, for a Π state, $J = 1, 2, 3, \ldots$ and there is no $J = 0$ level. Such a stack of rotational levels is shown in the upper part of Figure 7.28.

It was explained in Section 7.2.2 that all states with $\Lambda > 0$ are doubly degenerate, which can be thought of, classically, as being caused by the same energy being associated with clockwise or anticlockwise motion of the electrons about the internuclear axis. The degeneracy may be split, as it is in Figure 7.28, due to interaction of the orbital motion and the overall rotation. The splitting, $\Delta F(J)$, of the term values $F(J)$ due to this interaction is exaggerated in Figure 7.28 for clarity. It increases with the speed of overall rotation, i.e. with J, and is given by

$$\Delta F(J) = qJ(J + 1) \tag{7.94}$$

The effect is known as Λ-type doubling[†]. The quantity q is constant for a particular electronic state.

The rotational selection rule for a $^1\Pi - {}^1\Sigma$ transition is

$$\Delta J = 0, \pm 1 \tag{7.95}$$

giving a P, Q, and R branch as shown in Figure 7.28. This figure has been drawn for the more usual case, especially if the $^1\Sigma$ state is the ground state, in which $r'_e > r''_e$ resulting in $B'_e < B''_e$ and a more rapid divergence with increasing J of rotational levels in the $^1\Sigma$ state than the $^1\Pi$ state. The result is a convergence in the R branch and a divergence in the P branch, as for the $^1\Sigma - {}^1\Sigma$ example in Figure 7.25. There is also a small divergence of the Q branch to low wavenumber. The band is said to be degraded to low wavenumber or to the red. Figure 7.29 shows, as an example, the $A^1\Pi - X^1\Sigma^+$ electronic transition in emission at 424 nm of the short-lived molecule AlH. The band is degraded more strongly to the red than that in Figure 7.28, leading to an R-branch head and considerable overlap of the P and Q branches.

Whereas the $+$ and $-$ or e and f labels attached to the rotational levels for a $^1\Sigma - {}^1\Sigma$ transition in Figure 7.25 were superfluous, so far as rotational selection rules were concerned, they are essential for a $^1\Pi - {}^1\Sigma$ transition in

---

[†] This behaviour is very similar to that in a $\Pi - \Sigma$ vibrational transition in a linear polyatomic molecule (Section 6.2.3.1) in which the splitting is known as ℓ-type doubling. Quantitatively, though, Λ-type doubling is often a much larger effect.

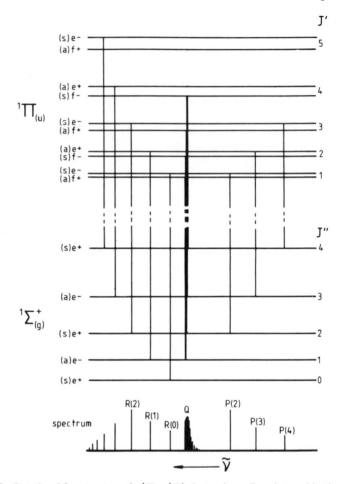

Figure 7.28 Rotational fine structure of a $^1\Pi - {}^1\Sigma^+$ electronic or vibronic transition in a diatomic molecule for which $r'_e > r''_e$. The g and u subscripts and s and a labels apply only to a homonuclear molecule.

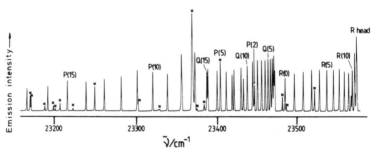

Figure 7.29 The $A^1\Pi - X^1\Sigma^+$ electronic transition of AlH in emission. Lines marked with a cross are not due to AlH.

order to tell us which of the split components in the $^1\Pi$ state is involved in a particular transition.

The $+$ or $-$ label indicates whether the wave function is symmetric or antisymmetric respectively to reflection across any plane containing the internuclear axis. Whether the $+$ component is below or above the $-$ component for, say, $J = 1$ depends on the sign of q in equation (7.94). The selection rules[†].

$$+ \leftrightarrow -, \; + \nleftrightarrow +, \; - \nleftrightarrow - \tag{7.96}$$

result in the P and R branches involving the upper components and the Q branch the lower components for the case in Figure 7.28. The selection rules

$$e \leftrightarrow f, \; e \nleftrightarrow e, \; f \nleftrightarrow f \text{ for } \Delta J = 0; \; e \nleftrightarrow f, \; e \leftrightarrow e, \; f \leftrightarrow f \text{ for } \Delta J = \pm 1 \tag{7.97}$$

involving the alternative e or f labels lead to exactly the same result.

The method of combination differences applied to the P and R branches gives the lower state rotational constants B'', or B'' and D'', just as in a $^1\Sigma - {}^1\Sigma$ transition, from equation (6.29) or equation (6.32). These branches also give rotational constants B'_u, or B'_u and D'_u, relating to the *upper* components of the $^1\Pi$ state, from equation (6.30) or equation (6.33). The constants B'_ℓ, or B'_ℓ and D'_ℓ, relating to the *lower* components of the $^1\Pi$ state, may be obtained from the Q branch. The value of q can be obtained from B'_u and B'_ℓ.

In the $A^1\Pi - X^1\Sigma^+$ electronic transition of AlH, shown in Figure 7.29, the considerable degradation of the band to the red is due not to an appreciable geometry change, since r_e is 1.6478 Å in the $X^1\Sigma^+$ state and 1.648 Å in the $A^1\Pi$ state, but to the relatively high value of q of 0.0080 cm^{-1} in the $A^1\Pi$ state.

The 'g' and 'u' subscripts in Figure 7.28 are appropriate only to a homonuclear diatomic molecule. This is the case also for the 's' and 'a' labels which may result in intensity alternations for J even or odd in the initial state of the transition. Figure 7.28 would apply equally to a $^1\Pi - {}^1\Sigma^-$ type of transition.

7.3 Electronic Spectroscopy of Polyatomic Molecules

7.3.1 Molecular Orbitals and Electronic States

Polyatomic molecules cover such a wide range of different types that it is not possible here to discuss the MOs and electron configurations of more than a very few. The molecules that we shall discuss are those of the general type AH_2, where A is a first-row element, formaldehyde (H_2CO), benzene, and some regular octahedral transition metal complexes.

For non-linear polyatomic molecules the use of symmetry arguments in discussing orbitals and electronic states becomes almost essential, whereas for atoms and diatomic molecules this could be largely avoided. For linear

---

† Note that these are the opposite of the electronic selection rules in equation (7.70).

polyatomic molecules the classification of orbitals and states and the description of electronic and associated rotational selection rules are very similar to those for diatomic molecules and symmetry arguments, again, may be largely avoided. However, it is possible that a polyatomic molecule which is linear in its ground state may be non-linear in some excited electronic states. For example, acetylene (HC≡CH) has a linear ground state ($^1\Sigma_g^+$) but a *trans* bent first excited singlet state necessitating the use of symmetry in discussing states, transitions, etc.

The total electron density contributed by all the electrons in any molecule is a property which can be visualized and it is possible to imagine an experiment in which it could be observed. It is when we try to break down this electron density into a contribution from each electron that problems arise. The methods employing hybrid orbitals or equivalent orbitals are useful in certain circumstances such as rationalizing properties of a localized part of the molecule. However the promotion of an electron from one orbital to another, in an electronic transition, or the complete removal of it, in an ionization process, both obey symmetry selection rules. For this reason the orbitals used to describe the difference between either two electronic states of the molecule or an electronic state of the molecule and an electronic state of the positive ion must be MOs which belong to symmetry species of the point group to which the molecule belongs. Such orbitals are called symmetry orbitals and are the only type we shall consider here.

7.3.1.1 AH_2 molecules

The valence atomic orbitals (AOs) of A and H are the most important, so far as the valence MOs of AH_2 are concerned, and these are $2s$ and $2p$ of the first-row element A (Li to Ne) and the $1s$ AO on H.

AH_2 molecules may have two possible extreme geometries: linear, belonging to the $D_{\infty h}$ point group, or bent with an angle of 90°, belonging to the C_{2v} point group. We shall construct MOs for these two extremes and see how they correlate as the angle changes smoothly from 90 to 180°.

7.3.1.1(a) $\angle HAH = 180°$ The $2s$ and $2p$ orbitals on A and $1s$ orbitals on H must be assigned to symmetry species of the $D_{\infty h}$ point group (see Table A.37 in the Appendix for the character table). The spherically symmetrical $2s$ orbital belongs to the species σ_g^+, the $2p_z$ AO, where z is the internuclear axis, belongs to σ_u^+, and the $2p_x$ and $2p_y$ AOs, which remain degenerate, belong to π_u. A $1s$ orbital on H cannot, by itself, be assigned to a symmetry species. We get over this problem by taking in-phase and out-of-phase combinations of both of them, as shown in Figure 7.30. The resulting orbitals belong to the σ_g^+ and σ_u^+ symmetry species respectively.

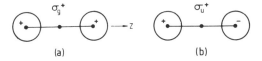

Figure 7.30 (a) In-phase and (b) out-of-phase $1s$ AOs on the hydrogen atoms of linear AH_2.

The way in which the MOs are formed is indicated on the right-hand side of Figure 7.31.

The requirements for formation of MOs are the same as for a diatomic molecule, namely that the orbitals from which they are formed must be of comparable energy and of the right symmetry. This allows the $1s + 1s$ orbital of Figure 7.30(a) to combine only with $2s$ on A. A σ_g^+ MO results, but the superscript '$+$' is usually omitted as there are no σ^- MOs. The MO is labelled $2\sigma_g$ according to the convention of numbering MOs of the same symmetry in order of increasing energy. If the $2s$ AO on A is out-of-phase with the $1s + 1s$ orbital the resulting MO, which is $3\sigma_g$, is antibonding with nodal planes between A and H. The $1\sigma_g$ MO is not shown in Figure 7.31: it is the $1s$ AO on A, which remains virtually unchanged in AH_2.

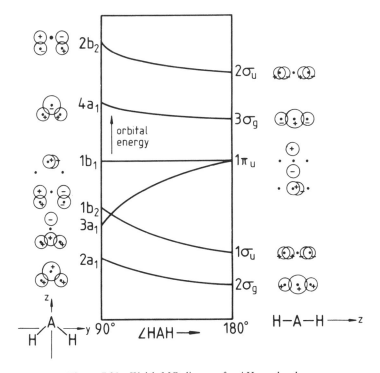

Figure 7.31 Walsh MO diagram for AH_2 molecules.

The $1s - 1s$ orbital, shown in Figure 7.30(b), combines with the $2p_z$ AO on A to form the $1\sigma_u$ and $2\sigma_u$ MOs which are bonding and antibonding respectively.

The $2p_x$ and $2p_y$ AOs on A cannot combine with either $1s + 1s$ or $1s - 1s$ on the hydrogen atoms because their symmetry does not allow it. They remain as doubly degenerate AOs on A classified as $1\pi_u$ in $D_{\infty h}$.

Arranging the MOs in order of increasing energy is achieved using the general principle that a decrease in the s character or an increase in the number of nodes in an MO increases the energy. The $2\sigma_g$ and $1\sigma_u$ MOs are both bonding between A and H but the nodal plane through A results in $1\sigma_u$ being higher in energy. A similar argument places $2\sigma_u$ above $3\sigma_g$.

7.3.1.1(b) ∠$HAH = 90°$ In bent AH_2 the $2s$, $2p_x$, $2p_y$, and $2p_z$ AOs on A can be assigned to a_1, b_1, b_2, and a_1 species respectively of the C_{2v} point group using the axis notation in Figure 7.31 and the character table in Table A.11 in the Appendix. As for linear AH_2, the $1s$ AOs on the H atoms must be delocalized to give the in-phase and out-of-phase combinations, which have the symmetry species a_1 and b_2 respectively, as shown in Figure 7.32.

The $1s + 1s$, a_1 orbital can combine with the $2s$ or $2p_z$ AO on A to give the $2a_1$, $3a_1$, and $4a_1$ MOs shown on the left-hand side of Figure 7.31. The $1a_1$ MO is the virtually unchanged $1s$ AO on A and is not shown in the figure. The $2p_y$ AO on A can combine only with the $1s - 1s$ orbital on the hydrogen atoms, but $2p_x$ cannot combine with any other orbital: it becomes the $1b_1$ lone pair orbital.

The ordering in Figure 7.31 of the MOs in terms of energy follows the rules that decreased s character and increased number of nodal planes increase the energy, but some details rely on experimental data.

Correlation of the MOs as the HAH angle changes from 90 to 180° is shown in Figure 7.31, which is known as a Walsh diagram after A. D. Walsh who devised similar diagrams for many other types of molecules (see the bibliography). The correlations should be obvious from their shapes, indicated on the left and right of the figure. It should be noted that the z axis in the linear molecule becomes the y axis in the bent molecule.

One particularly important correlation is between $3a_1$ and $1\pi_u$. Because of

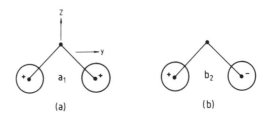

Figure 7.32 (a) In-phase and (b) out-of-phase $1s$ AOs on the hydrogen atoms of bent AH_2.

the relaxation of symmetry restrictions in the C_{2v} compared to the $D_{\infty h}$ point group, bending of the molecule results in some mixing between what become the $3a_1$ and $2a_1$ MOs. Since $2a_1$ is strongly bonding, with considerable $2s$ character, one effect of the mixing is to impart some $2s$ character to $3a_1$. This causes the very steep energy increase in $3a_1$ as the angle increases from 90 to 180°. A result of this is that some AH_2 molecules undergo not only a change of angle but a change of point group on going from the ground to an excited electronic state or on being ionized to AH_2^+. A change of point group can occur only in polyatomic molecules and this is only one of many such examples.

Using the building-up principle we can feed the electrons of any AH_2 molecule into the MOs of Figure 7.31 in pairs, except for the π orbitals which can accommodate four electrons, to give the ground or an excited configuration. The ground and first excited configurations are given in Table 7.8 for some first-row AH_2 molecules. The electronic states that arise from these configurations are given in the table and are obtained, as for diatomics, using equation (7.73) and taking account of electron spin. The $\tilde{X}, \tilde{A}, \tilde{B}, \ldots$ and $\tilde{a}, \tilde{b} \ldots$ system of labelling states is the same as in diatomics ($\tilde{A}, \tilde{B}, \ldots$ for excited states of the same multiplicity as the ground state $\tilde{X}$, and $\tilde{a}, \tilde{b}, \ldots$ for excited states of different multiplicity) except for the addition of a tilde above the label to avoid any confusion with symmetry species. The bond angles in Table 7.8 have been determined from electronic spectra, except for LiH_2 and BeH_2 which are unknown species.

The $1a_1$ or $1\sigma_g$ orbital is non-bonding and favours neither a bent nor a linear shape, but occupation of the $2\sigma_g$ and $1\sigma_u$ favours linearity since their energies are lowest for a 180° angle, as Figure 7.31 shows. Therefore, LiH_2 and BeH_2 are expected to have linear ground states. Promotion of an electron to the next

Table 7.8 Ground and excited configurations of some AH_2 molecules.

Molecule	Configuration	State	∠ HAH
LiH_2	$(1\sigma_g)^2(2\sigma_g)^2(1\sigma_u)^1$	$\tilde{X}^2\Sigma_u^+$	180°(?)
	$(1a_1)^2(2a_1)^2(3a_1)^1$	$\tilde{A}^2A_1$	<180°(?)
BeH_2	$(1\sigma_g)^2(2\sigma_g)^2(1\sigma_u)^2$	$\tilde{X}^1\Sigma_g^+$	180°(?)
	$(1a_1)^2(2a_1)^2(1b_2)^1(3a_1)^1$	$\begin{cases} \tilde{a}^3B_2 \\ \tilde{A}^1B_2 \end{cases}$	<180°(?) <180°(?)
BH_2	$(1a_1)^2(2a_1)^2(1b_2)^2(3a_1)^1$	$\tilde{X}^2A_1$	131°
	$(1\sigma_g)^2(2\sigma_g)^2(1\sigma_u)^2(1\pi_u)^1$	$\tilde{A}^2\Pi_u$	180°
CH_2	$(1a_1)^2(2a_1)^2(1b_2)^2(3a_1)^2$	$\tilde{a}^1A_1$	102.4°
	$(1a_1)^2(2a_1)^2(1b_2)^2(3a_1)^1(1b_1)^1$	$\begin{cases} \tilde{X}^3B_1 \\ \tilde{b}^1B_1 \end{cases}$	136° 140°
$NH_2(H_2O^+)$	$(1a_1)^2(2a_1)^2(1b_2)^2(3a_1)^2(1b_1)^1$	$\tilde{X}^2B_1$	103.4° (110.5°)
	$(1a_1)^2(2a_1)^2(1b_2)^2(3a_1)^1(1b_1)^2$	$\tilde{A}^2A_1$	144° (180°)
H_2O	$(1a_1)^2(2a_1)^2(1b_2)^2(3a_1)^2(1b_1)^2$	$\tilde{X}^1A_1$	104.5°

highest, $3a_1 - 1\pi_u$, orbital has a drastic effect because this orbital very much favours a bent molecule. It is anticipated, from molecules like BH_2 and CH_2 whose shapes are known, that one electron in the $3a_1 - 1\pi_u$ orbital more than counterbalances the four in the $2a_1 - 2\sigma_g$ and $1b_2 - 1\pi_u$ orbital, favouring linearity, and that molecules will be bent in these excited states.

In the $\tilde{X}^2A_1$ ground state of BH_2 the angle is known to be 131°. The fact that it is bent is due to the single electron in the $3a_1$ orbital. Promotion of this electron produces a linear molecule because the $1b_1 - 1\pi_u$ orbital favours no particular angle.

CH_2 has two electrons in the $3a_1$ orbital, resulting in the small angle of 102.4°. Promotion of an electron from $3a_1$ to $1b_1$ produces a singlet and a triplet state with, as expected, a larger angle, but the molecule is still bent. In fact the triplet state, $\tilde{X}^3B_1$, lies slightly (3165 cm^{-1} or 37.86 kJ mol^{-1}) below $\tilde{a}^1A_1$ so the former is the ground state and the latter a low-lying excited state. NH_2 shows similar geometry changes as it differs from CH_2 only by having an extra electron in the $1b_1$ orbital which favours no particular geometry. H_2O^+, having the same number of electrons as NH_2, is also quite similar.

In the ground configuration of H_2O there are two electrons in the $3a_1$ orbital strongly favouring a bent molecule. The only excited states known for H_2O are those in which an electron has been promoted from $1b_1$ to a so-called Rydberg orbital. Such an orbital is large compared to the size of the molecule and resembles an atomic orbital. Because it is so large it resembles the $1b_1$ orbital in that it does not influence the geometry. So H_2O, in such Rydberg states, has an angle similar to that in the ground state.

The MO diagram in Figure 7.31 has been derived using a number of approximations, not least of which is the assumption that it is the same for all A. In spite of this, there is remarkable agreement between what the diagram predicts and the angles that are observed. Table 7.8 shows that double occupancy of the $3a_1$ orbital gives angles in the range 102.4 to 110.5°, single occupancy gives angles in the range 131 to 144° (with the $\tilde{A}^2A_1$ state of H_2O^+ a surprising exception), and zero occupancy results in a linear molecule.

The H_3 and H_3^+ molecules are special cases of AH_2 molecules in that neither of them has the linear or bent shape already discussed. They are both cyclic molecules although H_3 is known only in excited electronic states since, in its ground state, it is unstable with respect to $H + H_2$.

7.3.1.2 Formaldehyde (H_2CO)

In this molecule, which has sixteen electrons, we shall not be concerned with twelve of them, six in σ bonding orbitals in C–O and C–H bonds, two in each of the $1s$ orbitals on C and O, and two in the $2s$ orbital on O. The remaining four valence electrons occupy higher energy MOs, shown in Figure 7.33. The orbitals shown are a bonding π orbital, an antibonding π^* orbital, both between

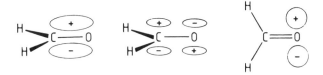

Figure 7.33 The $1b_1(\pi)$, and $2b_1(\pi^*)$, and $2b_2(n)$ MOs in formaldehyde.

C and O, and a non-bonding $2p_y$ orbital, n, on O (the z axis is the C–O direction and the x axis perpendicular to the plane of the molecule). Each is classified according to the C_{2v} point group (Table A.11 in the Appendix). The order of energies is $\pi^* > n > \pi$ and the ground configuration is

$$\ldots (1b_1)^2(2b_2)^2 \tag{7.98}$$

leading to an $\tilde{X}^1A_1$ ground state.

The lowest energy electron promotion is from the $2b_2$ non-bonding n orbital to the $2b_1$ antibonding π^* orbital, giving the configuration

$$\ldots (1b_1)^2(2b_2)^1(2b_1)^1 \tag{7.99}$$

resulting in $\tilde{a}^3A_2$ and $\tilde{A}^1A_2$ states.

Orbital promotions of this type give rise to states, such as the $\tilde{a}$ and $\tilde{A}$ states of formaldehyde, which are commonly referred to as $n\pi^*$ states. In addition, transitions to such states, for example the $\tilde{a} - \tilde{X}$ and $\tilde{A} - \tilde{X}$ transitions of formaldehyde, are referred to colloquially as $\pi^* - n$ or n-to-π^*, transitions.

There is a useful way of distinguishing a transition of the $\pi^* - n$ type from one of, say, the $\pi^* - \pi$ type. The former is blue-shifted, i.e. shifted to a lower wavelength, in a hydrogen-bonding solvent. The reason is that such a solvent, e.g. ethanol, forms a hydrogen bond by weak MO formation between the n orbital and $1s$ orbital on the hydrogen atom of the OH group of the solvent. This increases the binding energy of the n orbital and therefore increases the energy of the $\pi^* - n$ transition, shifting it to the blue.

However, an electron in the $2b_1$ π^* orbital favours a pyramidal shape for formaldehyde shown in Figure 7.34. The reason for this is that the π^* orbital can then overlap with the $1s + 1s$ orbital on the hydrogens, thereby gaining some C–H bonding character, and is also able to mix with the $2s$ orbital on C. The result is an increase in s character and a substantial lowering of the energy relative to the planar molecule. The angle ϕ (Figure 7.34) is $38°$ in the $\tilde{A}^1A_2$ state and $43°$ in the $\tilde{a}^3A_2$ state.

Figure 7.34 Non-planar formaldehyde in its $\tilde{a}$ and $\tilde{A}$ excited states.

Because of the pyramidal shape in these excited states the orbitals and states may be reclassified according to the C_s point group (Table A.1 in the Appendix).

7.3.1.3 Benzene

In a molecule with electrons in π orbitals, such as formaldehyde, ethylene, buta-1,3-diene and benzene, if we are concerned only with the ground state, or excited states obtained by electron promotion within π-type MOs, an approximate MO method due to Hückel may be useful.

The Hückel MO method is based on the LCAO method for diatomic molecules discussed in Section 7.2.1. Extension of the LCAO method to polyatomic molecules gives a secular determinant of the general type

$$\begin{vmatrix} H_{11} - E & H_{12} - ES_{12} & \cdots & H_{1n} - ES_{1n} \\ H_{12} - ES_{12} & H_{22} - E & \cdots & H_{2n} - ES_{2n} \\ \vdots & \vdots & & \vdots \\ H_{1n} - ES_{1n} & H_{2n} - ES_{2n} & \cdots & H_{nn} - E \end{vmatrix} = 0 \qquad (7.100)$$

analogous to equation (7.46) for a diatomic molecule, where n is the number of atoms and H_{nn} are Coulomb integrals, H_{mn} (for $m \neq n$) are resonance integrals, S_{mn} (for $m \neq n$) are overlap integrals, and E is the orbital energy, as in equation (7.46). This determinant may be abbreviated to

$$|H_{mn} - ES_{mn}| = 0 \qquad (7.101)$$

where $S_{mn} = 1$ when $m = n$.

For π-electron systems Hückel made the following approximations:

1. Only electrons in π orbitals are considered, those in σ orbitals being neglected. In a molecule such as ethylene, with sufficiently high symmetry, the σ–π separation is not an approximation since the σ and π MOs have different symmetry species and therefore cannot mix. However, in a molecule of low symmetry such as but-1-ene ($CH_3CH_2CH{=}CH_2$), the σ and π MOs have the *same* symmetry species; in the Hückel treatment, though, the σ MOs are still neglected because they are assumed to be much lower in energy than the π MOs.
2. It is assumed that, for $m \neq n$,

$$S_{mn} = 0 \qquad (7.102)$$

implying no overlap of atomic orbitals even for nearest-neighbour atoms.
3. When $m = n$ the Coulomb integral H_{nn} is assumed to be the same for each atom and is given the symbol α:

$$H_{nn} = \alpha \qquad (7.103)$$

4. When $m \neq n$ the resonance integral H_{mn} is assumed to be the same for any pair of directly bonded atoms and is given the symbol β:

$$H_{mn} = \beta \qquad (7.104)$$

5. When m and n are not directly bonded

$$H_{mn} = 0 \qquad (7.105)$$

The π-electron wave functions in the Hückel method are given by

$$\psi = \sum_i c_i \chi_i \qquad (7.106)$$

as in equation (7.36) for LCAO MOs of a diatomic molecule, but now the χ_i are only those AOs, very often $2p$ on C, N, or O, which are involved in the π MOs.

In the case of benzene, Hückel treatment of the six $2p$ orbitals on the carbon atoms and perpendicular to the plane of the ring leads to the secular determinant

$$\begin{vmatrix} x & 1 & 0 & 0 & 0 & 1 \\ 1 & x & 1 & 0 & 0 & 0 \\ 0 & 1 & x & 1 & 0 & 0 \\ 0 & 0 & 1 & x & 1 & 0 \\ 0 & 0 & 0 & 1 & x & 1 \\ 1 & 0 & 0 & 0 & 1 & x \end{vmatrix} = 0 \qquad (7.107)$$

derived from the secular equations of equation (7.100), making the substitutions in equations (7.102) to (7.105) and putting

$$\frac{\alpha - E}{\beta} = x \qquad (7.108)$$

By the method of solution of simultaneous equations or, much more easily, by solving the determinant of equation (7.107) we obtain the solutions

$$x = \pm 1, \ \pm 1, \ \text{or} \ \pm 2 \qquad (7.109)$$

from which equation (7.108) gives

$$E = \alpha \pm \beta, \ \alpha \pm \beta, \ \text{or} \ \alpha \pm 2\beta \qquad (7.110)$$

The fact that the $E = \alpha \pm \beta$ solution appears twice implies that the MOs with $E = \alpha + \beta$ are doubly degenerate, as are those with $E = \alpha - \beta$. Figure 7.35 is an energy level diagram illustrating this. As usual β, the resonance integral, is a negative quantity. The corresponding six MO wave functions may be obtained from equation (7.106) in a similar way to that described for diatomic molecules in Section 7.2.1.1.

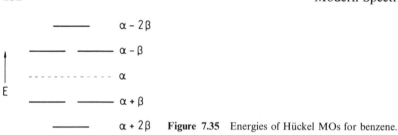

$\alpha - 2\beta$

$\alpha - \beta$

α

E

$\alpha + \beta$

$\alpha + 2\beta$ **Figure 7.35** Energies of Hückel MOs for benzene.

Figure 7.36 illustrates the MO wave functions and gives the symmetry species, according to the D_{6h} point group, which may be confirmed using the character table in Table A.36 in the Appendix. In this figure only the parts of the wave functions above the plane of the carbon ring are illustrated. The parts below are identical in form but opposite in sign, which means that, like all π orbitals in planar molecules, they are antisymmetric to reflection in the plane of the molecule. The energy increases with the number of nodal planes perpendicular to the ring.

The ground configuration of benzene is obtained by feeding the six electrons, which were originally in $2p_z$ AOs on the carbon atoms (z axis perpendicular to the ring), into the lower energy MOs giving

$$\dots (1a_{2u})^2 (1e_{1g})^4 \tag{7.111}$$

taking account of the fact that an e orbital is doubly degenerate and can accommodate four electrons. As for all molecules with occupied orbitals that are all filled, the ground state is a totally symmetric singlet state, that is $\tilde{X}^1 A_{1g}$.

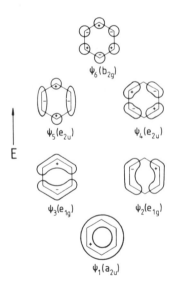

$\psi_6 (b_{2g})$

$\psi_5 (e_{2u})$ $\psi_4 (e_{2u})$

E

$\psi_3 (e_{1g})$ $\psi_2 (e_{1g})$

$\psi_1 (a_{2u})$ **Figure 7.36** Hückel MOs in benzene.

The first excited configuration is obtained by promoting an electron from an e_{1g} to an e_{2u} orbital, resulting in

$$\ldots (1a_{2u})^2 (1e_{1g})^3 (1e_{2u})^1 \qquad (7.112)$$

The states arising from this configuration are the same as those from ... $(1a_{2u})^2 (1e_{1g})^1 (1e_{2u})^1$, as a single vacancy in e_{1g} can be treated like an electron. The states can be obtained in a similar way to those for the excited configuration of N_2 in equation (7.76). The symmetry species $\Gamma(\psi_e^o)$ of the orbital part of the electronic wave function is obtained from

$$\Gamma(\psi_e^o) = e_{1g} \times e_{2u} = B_{1u} + B_{2u} + E_{1u} \qquad (7.113)$$

This result is similar to that for $e \times e$, in equation (4.29), in the C_{3v} point group and can be verified using the D_{6h} character table in Table A.36 in the Appendix. As the two electrons (or one electron and one vacancy) in the partially occupied orbitals may have parallel ($S = 0$) or antiparallel ($S = 1$) spins there are six states arising from the configuration in equation (7.112), namely $^{1,3}B_{1u}$, $^{1,3}B_{2u}$, $^{1,3}E_{1u}$. The singlet states are, in order of increasing energy, $\tilde{A}^1 B_{2u}$, $\tilde{B}^1 B_{1u}$, and $\tilde{C}^1 E_{1u}$, although there is some doubt about the identification of the $\tilde{B}$ state. The triplet states are $\tilde{a}^3 B_{1u}$, $\tilde{b}^3 E_{1u}$, and $\tilde{c}^3 B_{2u}$ in order of increasing energy.

7.3.1.4 Crystal field and ligand field molecular orbitals

Transition metal atoms are distinguished from other atoms by their having partially filled $3d$, $4d$, or $5d$ orbitals. Here we consider only metals of the first transition series, Sc, Ti, V, Cr, Mn, Fe, Co, Ni, Cu, and Zn, in which the $3d$ orbital is involved.

Transition metals readily form complexes, such as $[Fe(CN)_6]^{4-}$, the ferrocyanide ion, $Ni(CO)_4$, nickel tetracarbonyl, and $[CuCl_4]^{2-}$, the copper tetrachloride ion. MO theory applied to such species has tended to be developed independently. It is for this reason that the terms 'crystal field theory' and 'ligand field theory' have arisen which tend to disguise the fact that they are both aspects of MO theory.

The word 'ligand' to describe an atom, or group of atoms, attached to a central metal atom can also be confusing. This has arisen because the type of bonding in complexes tends to be different from that in, say, H_2O. However the difference is quantitative rather than qualitative and there is no reason why we should not refer to the hydrogen atoms in H_2O as ligands—it just happens that we rarely do.

Ligands in a transition metal complex are usually arranged in a highly symmetrical way. For example, six ligands often take up an octahedral configuration, as in $[Fe(CN)_6]^{4-}$, and four ligands a tetrahedral, as in $Ni(CO)_4$, or

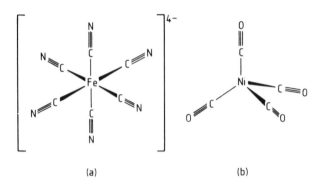

Figure 7.37 (a) Octahedral $[Fe(CN)_6]^{4-}$ and (b) tetrahedral $Ni(CO)_4$.

a square planar configuration, as in $[CuCl_4]^{2-}$. Figure 7.37 shows $[Fe(CN)_6]^{4-}$ and $Ni(CO)_4$ but we shall consider only the regular octahedral case in detail.

In a transition metal complex the higher energy occupied MOs can be regarded as perturbed d orbitals of the metal atom. In an octahedral complex, if the perturbation is weak, the ligands can be treated as point charges at the corners of a regular octahedron. This is reminiscent of the perturbation of the Na^+ orbitals by six octahedrally arranged nearest-neighbour Cl^- ions in a sodium chloride crystal, and it is for this reason that this aspect of MO theory is known as crystal field theory.

When the ligands interact more strongly the MOs of the ligands must be taken into account. This type of MO theory is referred to as ligand field theory.

7.3.1.4(a) Crystal field theory   In the presence of six point charges arranged octahedrally on the cartesian axes the five d orbitals of Figure 1.8 are perturbed and must be classified according to the O_h point group (see Section 4.2.9 and also Table A.43 in the Appendix). Table 7.9 gives the symmetry species of the d orbitals in octahedra (O_h), as well as tetrahedral (T_d) and other crystal fields.

The results in Table 7.9 will not be derived here but they show that a set of five d orbitals breaks down into a doubly degenerate e_g orbital and a triply degenerate t_{2g} orbital in a regular octahedral crystal field. Since the d_{z^2} and $d_{x^2-y^2}$ orbitals have much of their electron density along metal–ligand bonds, electrons in them experience more repulsion by the ligand electrons than those in d_{xy}, d_{yz}, or d_{xz} orbitals. The result is that the e_g orbitals, derived from d_{z^2} and $d_{x^2-y^2}$, are pushed up in energy by $\frac{3}{5}\Delta_0$ and the t_{2g} orbitals are pushed down by $\frac{2}{5}\Delta_0$, where Δ_0 is the $e_g - t_{2g}$ splitting, as shown in Figure 7.38. The value of Δ_0 is typically such that promotion of an electron from the t_{2g} to the e_g

Table 7.9 Symmetry species of orbitals resulting from the splitting of d orbitals by various ligand arrangements.

Point group	d_{z^2}	$d_{x^2-y^2}$	d_{xy}	d_{yz}	d_{xz}
O_h	← e_g →		← t_{2g} →		
T_d	← e →		← t_2 →		
D_{3h}	a'_1	← e' →		← e'' →	
D_{4h}	a_{1g}	b_{1g}	b_{2g}	← e_g →	
$D_{\infty h}$	σ_g^+	← δ_g →		← π_g →	
C_{2v}	a_1	a_1	a_2	b_2	b_1
C_{3v}	a_1	a_1	a_2	← e →	
C_{4v}	a_1	b_1	b_2	← e →	
D_{2d}	a_1	b_1	b_2	← e →	
D_{4h}	a_1	← e_2 →		← e_3 →	

orbital leads to an absorption in the visible region of the spectrum and the characteristic property that such complexes are usually coloured.

Figure 7.39 shows how the d electrons are fed into the t_{2g} and e_g orbitals in an octahedral complex when the metal atom or ion has a d^1, d^2, d^3, d^8, d^9, or d^{10} configuration. Just as in the ground configuration of O_2 in equation (7.59), electrons in degenerate orbitals prefer to have parallel spins for minimum energy. For example, in $[Cr(H_2O)_6]^{3+}$ the Cr^{3+} has a d^3 configuration (having lost one $3d$ and two $4s$ electrons) and they each go into different t_{2g} orbitals with parallel spins to give a quartet ground state. (In fact it is $\tilde{X}^4A_{2g}$, but we shall not derive the states arising from a $(t_{2g})^3$ configuration here.)

Taking into account the preference for parallel spins the configurations illustrated in Figure 7.39 are obtained unambiguously. However this is not the case for d^4, d^5, d^6, and d^7 configurations. Figure 7.40 shows that the way in which the electrons are fed into the orbitals to give the minimum energy depends on the magnitude of the crystal field splitting Δ_o. If it is small, as in

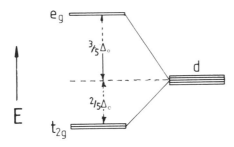

Figure 7.38 Splitting of d orbitals in a regular octahedral field.

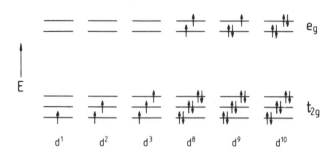

Figure 7.39 Electron configurations in d^1, d^2, d^3, d^8, d^9, and d^{10} octahedral complexes.

Figure 7.40(a), electrons prefer to go into e_g orbitals with parallel spins rather than, with antiparallel spins, to fill up the t_{2g} orbitals. Figure 7.40(b) shows that a larger Δ_o results in electrons preferring to have antiparallel spins in t_{2g} orbitals rather than be promoted to e_g orbitals. Complexes behaving as in Figure 7.40(a) and (b) are known as weak field, high spin and strong field, low spin complexes respectively.

As an example $[\mathrm{Cr(H_2O)_6}]^{2+}$ has a d^4 weak field, high spin configuration giving a quintet (in fact $\tilde{X}^5E_g$) ground state. In $[\mathrm{Fe(CN)_6}]^{4-}$, Fe^{2+} has a d^6 configuration and the complex is of the strong field, low spin type having a $(t_{2g})^6$ configuration. The ground state is therefore singlet (and is, in fact, $\tilde{X}^1A_{1g}$ since all occupied orbitals are filled).

7.3.1.4(b)   Ligand field theory   When ligands interact so strongly with the central metal atom that they can no longer be treated as negative point charges the crystal field approximation breaks down and the MOs of the ligand L must be considered.

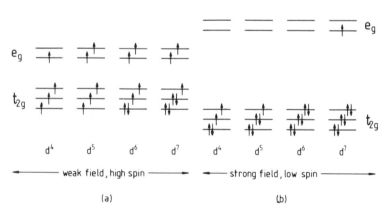

Figure 7.40 (a) Weak field, high spin and (b) strong field, low spin configurations in d^4, d^5, d^6, and d^7, octahedral complexes.

Table 7.10 Classification of σ ligand orbitals in various point groups.

Point group	σ orbital symmetry species
O_h	$a_{1g} + e_g + t_{1u}$ (in octahedral ML_6)
T_d	$a_1 + t_2$ (in tetrahedral ML_4)
D_{3h}	$2a'_1 + a''_2 + e'$ (in trigonal bipyramidal ML_5)
D_{4h}	$a_{1g} + b_{1g} + e_u$ (in square planar ML_4)
	$2a_{1g} + a_{2u} + b_{1g} + e_u$ (in *trans*-octahedral $ML_4 L'_2$)
$D_{\infty h}$	$\sigma_g + \sigma_u$ (in linear ML_2)
C_{2v}	$a_1 + b_2$ (in non-linear ML_2)
	$2a_1 + b_1 + b_2$ (in tetrahedral $ML_2 L'_2$)
	$3a_1 + a_2 + b_1 + b_2$ (in *cis*-octahedral $ML_4 L'_2$)
C_{3v}	$2a_1 + e$ (in tetrahedral $ML_3 L'$)
	$2a_1 + 2e$ (in all-*cis*-octahedral $ML_3 L'_3$)
C_{4v}	$2a_1 + b_1 + e$ (in square pyramidal $ML_4 L'$)
	$3a_1 + b_1 + e$ (in octahedral $ML_5 L'$)
D_{2d}	$2a_1 + 2b_2 + 2e$ (in dodecahedral ML_8)
D_{4d}	$a_1 + b_2 + e_1 + e_2 + e_3$ (in square antiprism ML_8)

The ligand MOs are of two types: σ MOs, which are cylindrically symmetrical about the metal–ligand bond, and π MOs which are not. The σ type of metal–ligand bonding is usually stronger as, for example, is provided by the lone pair orbital on CO in metal carbonyls. We shall neglect π-type bonding and consider in detail only octahedral cases.

Table 7.10 shows how σ ligand orbitals are classified in various point groups with different ligand arrangements. It shows that, in octahedral ML_6, the six σ ligand orbitals are split into a_{1g}, e_g, and t_{1u} orbitals. These are shown on the right-hand side of Figure 7.41. The effect of these on the e_g and t_{2g} orbitals, derived in the crystal field approximation, is for the e_g orbitals to interact, the crystal field orbital being pushed up and the ligand orbital being pushed down. The result, as shown in Figure 7.41, is to increase Δ_o compared to its value

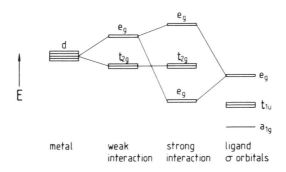

Figure 7.41 Perturbation of crystal field MOs by ligand MOs.

in the crystal field approximation. This increase in Δ_o leads, in turn, to a tendency towards low rather than high spin complexes, as in $[Fe(CN)_6]^{4-}$.

7.3.1.4(c) Electronic transitions Both ground and excited electron configurations of metal complexes often give rise to quite complex manifolds of states which we shall not derive here. However, there is one simplifying factor which is extremely useful in considering electronic spectra of transition metal complexes. All the higher energy orbitals which may be occupied, in regular octahedral complexes, are t_{2g} or e_g, whether we use the crystal field or ligand field approach. Since the subscript g means 'symmetric to inversion through the centre of the molecule' all states must be g also, since g multiplied by itself any number of times (actually the number of times g orbitals are occupied) always gives g. So all ground and excited states arising from t_{2g} and e_g occupancy are all g states. Just as in a homonuclear diatomic molecule (see equation 7.71) $g-g$ transitions are forbidden.

However transition metal complexes *do* absorb in the visible region, giving them a characteristic colour. How can this happen if the transitions are forbidden? The answer is that interaction may occur between the motion of the electrons and vibrational motions so that some vibronic transitions are allowed (see Section 7.3.4.2b).

7.3.2   Electronic and Vibronic Selection Rules

In the case of atoms (Section 7.1) a sufficient number of quantum numbers is available for us to be able to express electronic selection rules entirely in terms of these quantum numbers. For diatomic molecules (Section 7.2.3) we require, in addition to the quantum numbers available, one or, for homonuclear diatomics, two symmetry properties $(+, -$ and $g, u)$ of the electronic wave function to obtain selection rules.

In non-linear polyatomic molecules the process of deterioration of quantum numbers continues to such an extent that only the total electron spin quantum number S remains. The selection rule

$$\Delta S = 0 \qquad (7.114)$$

still applies, unless there is an atom with a high nuclear charge in the molecule. For example, triplet–singlet transitions are extremely weak in benzene but much more intense in iodobenzene.

For the orbital parts ψ_e^o of the electronic wave functions of two electronic states the selection rules depend entirely on symmetry properties. (In fact the electronic selection rules can also be obtained, from symmetry arguments only, for diatomic molecules and atoms, using the $D_{\infty h}$ (or $C_{\infty v}$) and K_h point groups respectively; but it is more straightforward to use quantum numbers when these are available.)

Electronic transitions mostly involve interaction between the molecule and the electric component of the electromagnetic radiation (Section 2.1). The selection rules are, therefore, electron dipole selection rules analogous to those derived in Section 6.2.2.1 for infrared vibrational transitions in polyatomic molecules.

The electronic transition intensity is proportional to $|R_e|^2$, the square of the electronic transition moment R_e, where

$$R_e = \int \psi_e'^* \mu \psi_e'' \, d\tau_e \tag{7.115}$$

which is similar to equation (6.44) for an infrared vibrational transition. For an allowed electronic transition $|R_e| \neq 0$ and the symmetry requirement for this is

$$\Gamma(\psi_e') \times \Gamma(\mu) \times \Gamma(\psi_e'') = A \tag{7.116}$$

for a transition between non-degenerate states or

$$\Gamma(\psi_e') \times \Gamma(\mu) \times \Gamma(\psi_e'') \supset A \tag{7.117}$$

where $\supset$ means 'contains', for transitions between states where at least one of them is degenerate. The symbol A stands for the totally symmetric species of the point group concerned.

The components of R_e along the cartesian axes are given by

$$R_{e,x} = \int \psi_e'^* \mu_x \psi_e'' \, d\tau_e$$

$$R_{e,y} = \int \psi_e'^* \mu_y \psi_e'' \, d\tau_e \tag{7.118}$$

$$R_{e,z} = \int \psi_e'^* \mu_z \psi_e'' \, d\tau_e$$

and, since

$$|R_e|^2 = (R_{e,x})^2 + (R_{e,y})^2 + (R_{e,z})^2 \tag{7.119}$$

the electronic transition is allowed if any of $R_{e,x}$, $R_{e,y}$, or $R_{e,z}$ is non-zero. It follows that, analogous to equation (6.53), for a transition to be allowed

$$\Gamma(\psi_e') \times \Gamma(T_x) \times \Gamma(\psi_e'') = A$$

and/or

$$\Gamma(\psi_e') \times \Gamma(T_y) \times \Gamma(\psi_e'') = A \tag{7.120}$$

and/or

$$\Gamma(\psi_e') \times \Gamma(T_z) \times \Gamma(\psi_e'') = A$$

for transitions between non-degenerate states (replace = by $\supset$ if a degenerate state is involved), where T_x, etc., are translations along the corresponding axes.

If the product of two symmetry species is totally symmetric those species must be the same. Therefore we can rewrite equation (7.120) as

$$\Gamma(\psi'_e) \times \Gamma(\psi''_e) = \Gamma(T_x) \text{ and/or } \Gamma(T_y) \text{ and/or } \Gamma(T_z) \qquad (7.121)$$

or, if a degenerate state is involved, = is replaced by $\supset$. This is the general selection rule for a transition between two electronic states. If the lower state is the ground state of a molecule with only filled orbitals, a so-called closed shell molecule, ψ''_e is totally symmetric and equation (7.121) simplifies to

$$\Gamma(\psi'_e) = \Gamma(T_x) \text{ and/or } \Gamma(T_y) \text{ and/or } \Gamma(T_z) \qquad (7.122)$$

This result is the same as the infrared vibrational selection rule in equation (6.55).

If vibrations are excited in either the lower or the upper electronic state, or both, the vibronic transition moment $\boldsymbol{R}_{ev}$, corresponding to the electronic transition moment $\boldsymbol{R}_e$ in equation (7.115), is given by

$$\boldsymbol{R}_{ev} = \int \psi'^*_{ev} \boldsymbol{\mu} \psi''_{ev} \, \mathrm{d}\tau_{ev} \qquad (7.123)$$

where ψ_{ev} is a vibronic wave function. Following the same arguments as for electronic transitions the selection rule analogous to that in equation (7.121) is

$$\Gamma(\psi'_{ev}) \times \Gamma(\psi''_{ev}) = \Gamma(T_x) \text{ and/or } \Gamma(T_y) \text{ and/or } \Gamma(T_z) \qquad (7.124)$$

or, since

$$\Gamma(\psi_{ev}) = \Gamma(\psi_e) \times \Gamma(\psi_v) \qquad (7.125)$$

we obtain

$$\Gamma(\psi'_e) \times \Gamma(\psi'_v) \times \Gamma(\psi''_e) \times \Gamma(\psi''_v) = \Gamma(T_x) \text{ and/or } \Gamma(T_y) \text{ and/or } \Gamma(T_z) \quad (7.126)$$

If a degenerate state is involved the = is replaced by $\supset$. Very often either the same vibration is excited in both states, in which case $\Gamma(\psi'_v) = \Gamma(\psi''_v)$ and the selection rule is the same as the electronic selection rule, or no vibration is excited in the upper or lower state, resulting in $\Gamma(\psi'_v)$ or $\Gamma(\psi''_v)$ being totally symmetric.

7.3.3 Chromophores

The concept of a chromophore is analogous to that of a group vibration discussed in Section 6.2.1. Just as the wavenumber of a group vibration is treated as transferable from one molecule to another so is the wavenumber, or wavelength, at which an electronic transition occurs in a particular group. Such a group is called a chromophore since it results in a characteristic colour of the compound due to absorption of visible or, broadening the use of the word 'colour', ultraviolet radiation.

The ethylenic group, $\diagdown$C$=$C$\diagdown$, is an example. Whatever molecule contains the group, such as $H_2C{=}CH_2$, $XHC{=}CH_2$, $X_2C{=}CH_2$, or cyclohexene, shows an intense absorption system with a maximum intensity at about 180 nm. However the group can act as a chromophore only if it is not conjugated with any other π-electron system: for example, buta-1,3-diene and benzene absorb at much longer wavelengths. On the other hand, the benzene ring itself can be treated as a chromophore showing a characteristic, fairly weak absorption at about 260 nm, like benzene itself (Section 7.3.1.3) and, say, phenyl-cyclohexane.

Similarly, the acetylenic group, $-$C$\equiv$C$-$, shows an intense absorption system at about 190 nm and the allylic group, $\diagdown$C$=$C$=$C$\diagdown$, absorbs strongly at about 225 nm.

A transition involving a $\pi^* - n$ promotion is useful in identifying a chromophore as it gives a characteristically weak absorption system which is usually to high wavelength of systems due to $\pi^* - \pi$ promotions and may be interfered with by them. The aldehyde group, $-$CHO, is a useful chromophore showing a weak π^*-n absorption system at about 280 nm, like formaldehyde itself (Section 7.3.1.2). However, in a molecule such as benzaldehyde (C_6H_5CHO), the aldehyde group is part of a conjugated π-electron system and can no longer be treated as a chromophore.

Like group vibrations, the wavelength at which a chromophore absorbs can be employed as an analytical tool, but a rather less useful one.

7.3.4 Vibrational Coarse Structure

As for diatomic molecules (Section 7.2.5.2) the vibrational (vibronic) transitions accompanying an electronic transition fall into the general categories of progressions and sequences as illustrated in Figure 7.18. The main differences in a polyatomic molecule are that there are $3N - 6$ (or $3N - 5$ for a linear molecule) vibrations—not just one—and that some of these lower the symmetry of the molecule as they are non-totally symmetric.

7.3.4.1 Sequences

The most common type of sequence, one with $\Delta v = 0$ shown in Figure 7.18, is always allowed by symmetry because, whatever the symmetry of the vibration involved, $\Gamma(\psi'_v) = \Gamma(\psi''_v)$ in equation (7.126) and the product $\Gamma(\psi'_v) \times \Gamma(\psi''_v)$, is totally symmetric. There may be several $\Delta v_i = 0$ sequences in various vibrations i but only those vibrations with levels of a sufficiently low wavenumber to be appreciably populated (see equation 7.83) will form sequences. In a planar molecule the lowest wavenumber vibrations are usually out-of-plane vibrations.

7.3.4.2   Progressions

7.3.4.2(a)   Totally symmetric vibrations   When the vibrations involved are totally symmetric, progressions are formed for very much the same reasons as in a diatomic. The Franck–Condon principle, described in Section 7.2.5.3, applies to each vibration separately. If there is a geometry change from the lower to the upper electronic state in the direction of one of the normal coordinates then the corresponding totally symmetric vibration is excited and forms a progression. The length of the progression depends on the size of the geometry change, as illustrated for a diatomic in Figure 7.22. Intensities along each progression are given by equation (7.88) where the vibrational overlap integral applies to the vibration concerned.

One example of a long progression involves the CO-stretching vibration in the $\tilde{A}^1 A_2 - \tilde{X}^1 A_1$ system of formaldehyde (see Section 7.3.1.2). Because the π^*–n electron promotion involves an electron going from a non-bonding orbital to one which is antibonding between the C and O atoms, there is an increase of the C=O bond length (from 1.21 to 1.32 Å) in the excited state and a long progression in the C=O stretching vibration results.

In the $\tilde{A}^1 B_{2u} - \tilde{X}^1 A_{1g}$ system of benzene (see Section 7.3.1.3) an electron is promoted from an e_{1g} orbital to a more antibonding e_{2u} orbital, shown in Figure 7.36. The result is an increase in all the C–C bond lengths (from 1.397 to 1.434 Å) in the excited state. This gives rise to the long progression shown in the low resolution absorption spectrum in Figure 7.42, involving $v_1{}^\dagger$, the symmetrical ring-breathing vibration shown in Figure 6.13(f).

In electronic spectroscopy of polyatomic molecules the system used for labelling vibronic transitions employs $N_{v''}^{v'}$ to indicate a transition in which vibration N is excited with v'' quanta in the lower state and v' quanta in the upper state. The pure electronic transition is labelled 0_0^0. The system is very similar to the rather less often used system for pure vibrational transitions described in Section 6.2.2.1.

In Figure 7.42 it is seen that the progression is built not on the 0_0^0 but on the 6_0^1 band. The reason for this will become clear when we have seen, in the following section, how non-totally symmetric vibrations may be active in an electronic band system.

7.3.4.2(b)   Non-totally symmetric vibrations   The general vibronic selection rule in equation (7.126) shows that many vibronic transitions involving one

---

† Using the Wilson numbering (see Bibliography to Chapter 6).

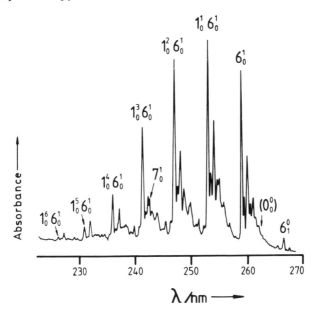

Figure 7.42   Low resolution $\tilde{A}^1B_{2u} - \tilde{X}^1A_{1g}$ absorption spectrum of benzene.

quantum of a non-totally symmetric vibration are allowed. For example, consider a molecule such as chlorobenzene in the C_{2v} point group for which the ground state is $^1A_1$ and an excited electronic state is $^1B_2$. If one quantum of a b_2 vibration X is excited in the upper electronic state and no vibration at all is excited in the lower electronic state equation (7.126) becomes

$$\Gamma(\psi'_e) \times \Gamma(\psi'_v) \times \Gamma(\psi''_e) \times \Gamma(\psi''_v) = B_2 \times B_2 \times A_1 \times A_1$$
$$= A_1 = \Gamma(T_z) \qquad (7.127)$$

The vibronic transition X^1_0, together with X^0_1 in which the same b_2 vibration is excited only in the lower state, are illustrated in Figure 7.43. Both vibronic

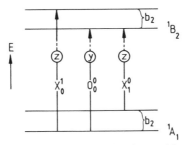

Figure 7.43   Some allowed electronic and vibronic transitions in a C_{2v} molecule.

transitions are allowed and are polarized along the z axis since the product of equation (7.127) is A_1 for both transitions. The electronic transition is also allowed but is polarized along the y axis. How do the vibronic transitions obtain their intensity?

The answer, very often, is that they do not obtain any intensity. Many such vibronic transitions, involving non-totally symmetric vibrations but which are allowed by symmetry, can be devised in many electronic band systems but, in practice, few have sufficient intensity to be observed. For those which do have sufficient intensity the explanation first put forward as to how it is derived was due to Herzberg and Teller.

The Franck–Condon approximation (see Section 7.2.5.3) assumes that an electronic transition is very rapid compared to the motion of the nuclei. One important result is that the transition moment R_{ev} for a vibronic transition is given by

$$R_{ev} = R_e \int \psi_v'^* \psi_v'' \, dQ \tag{7.128}$$

i.e. the electronic transition moment multiplied by the vibrational overlap integral, as for a diatomic molecule (equation 7.88) but replacing r by the general vibrational coordinate Q. Herzberg and Teller said that, if a non-totally symmetric vibration is excited, the Franck–Condon approximation breaks down. The breakdown can be represented by expanding the electronic transition moment as a Taylor series. Including only the first two terms of the series, this gives

$$R_e \simeq (R_e)_{eq} + \sum_i \left(\frac{\partial R_e}{\partial Q_i} \right)_{eq} Q_i \tag{7.129}$$

where the subscript 'eq' refers to the equilibrium configuration of the molecule. We have, up to now, neglected the second term on the right-hand side and it is this term which allows for R_e changing as vibration i, with coordinate Q_i, is excited. Inserting this expression for R_e into equation (7.128) gives

$$R_{ev} = \int \psi_v'^* \left[(R_e)_{eq} + \sum_i \left(\frac{\partial R_e}{\partial Q_i} \right)_{eq} Q_i \right] \psi_v'' \, dQ_i \tag{7.130}$$

and, integrating the two terms separately, we get

$$R_{ev} = (R_e)_{eq} \int \psi_v'^* \psi_v'' \, dQ_i + \sum_i \left(\frac{\partial R_e}{\partial Q_i} \right)_{eq} \int \psi_v'^* Q_i \psi_v'' \, dQ_i \tag{7.131}$$

The first term on the right-hand side is the same as in equation (7.128). Herzberg and Teller suggested that the second term, in particular $(\partial R_e / \partial Q_i)_{eq}$, may be non-zero for certain non-totally symmetric vibrations. As the intensity is proportional to $|R_{ev}|^2$ this term is the source of intensity of such vibronic transitions.

Examples of vibronic transitions involving non-totally symmetric vibrations are in the $\tilde{A}^1B_2 - \tilde{X}^1A_1$ system of chlorobenzene, a C_{2v} molecule. One b_2 vibration ν_{29}, with a wavenumber of 615 cm^{-1} in the $\tilde{X}$ state and 523 cm^{-1} in the $\tilde{A}$ state, is active in 29^1_0 and 29^0_1 bands similar to the case shown in Figure 7.43. There are ten b_2 vibrations in chlorobenzene but the others are much less strongly active. The reason is that $(\partial R_e/\partial Q_{29})_{eq}$ is much greater than the corresponding terms for all the other b_2 vibrations.

The $\tilde{A}^1B_2 - \tilde{X}^1A_1$ system of chlorobenzene is electronically allowed, since $B_2 = \Gamma(T_z)$, which satisfies equation (7.122). The 0^0_0 band, and progressions in totally symmetric vibrations built on it, obtain their intensity in the usual way, through the first term on the right-hand side of equation (7.131).

The $\tilde{A}^1B_{2u} - \tilde{X}^1A_{1g}$ system of benzene (see Section 7.3.1.3), shown in Figure 7.42, is a particularly interesting one. It is electronically forbidden, since B_{2u} is not the symmetry species of a translation (see Table A.36 in the Appendix); therefore $(R_e)_{eq}$ is zero and the 0^0_0 band is not observed. All the intensity derives from the second term on the right-hand side of equation (7.131) and most of it through the e_{2g} vibration ν_6. Since

$$B_{2u} \times e_{2g} = E_{1u} = \Gamma(T_x, T_y) \qquad (7.132)$$

the 6^1_0 vibronic transition is allowed and is shown in Figure 7.42. The large value of $(\partial R_e/\partial Q_6)_{eq}$ is responsible for the appreciable intensity of this band whereas the other three e_{2g} vibrations, ν_7, ν_8, and ν_9, are much less strongly active. The very weak 7^1_0 band is shown in Figure 7.42 which shows also the 6^0_1 band, weakened by the Boltzmann factor, and the position where the forbidden 0^0_0 band would be.

The $\tilde{A}^1A_2 - \tilde{X}^1A_1$, π^*n system of formaldehyde (see Section 7.3.1.2) is also electronically forbidden since A_2 is not a symmetry species of a translation (see Table A.11 in the Appendix). The main non-totally symmetric vibration which is active is ν_4, the b_1 out-of-plane bending vibration (see question 3 in Chapter 4) in 4^1_0 and 4^0_1 transitions.

There are two further points of interest about this system of formaldehyde. Firstly, as mentioned in Section 7.3.1.2, the molecule is pyramidal in the $\tilde{A}$ state so that the potential function for ν_4 is anharmonic in that state. It is W-shaped, rather like that for ν_2 in the ground electronic state of ammonia in Figure 6.38, and the change of geometry from the $\tilde{X}$ to the $\tilde{A}$ state leads to a progression in ν_4 in accordance with the Franck–Condon principle. The second point of interest is that the 0^0_0 band is observed very weakly. The reason for this is that, although the transition is forbidden by electric dipole selection rules, it is allowed by magnetic dipole selection rules. A transition is magnetic dipole allowed if the excited state symmetry species is that of a *rotation* of the molecule: in this case $A_2 = \Gamma(R_z)$, as shown in Table A.11 in the Appendix.

All the forbidden electronic transitions of regular octahedral transition metal complexes, mentioned in Section 7.3.1.4, are induced by non-totally symmetric vibrations.

Although we have considered cases where $(\partial \mathbf{R}_e/\partial Q_i)_{eq}$ in equation (7.131) may be quite large for a non-totally symmetric vibration, a few cases are known where $(\partial \mathbf{R}_e/\partial Q_i)_{eq}$ is appreciable for totally symmetric vibrations. In such cases the second term on the right-hand side of equation (7.131) provides an additional source of intensity for X_0^1 or X_1^0 vibronic transitions when ν_X is totally symmetric.

7.3.5 Rotational Fine Structure

As is the case for diatomic molecules, rotational fine structure of electronic spectra of polyatomic molecules is very similar, in principle, to that of their infrared vibrational spectra. For linear, symmetric rotor, spherical rotor, and asymmetric rotor molecules the selection rules are the same as those discussed in Sections 6.2.3.1 to 6.2.3.4. The major difference, in practice, is that, as for diatomics, there is likely to be a much larger change of geometry, and therefore of rotational constants, from one electronic state to another than from one vibrational state to another.

From the ground to an excited electronic state the electron promotion involved is likely to be to a less strongly bonding orbital, leading to an increase in molecular size and a decrease in rotational constants. The effect on the rotational fine structure is to degrade it to low wavenumber to give a strongly asymmetrical structure, unlike the symmetrical structure typical of vibrational transitions.

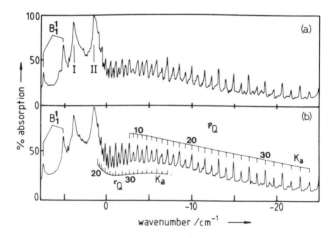

Figure 7.44 (a) Observed and (b) best computed rotational contour of the type B 0_0^0 band of the $\tilde{A}^1 B_{2u} - \tilde{X}^1 A_g$ system of 1,4-difluorobenzene, with a weak overlapping sequence band labelled B_1^1. (Reproduced, with permission, from Cvitaš, T., and Hollas, J. M., *Mol. Phys.*, **18**, 793, 1970.)

Examples of this degradation of bands are shown in Figures 7.44 and 7.45. Figure 7.44(a) shows the rotational fine structure of the 0^0_0 band of the $\tilde{A}^1B_{2u} - \tilde{X}^1A_g$ system of 1,4-difluorobenzene, belonging to the D_{2h} point group. The fine structure is in the form of a contour of tens of thousands of unresolved rotational transitions which, nevertheless, shows well-defined features (B^1_1 is an overlapping weaker band of a similar type). Since $B_{2u} = \Gamma(T_y)$, as given by Table A.32 in the Appendix, the electronic transition is allowed and is polarized along the y axis (in-plane, perpendicular to the F–C---C–F line). 1,4-Difluorobenzene is a prolate asymmetric rotor and, because the y axis is the b inertial axis, type B rotational selection rules apply. In Figure 7.44(b) is a computer simulation of the observed bands with the best set of trial rotational constants for the $\tilde{A}$ state. Type A and type C bands have not been observed in the spectrum but rotational contours, computed with the same rotational constants as for Figure 7.44(b) but with type A or type C selection rules, are shown in Figure 7.45.

Typically the type A, B, and C bands in Figures 7.44 and 7.45 are all strongly degraded to low wavenumber and are easily distinguishable from each other. However, it is typical of electronic spectra that type A, B, and C bands are, in general, quite different in appearance for different molecules because of the sensitivity of the contours to wide differences in changes of rotational constants from the lower to the upper state. It follows that, although we can use the concept of type A, B or C vibrational bands having a typical rotational contour,

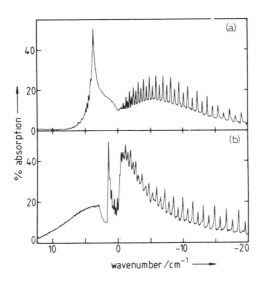

Figure 7.45 Computed (a) type A and (b) type C rotational contours for 1,4-difluorobenzene using the same rotational constants as for Figure 7.44(b). (Reproduced, with permission, from Cvitaš, T., and Hollas, J. M., *Mol. Phys.*, **18**, 793, 1970.)

as in ethylene (Figures 6.28 to 6.30) or perdeuteronaphthalene (Figure 6.32), it is not possible to do so for electronic or vibronic bands.

Similarly, parallel or perpendicular electronic or vibronic bands of a symmetric rotor (see Section 6.2.3.2) show widely varying shapes, unlike the parallel or perpendicular vibrational bands illustrated in Figure 6.26 and 6.27 respectively.

7.3.6 Diffuse Spectra

By comparison with diatomics, polyatomic molecules are more likely to show diffuseness in the rotational, and even the vibrational, structure of electronic transitions. There is also a tendency to increased diffuseness with increasing vibrational energy in an excited electronic state. These observations indicate that we must look for a different explanation in polyatomic molecules in which, for example, repulsive states, such as those described for diatomics in Section 7.2.5.6, are very unlikely.

One of the simplest reasons for this tendency to diffuseness in electronic spectra of large molecules is that of spectral congestion—too many transitions crowded into a small wavenumber range. In anthracene, $C_{14}H_{10}$, it has been estimated that there are, on average, about a hundred rotational transitions, accompanying the $\tilde{A}^1 B_{1u} - \tilde{X}^1 A_g$ electronic transition, within the rotational (Doppler-limited) line width. In addition there is congestion of vibronic transitions, particularly of sequence bands, due to the large number of vibrations—anthracene, for example, has sixty-six. The result is that the $\tilde{A} - \tilde{X}$ absorption spectrum appears as a quasi-continuum, even in the gas phase at low pressure.

The line width $\Delta\tilde{\nu}$ of a rotational transition accompanying an electronic or vibronic transition is related to the lifetime τ of the excited state and the first-order rate constant k for decay by

$$\Delta\tilde{\nu} = \frac{1}{2\pi c \tau} = \frac{k}{2\pi c} \tag{7.133}$$

derived from equations (2.25) and (2.23). We consider an absorption process from the ground (singlet) state S_0 of a molecule to the lowest excited singlet state S_1, shown in Figure 7.46. When a molecule arrives in S_1 it may decay by fluorescence to the ground state in a radiative transition or it may decay by various non-radiative processes. Then the first-order rate constant contains a radiative (k_r) and a non-radiative (k_{nr}) contribution so that

$$\Delta\tilde{\nu} = \frac{1}{2\pi c}\left(\frac{1}{\tau_r} + \frac{1}{\tau_{nr}}\right) = \frac{k_r + k_{nr}}{2\pi c} \tag{7.134}$$

In a large molecule the vibrational and rotational levels associated with any electronic state become so extremely congested at high vibrational energies that

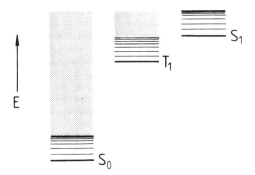

Figure 7.46 States S_0, S_1, and T_1 of a polyatomic molecule showing regions of low density of vibrational states and, at higher energy, pseudo-continua.

they form a pseudo-continuum. This is illustrated for S_0, S_1, and the lowest excited triplet state T_1, lying below S_1, in Figure 7.46. It is apparent that even the zero-point level of S_1 may be degenerate with pseudo-continua associated with S_0 and T_1, thereby increasing the efficiency of crossing from S_1 to S_0, known as internal conversion, and from S_1 to T_1, known as inter-system crossing. Either of these non-radiative processes decreases the quantum yield Φ_F for fluorescence from S_1 where

$$\Phi_F = \frac{\text{number of molecules fluorescing}}{\text{number of quanta absorbed}} \qquad (7.135)$$

For some $S_1 - S_0$ transitions, Φ_F is so low that fluorescence has not been detected.

The fluorescence lifetime τ_F can be measured directly and is the lifetime of the S_1 state, taking into account all decay processes. It is related to k_r and k_{nr} by

$$\tau_F = \frac{1}{k_r + k_{nr}} \qquad (7.136)$$

Since Φ_F is related to k_r and k_{nr} by

$$\Phi_F = \frac{k_r}{k_r + k_{nr}} \qquad (7.137)$$

measurement of τ_F and Φ_F can be translated into values of k_r and k_{nr}. Table 7.11 gives some values obtained for fluorescence from the zero-point level (0^0) and the 6^1 and $1^1 6^1$ vibronic levels in the S_1 ($\tilde{A}^1 B_{2u}$) state of benzene by observing fluorescence, at low pressure to avoid collisions, following selective population of these levels: the technique is referred to as single vibronic level fluorescence (see Section 9.3.7). The results show that the rate of non-radiative relaxation of S_1 is greater than that for radiative relaxation for all three levels

Table 7.11 Fluorescence quantum yield Φ_F, fluorescence lifetime τ_F, radiative k_r, and non-radiative k_{nr} rate constants for the S_1 state of benzene.

Vibronic level	Φ_F	τ_F/ns	k_r/s^{-1}	k_{nr}/s^{-1}
0^0	0.22	90	2.4×10^{-6}	8.7×10^{-6}
6^1	0.27	80	3.4×10^{-6}	9.1×10^{-6}
$1^1 6^1$	0.25	68	3.7×10^{-6}	11.0×10^{-6}

and that there is some vibrational dependence. This dependence is reflected also in the values for τ_F which decrease with increasing vibrational energy due to increasing density of the degenerate pseudo-continuum. S_1 to T_1 relaxation has been shown to be the dominant mechanism.

In the $S_1 - S_0$ system of benzene there is a dramatic fall of Φ_F to almost zero when the energy of the vibronic levels populated prior to emission is in the range 2800 to 3300 cm^{-1} (33.5 to 39.5 kJ mol^{-1}) above the zero-point level of S_1. This is due to a third channel for decay opening up (the other two being to T_1 or S_0) but the nature of this decay is the subject of speculation.

Questions

1. Calculate the wavelength of the first member of each of the series in the spectrum of atomic hydrogen with $n'' = 90$ and 166 and state in which region of the electromagnetic spectrum they occur.

2. From the ground configuration of titanium derive the ground state saying which rules enable you to do this. Derive the states arising from the $KL3s^2 3p^6 4s^2 3d^1 4f^1$ excited configuration and discuss the transitions which may occur between states arising from the two configurations.

3. Indicate which of the following electronic transitions are forbidden in a diatomic molecule stating which selection rules result in the forbidden character:

$$^1\Pi_g - {}^1\Pi_u, \ {}^1\Delta_u - {}^1\Sigma_g^+, \ {}^3\Phi_g - {}^1\Pi_g, \ {}^4\Sigma_g^+ - {}^2\Sigma_u^+.$$

4. How would the components of a $^3\Delta_g$ electronic state be described in the case (c) coupling approximation?

5. Derive the states which arise from the following electron configurations:

$$\begin{aligned}
C_2 \quad &\ldots \quad (\sigma_u^* 2s)^2 \, (\pi_u 2p)^3 \, (\sigma_g 2p)^1 \\
NO \quad &\ldots \quad (\sigma 2p)^1 \, (\pi 2p)^4 \, (\pi^* 2p)^2 \\
CO \quad &\ldots \quad (\sigma^* 2s)^1 \, (\pi 2p)^4 \, (\sigma 2p)^2 \, (\pi^* 2p)^1 \\
B_2 \quad &\ldots \quad (\sigma_u^* 2s)^2 \, (\pi_u 2p)^1 \, (\pi_g^* 2p)^1
\end{aligned}$$

What is the ground state of B_2?

6. From the following separations of vibrational levels in the $A^1\Pi$ excited electronic state of CO obtain values of ω_e and $\omega_e x_e$ and also for the dissociation energy D_e:

$(v+1) - v$	1–0	2–1	3–2	4–3	5–4	6–5
$[G(v+1) - G(v)]/\text{cm}^{-1}$	1484	1444	1411	1377	1342	1304

7. Calculate the ratio of molecules in the $v = 1$ compared to the $v = 0$ level of the ground electronic state at 293 K for H_2 ($\omega = 4401 \text{ cm}^{-1}$), F_2 ($\omega = 917 \text{ cm}^{-1}$), and I_2 ($\omega = 215 \text{ cm}^{-1}$). At what temperature, for each molecule, would the ratio be 0.500?

8. Sketch potential energy curves for the following states of CdH, Br_2, and CH, given their internuclear distances r_e, and suggest qualitative intensity distributions in the $v'' = 0$ progressions for transitions between the states observed in absorption:

Molecule	State	$r_e/\text{Å}$
CdH	$X^2\Sigma^+$	1.781
	$A^2\Pi$	1.669
Br_2	$X^1\Sigma_g^+$	2.281
	$B^3\Pi_{0_u^+}$	2.678
CH	$X^2\Pi$	1.120
	$C^2\Sigma^+$	1.114

9. Measure, approximately, the wavenumbers of the P- and R-branch rotational transitions in the 0–0 band of the $A^1\Sigma^+ - X^1\Sigma^+$ electronic transition of CuH in Figure 7.26 and hence obtain values for r_0 in the A and X states, neglecting centrifugal distortion.

10. Discuss, briefly, the valence molecular orbitals of AlH_2 and the shape of the molecule in the ground and first excited singlet states.

11. For formaldehyde give the lowest MO configuration resulting from a $\pi^*-\pi$ promotion and deduce the resulting states.

12. Show how the determinant equation (7.107) gives the results in equation (7.109).

13. Write down the crystal field orbital configurations of the following transition metal complexes: $[Cu(H_2O)_6]^{2+}$, $[V(H_2O)_6]^{3+}$, $[Mn(H_2O)_6]^{2+}$–high spin, $[Co(NH_3)_6]^{3+}$–low spin. Explain why $[Mn(H_2O)_6]^{2+}$ is colourless.

14. A $^1B_{3g} - {}^1A_g$ electronic transition in a molecule belonging to the D_{2h} point group is forbidden. What are the possible symmetry species of a vibration X which would result in the X_0^1 and X_1^0 transitions being allowed?

15. Show that the following electronic transitions:

(a) $^1E - {}^1A_1$ in methyl fluoride;

(b) $^1A_2'' - {}^1A_1'$ in 1,3,5-trichlorobenzene;

(c) $^1B_2 - {}^1A_1$ in allene ($CH_2 = C = CH_2$);

are allowed, determine the direction of the transition moment and state the rotational selection rules which apply.

Bibliography

Candler, C. (1964). *Atomic Spectra*, Hilger and Watts, London.

Condon, E. V., and Shortley, G. H. (1953). *The Theory of Atomic Spectra*, Cambridge University Press, London.

Coulson, C. A. (1961). *Valence*, Oxford University Press, Oxford.

Coulson, C. A., and McWeeney, R. (1979). *Coulson's Valence*, Oxford University Press, Oxford.

Herzberg, G. (1944). *Atomic Spectra and Atomic Structure*, Dover, New York.

Herzberg, G. (1950). *Spectra of Diatomic Molecules*, Van Nostrand, New York.

Herzberg, G. (1966). *Electronic Spectra of Polyatomic Molecules*, Van Nostrand, New York.

Huber, K. P., and Herzberg, G. (1979). *Constants of Diatomic Molecules*, Van Nostrand Reinhold, New York.

Kettle, S. F. A. (1985). *Symmetry and Structure*, Wiley, London.

King, G. W. (1964). *Spectroscopy and Molecular Structure*, Holt, Rinehart and Winston, New York.

Kuhn, H. G. (1969). *Atomic Spectra*, Longman, London.

Murrell, J. N., Kettle, S. F. A., and Tedder, J. M. (1965). *Valence Theory*, Wiley, London.

Murrell, J. N. , Kettle, S. F. A., and Tedder, J. M. (1978). *The Chemical Bond*, Wiley, London.

Rosen, B. (Ed.) (1970). *Spectroscopic Data Relative to Diatomic Molecules*, Pergamon, Oxford.

Steinfeld, J. I. (1974). *Molecules and Radiation*, Harper and Row, New York.

Walsh, A. D. (1953). *J. Chem. Soc.*, **1953**, 2260–2317.

CHAPTER 8 ———————————————————————————

Photoelectron and Related Spectroscopies

8.1 Photoelectron Spectroscopy

Photoelectron spectroscopy involves the ejection of electrons from atoms or molecules following bombardment by monochromatic photons. The ejected electrons are called photoelectrons and were mentioned, in the context of the photoelectric effect, in Section 1.2. The effect was observed originally on surfaces of easily ionizable metals, such as the alkali metals. Bombardment of the surface with photons of tunable frequency does not produce any photoelectrons until the threshold frequency is reached (see Figure 1.2). At this frequency, v_t, the photon energy is just sufficient to overcome the work function Φ of the metal, so that

$$hv_t = \Phi \tag{8.1}$$

At higher frequencies the excess energy of the photons is converted into kinetic energy of the photoelectrons

$$hv = \Phi + \tfrac{1}{2}m_e v^2 \tag{8.2}$$

where m_e and v are their mass and velocity.

Work functions of alkali metal surfaces are only a few electronvolts[†] so that the energy of near ultraviolet radiation is sufficient to produce ionization.

Photoelectron spectroscopy is a simple extension of the photoelectric effect involving the use of higher energy incident photons and applied to the study not only of solid surfaces but also of samples in the gas phase. Equations (8.1) and (8.2) still apply but, for gas phase measurements in particular, the work function is usually replaced by the ionization energy I[‡], so that equation (8.2) becomes

$$hv = I + \tfrac{1}{2}m_e v^2 \tag{8.3}$$

———————————————————————————

[†] 1 eV = 96.485 kJ mol^{-1} = 8065.54 cm^{-1}.
[‡] This is often referred to as the ionization potential but, since equation (8.3) shows that I has dimensions of energy, the term ionization energy is to be preferred.

273

Even though Einstein developed the theory of the photoelectric effect in 1906 photoelectron spectroscopy, as we now know it, was not developed until the early 1960s, particularly by Siegbahn, Turner, and Price.

For an atom or molecule in the gas phase Figure 8.1 schematically divides the orbitals (AOs or MOs) into core orbitals and valence orbitals. Each orbital is taken to be non-degenerate and can accommodate two electrons with antiparallel spins. The orbital energy, always negative, is measured relative to a zero of energy corresponding to removal of an electron in that orbital to infinity. The valence, or outer shell, electrons have higher orbital energies than the core, or inner shell, electrons. A monochromatic source of soft (low energy) X-rays may be used to remove core electrons and the technique is often referred to as X-ray photoelectron spectroscopy, sometimes, as here, abbreviated to XPS. On the other hand, far-ultraviolet radiation has sufficient energy to remove only valence electrons and such a source is used in ultraviolet photoelectron spectroscopy, or UPS.

Although the division into XPS and UPS is conceptually artificial it is often a practically useful one because of the different experimental techniques used.

Acronyms abound in photoelectron and related spectroscopies but we shall use only XPS, UPS, and, in Sections 8.2 and 8.3, AES, XRF and EXAFS. In addition, ESCA is worth mentioning, briefly. It stands for 'electron spectroscopy for chemical analysis' in which electron spectroscopy refers to the various branches of spectroscopy which involve the ejection of an electron from an atom or molecule. However, because ESCA was an acronym introduced by workers in the field of XPS it is most often used to refer to XPS rather than to electron spectroscopy in general.

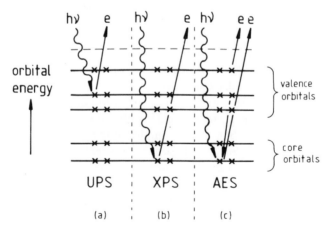

Figure 8.1 Processes occurring in (a) ultraviolet photoelectron spectroscopy (UPS), (b) X-ray photoelectron spectroscopy (XPS), (c) Auger electron spectroscopy (AES).

Figures 8.1(a) and (b) illustrates the processes involved in UPS and XPS. Both result in the ejection of a photoelectron following interaction of the atom or molecule M which is ionized to produce the singly charged M^+.

$$M + h\nu \rightarrow M^+ + e \tag{8.4}$$

8.1.1 Experimental Methods

Figure 8.2 illustrates the important components of an ultraviolet or X-ray photoelectron spectrometer. When the sample in the target chamber is bombarded with photons, photoelectrons are ejected in all directions. Some pass through the exit slit and into the electron energy analyser which separates the electrons according to their kinetic energy, rather in the way that ions are separated in a mass spectrometer. The electrons pass through the exit slit of the analyser onto an electron detector and the spectrum recorded is the number of electrons per unit time (often counts s^{-1}) as a function of *either* ionization energy *or* kinetic energy of the photoelectrons (care is sometimes needed to deduce which energy is plotted: they increase in opposite directions for a particular spectrum).

8.1.1.1 Sources of monochromatic ionizing radiation

For UPS a source providing at least 20 eV, and preferably up to 50 eV, of energy is required since the lowest ionization energy of an atom or molecule is typically 10 eV and it is desirable to detect higher energy ionization processes as well. Such sources are mostly produced by a discharge in He or Ne gas resulting in emission of far-ultraviolet radiation from the atom or positive ion. Of these the most often used is the helium discharge which produces, predominantly, 21.22 eV radiation due to the 2^1P_1 ($1s^12p^1$) $-$ 1^1S_0 ($1s^2$) transition of the atom at 58.4 nm (see Figure 7.9): this is referred to as He Iα, or simply He I, radiation.

It is possible to change the conditions in the helium discharge lamp so that

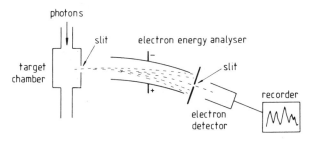

Figure 8.2   The principal components of a photoelectron spectrometer.

the helium is ionized predominantly to He$^+$ (He II). The radiation is due mainly to the $n = 2 - n = 1$ transition of He II (analogous to the first member of the Lyman series of the hydrogen atom in Figure 1.1) at 30.4 nm with an energy of 40.81 eV. A thin aluminium foil filter can be used to remove any He I radiation.

When neon is used in a discharge lamp radiation is produced predominantly with *two* close wavelengths, 74.4 and 73.6 nm corresponding to energies of 16.67 and 16.85 eV, making this source rather less useful than the more truly monochromatic, and more highly energetic, He I source.

Commonly used sources of X-ray radiation are Mg$K\alpha$ and Al$K\alpha$, where $K\alpha$ indicates that an electron has been ejected, by electrons falling on a Mg or Al surface, from the K ($n = 1$) shell and the radiation is due to the energy emitted when an electron falls back from the next highest energy shell, here the L ($n = 2$) shell, to fill the vacancy. The Mg$K\alpha$ radiation consists primarily of a 1253.7 and 1253.4 eV doublet while the Al$K\alpha$ radiation also consists primarily of a doublet, with energies of 1486.7 and 1486.3 eV. In addition there is a weak, continuous background, known as *bremsstrahlung*, and also several satellite lines accompanying both doublets. The *bremsstrahlung* and the satellites may be removed with a monochromator illustrated in Figure 8.3. A quartz crystal Q is bent to form a concave X-ray diffraction grating. The X-ray source and the target chamber T of the photoelectron spectrometer are then placed in positions on the Rowland circle. This is a circle whose diameter is equal to the radius of curvature of the grating and has the property that radiation from a source lying on the circle is diffracted by the grating and refocused at some other point on the circle.

A monochromator is useful not only for removing unwanted lines from the X-ray source but also for narrowing the otherwise broad lines. For example, each of the Mg$K\alpha$ and Al$K\alpha$ doublets is unresolved and about 1 eV wide at half-intensity. A monochromator can reduce this to about 0.2 eV. This reduction

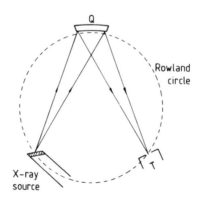

Figure 8.3 An X-ray monochromator using a bent quartz crystal Q.

of the line width is very important because in an XPS spectrum, unlike a UPS spectrum, the resolution is limited by the line width of the ionizing radiation. Unfortunately, even after line narrowing to 0.2 eV (1600 cm^{-1}), the line width is excessive compared to that of a helium or neon ultraviolet source. Because of this, although valence electrons as well as core electrons are removed by X-ray radiation, much higher resolution is achieved when the valence shells are investigated by UPS.

An important source of both far-ultraviolet and X-ray radiation is the storage ring or synchrotron radiation source (SRS). The one illustrated in Figure 8.4 is at Daresbury (England). An electron beam is generated in a small linear accelerator. Pulses of these electrons are injected tangentially into a booster synchroton where they are accelerated by 500 MHz radiation which also causes them to form bunches with a time interval of 2.0 ns between bunches. The electrons are restricted to a circular orbit by dipole bending magnets. When the electrons have achieved an energy of 600 MeV they are taken off tangentially into the storage ring where they are further accelerated to 2 GeV. Subsequently only a small amount of power is necessary to keep the electrons circulating.

The stored orbiting electrons lose energy continuously, with a half-life of about 8 h, in the form of electromagnetic radiation which is predominantly plane polarized and emerges tangentially in pulses of length 0.17 ns and spacing 2.0 ns. Several radiation ports are arranged round the ring to serve as experimental stations. The radiation is continuous from the far-infrared to the X-ray region but the gain in intensity, compared to more conventional sources, is

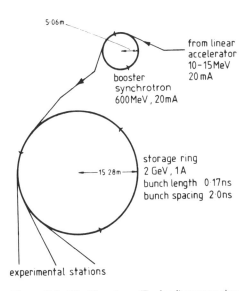

Figure 8.4 The Daresbury (England) storage ring.

most marked in the far-ultraviolet and, particularly, the X-ray region where the gain is a factor of 10^5 to 10^6. A monochromator must be used to select a narrow band of wavelengths.

8.1.1.2 Electron velocity analysers

Measurement of the kinetic energy of the photoelectrons (equation 8.3) involves measurement of their velocity. This has been achieved by various methods and Figure 8.5 shows four of them.

In the slotted grid analyser, in Figure 8.5(a), the photoelectrons are generated along the axis of a cylindrical electron collector. A retarding potential, applied to the cylindrical retarding grid, is smoothly changed and, when it corresponds to the energy of some of the emitted photoelectrons, there is a decrease in the electron current I detected. This current shows a series of steps when plotted against retarding potential V, each step corresponding to an ionization energy of the sample. A simple differentiating circuit converts the steps into peaks in the photoelectron spectrum which is a plot of dI/dV versus V. The purpose of the slotted grid is to restrict the photoelectrons travelling towards the retarding grid to those emitted normal to the axis of the cylinder.

The spherical grid analyser, in Figure 8.5(b), has the advantage of collecting all the photoelectrons generated at the centre, in whatever direction they are travelling. There are two spherical retarding grids in this design.

The analysers in Figure 8.5(a) and (b) are useful mainly for gaseous samples. Those in Figure 8.5(c) and (d) may be used for solid samples also. In the 127° cylindrical analyser in Figure 8.5(c) some of the generated photoelectrons enter the slit and traverse a path between two plates which are 127.28° sections of concentric cylinders. A variable electrostatic field is applied across the plates and the electrons emerge from the exit slit only if the field is such that they follow the circular path shown. The electrons emerging are counted as the field is smoothly varied.

The hemispherical analyser in Figure 8.5(d) works on a similar principle but has the advantage of collecting more photoelectrons. An analyser consisting of two concentric plates which are parts of hemispheres, so-called spherical sector plates, is often used in a spectrometer which operates for both UPS and XPS.

8.1.1.3 Electron detectors

The detector may be a simple electrometer when using a cylindrical or spherical grid analyser. With the other types, fewer electrons are being collected and an electron multiplier, having much greater sensitivity, is necessary. This consists of a number of dynodes, each of which produces more electrons than it receives. For a measurable current, about ten to twenty dynodes are required. Alternatively, a multichannel electron multiplier in the focal plane of the analyser can be used to collect simultaneously electrons with a range of energies.

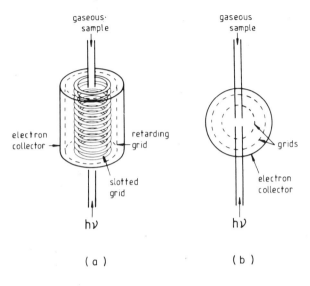

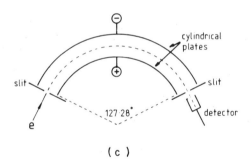

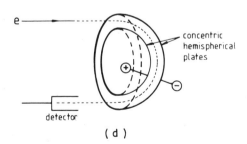

Figure 8.5 (a) Slotted grid, (b) spherical grid, (c) 127° cylindrical, and (d) hemispherical analysers.

8.1.1.4   Resolution

As already mentioned, the resolution in XPS is usually limited by the line width of the ionizing radiation. In UPS the obtainable resolution depends on such factors as the efficiency of shielding of the spectrometer from stray (including the earth's) magnetic fields and the cleanliness of the analyser surfaces. Resolution decreases when the kinetic energy of the photoelectrons is below about 5 eV. The highest resolution obtained in UPS is about 4 meV (32 cm^{-1}), which means that the technique is basically a low resolution one compared to the spectroscopic methods discussed previously. For this reason rotational fine structure accompanying ionization processes in UPS of gases is very difficult to observe. Only for a few molecules with small moments of inertia, such as H_2 and H_2O, has this been achieved and we shall not discuss it further.

The resolution of 0.2 eV (1600 cm^{-1}) of which XPS is typically capable is too low, in most cases, for even the vibrational coarse structure accompanying ionization to be observed.

8.1.2   Ionization Processes and Koopmans' Theorem

The photoionization process with which we shall be concerned in both UPS and XPS is that in equation (8.4) in which only the singly charged M^+ is produced. The selection rule for such a process is trivial—*all* ionizations are allowed.

When M is an atom the total change in angular momentum for the process $M + h\nu \rightarrow M^+ + e$ must obey the electric dipole selection rule $\Delta \ell = \pm 1$ (see equation 7.21), but the photoelectron can take away *any* amount of momentum. If, for example, the electron removed is from a d orbital ($\ell = 2$) of M it carries away one or three quanta of angular momentum depending on whether $\Delta \ell = -1$ or $+1$ respectively. The wave function of a free electron can be described, in general, as a mixture of $s, p, d, f, \ldots$ wave functions but, in this case, the ejected electron has just p and f character.

In molecules, also, there is no restriction on the removal of an electron. The main difference from atoms is that, since the symmetry is lower, the MOs themselves are mixtures of $s, p, d, f, \ldots$ AOs and the ejected electron is described by a more complex mixture of $s, p, d, f, \ldots$ character.

From Figure 8.1(a) and (b) it would appear that the energy required to eject an electron from an orbital (atomic or molecular) is a direct measure of the orbital energy. This is approximately true and was originally proposed by Koopmans whose theorem can be stated as follows: 'For a closed shell molecule the ionization energy of an electron in a particular orbital is approximately equal to the negative of the orbital energy calculated by a self-consistent field (SCF—see Section 7.1.1) method', or, for orbital i,

$$I_i \simeq -\varepsilon_i^{SCF} \tag{8.5}$$

The negative sign is due to the convention that orbital energies ε_i are negative.

At the level of simple valence theory Koopmans' theorem seems to be so self-evident as to be scarcely worth stating. However, with more accurate theory, this is no longer so and great interest attaches to why equation (8.5) is only approximately true.

The measured ionization energy I_i is the difference in energy between M and M^+, but the approximation arises in equating this with the orbital energy. Orbitals are an entirely theoretical concept and their energies can be obtained exactly *only* by calculation and then, for a many-electron system, only with difficulty. Experimental ionization energies are the measurable quantities which correspond most closely to the orbital energies.

The three most important factors which may contribute to the approximation in equation (8.5) relate to the main deficiencies in SCF calculations and are:

1. Electron reorganization. The orbitals in M^+ are not quite the same as in M because there is one electron fewer. The electrons in M^+ are, therefore, in orbitals which are slightly reorganized relative to those calculated for M, that is

$$\varepsilon_i^{SCF}(M^+) \neq \varepsilon_i^{SCF}(M) \tag{8.6}$$

2. Electron correlation. Electrons in an atom or molecule do not move entirely independently of each other but their movements are correlated. The associated correlation energy is often neglected in SCF calculations.
3. Relativistic effects. The neglect of the effects of relativity on orbital energies is another common defect of SCF calculations and is particularly important for core orbital energies which can be very large.

In most cases of closed-shell molecules Koopmans' theorem is a reasonable approximation but N_2 (see Section 8.1.3.2b) is a notable exception. For open-shell molecules, such as O_2 and NO, the theorem does not apply.

8.1.3 Photoelectron Spectra and their Interpretation

The simplest, and perhaps the most important, information derived from photoelectron spectra is the ionization energies for valence and core electrons. Before the development of photoelectron spectroscopy very few of these were known, especially for polyatomic molecules. For core electrons ionization energies were previously unobtainable and illustrate the extent to which core orbitals differ from the pure atomic orbitals pictured in simple valence theory.

Because of the general validity of Koopmans' theorem for closed-shell molecules ionization energies and, as we shall see, the associated vibrational structure represent a vivid illustration of the validity of quite simple-minded MO theory of valence electrons.

8.1.3.1 Ultraviolet photoelectron spectra of atoms

In the case of atoms UPS is unlikely to produce information which is not available from other sources. In addition many materials have such low vapour pressures that their UPS spectra may be recorded only at high temperatures. The noble gases, mercury, and, to some extent, the alkali metals are exceptions but we will consider here only the spectrum of argon.

The He I (21.22 eV) spectrum of argon is shown in Figure 8.6. The two peaks both result from the removal of an electron from the $3p$ orbital in the process:

$$Ar(KL3s^2 3p^6) \rightarrow Ar^+(KL3s^2 3p^5) \tag{8.7}$$

From the discussion in Section 7.1.2.3 we can deduce that the state arising from the ground configuration of Ar is $^1S_0$. However, the ground configuration of Ar^+ gives rise to two states, $^2P_{1/2}$ and $^2P_{3/2}$, since $L = 1$ and $S = \frac{1}{2}$, giving a ^2P term, and $J = \frac{1}{2}$ or $\frac{3}{2}$. These two states of Ar^+ are split by spin–orbit coupling. Since the unfilled $3p$ orbital is more than half-filled the multiplet is inverted, the $^2P_{3/2}$ state being below $^2P_{1/2}$ (rule 4 on page 198). The splitting

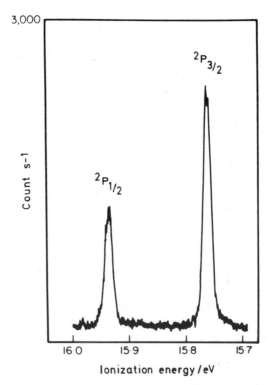

Figure 8.6 The He I UPS spectrum of argon. (Reproduced from Turner, D. W., Baker, C., Baker, A. D. and Brundle, C. R., *Molecular Photoelectron Spectroscopy*, p. 41, Wiley, London, 1970.)

Table 8.1 Lowest ionization energies of some noble gases.

	I/eV	
	$^2P_{3/2}$	$^2P_{1/2}$
Ar	15.759	15.937
Kr	14.000	14.665
Xe	12.130	13.436

by 0.178 eV of the two peaks in Figure 8.6 reflects the spin–orbit coupling. The ionization process $\text{Ar}^+(^2P_{3/2}) - \text{Ar}(^1S_0)$ is approximately twice as intense as $\text{Ar}^+(^2P_{1/2}) - \text{Ar}(^1S_0)$. This is due to the four-fold degeneracy of the $^2P_{3/2}$ state, resulting from $M_J = \frac{3}{2}, \frac{1}{2}, -\frac{1}{2}, -\frac{3}{2}$, compared to the two-fold degeneracy of the $^2P_{1/2}$ state, for which $M_J = +\frac{1}{2}, -\frac{1}{2}$.

The He I UPS spectra of Kr and Xe appear similar to that of Ar but the ionization energy decreases and the spin–orbit coupling increases with increasing atomic number, as illustrated by the data in Table 8.1.

8.1.3.2 Ultraviolet photoelectron spectra of molecules

8.1.3.2(a) Hydrogen The MO configuration of H_2 is $(\sigma_g 1s)^2$, as shown in Figure 7.14. The He I UPS spectrum in Figure 8.7 is caused by removing an electron from this MO to give the $X^2\Sigma_g^+$ ground state of H_2^+.

An obvious feature of the spectrum is that it does not consist of a single peak corresponding to the ionization but a vibronic band system, very similar to what we might observe for an electronic transition in a diatomic molecule (see Section 7.2.5.2). The Franck–Condon principle (Section 7.2.5.3) applies also to an ionization process so that the most probable ionization is to a vibronic state of the ion in which the nuclear positions and velocities are the same as in the molecule. This is illustrated for the process $H_2^+(X^2\Sigma_g^+) - H_2(X^1\Sigma_g^+)$ in Figure 8.8 which is similar to Figure 7.21 for an electronic transition. Removal of an electron from the $\sigma_g 1s$ bonding MO results in an increase in the equilibrium internuclear distance r_e in the ion and the most probable transition is to $v' = 2$. Figure 8.7 shows that there is a long $v'' = 0$ vibrational progression which just reaches the dissociation limit for H_2^+ at $v' \simeq 18$.

From the separations of the progression members the vibrational constants $\omega_e, \omega_e x_e, \omega_e y_e, \ldots$ (equation 6.16) for H_2^+ can be obtained. Comparison of $\omega_e = 2322$ cm^{-1} for the $X^2\Sigma_g^+$ state of H_2^+ with $\omega_e = 3115$ cm^{-1} for the $X^1\Sigma_g^+$ state of H_2 shows that the vibrational force constant (see equation (6.2)) is much smaller in the ground state of H_2^+ than in H_2. This, in turn, implies a weaker bond in H_2^+ consistent with the removal of a bonding electron.

Figures 8.7 and 8.8 illustrate the point that there are two ways in which we can define the ionization energy. One is the adiabatic ionization energy which

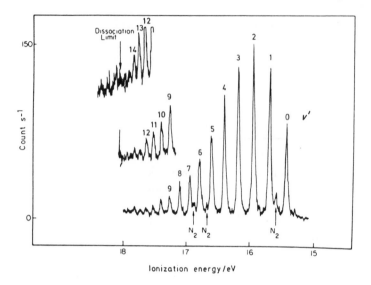

Figure 8.7 The He I UPS spectrum of H_2. (Reproduced from Turner, D. W., Baker, C., Baker, A. D., and Brundle, C. R., *Molecular Photoelectron Spectroscopy*, p. 44, Wiley, London, 1970.)

is defined as the enegy of the $v' = 0 - v'' = 0$ ionization. This quantity can be subject to appreciable uncertainty if the progression is so long that its first members are observed only weakly, or not at all. The second is the vertical ionization energy, which is the energy corresponding to the intensity maximum in the $v'' = 0$ progression (this may lie *between* bands). Alternatively, the vertical ionization energy has been defined as that corresponding to the

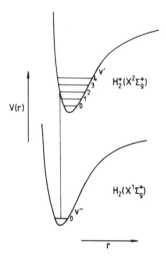

Figure 8.8 The Franck–Condon principle applied to the ionization of H_2.

transition at the centre of gravity of the band system, which may not be easy to determine when band systems overlap.

It is the vertical ionization energy to which Koopmans' theorem refers (equation 8.5).

8.1.3.2(b) Nitrogen The ground MO configuration of N_2 is given in equation (7.57) and the He I UPS spectrum shown in Figure 8.9.

According to the order of occupied MOs in equation (7.57) the lowest ionization energy (adiabatic value 15.58 eV) corresponds to the removal of an electron from the outermost, $\sigma_g 2p$, MO. The second (16.69 eV) and third (18.76 eV) lowest correspond to removal from $\pi_u 2p$ and $\sigma_u^* 2s$ respectively.

In the UPS of H_2 we saw how the extent of the vibrational progression accompanying the ionization is consistent, through the Franck–Condon principle, with the removal of an electron from a strongly bonding MO. Similarly, removal from a strongly antibonding MO would lead to a shortening of the bond in the ion and a long vibrational progression. However, removal from a weakly bonding or non-bonding MO would result in very little change of bond length and a very short progression (see Figure 7.22a).

Using the Franck–Condon principle in this way we can see that the band system associated with the second lowest ionization energy, showing a long progression, is consistent with the removal of an electron from a bonding $\pi_u 2p$ MO. The progressions associated with both the lowest and third lowest ionization energies are short, indicating that the bonding and antibonding

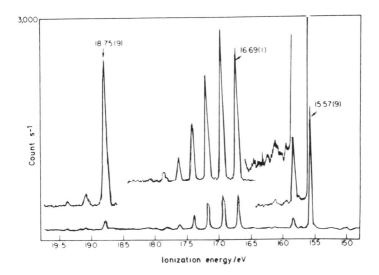

Figure 8.9 The He I UPS spectrum of N_2. (Reproduced from Turner, D. W., Baker, C., Baker, A. D., and Brundle, C. R., *Molecular Photoelectron Spectroscopy*, p. 46, Wiley, London, 1970.)

Table 8.2 Bond lengths of N_2^+ and N_2 in various electronic states.

Molecule	MO configuration	State	$r_e/\text{Å}$
N_2	$\ldots(\sigma_u^*2s)^2(\pi_u2p)^4(\sigma_g2p)^2$	$X^1\Sigma_g^+$	1.097 69
N_2^+	$\ldots(\sigma_u^*2s)^2(\pi_u2p)^4(\sigma_g2p)^1$	$X^2\Sigma_g^+$	1.116 42
N_2^+	$\ldots(\sigma_u^*2s)^2(\pi_u2p)^3(\sigma_g2p)^2$	$A^2\Pi_u$	1.174 9
N_2^+	$\ldots(\sigma_u^*2s)^1(\pi_u2p)^4(\sigma_g2p)^2$	$B^2\Sigma_u^+$	1.074

characteristics of the σ_g2p and σ_u^*2s MOs respectively are not pronounced. These observations are also consistent with the bond length of N_2^+ in various excited states known from the high resolution electronic emission spectrum. These bond lengths are compared with that of the ground state of N_2 in Table 8.2. In the X, A, and B states of N_2^+ there is a very small increase, a large increase, and a very small decrease respectively, compared to the X state of N_2.

There is a degeneracy factor of two associated with a π orbital compared to the non-degeneracy of a σ orbital, so that it might be expected that the integrated intensity of the second band system would be twice that of each of the other two. However, although the second band system is the most intense, other factors affect the relative intensities so that they are only an approximate guide to orbital degeneracies.

MO calculations of the SCF type for N_2^+ place the $A^2\Pi_u$ state *below* the $X^2\Sigma_g^+$ state. This discrepancy is an example of the breakdown of Koopmans' theorem due to deficiencies in the calculations.

8.1.3.2(c) Hydrogen bromide The MOs of the hydrogen halides are conceptually very simple. In HCl, for example, the occupied valence orbitals consist of the $3p_x$ and $3p_y$ lone pair AOs on the chlorine atom and a σ-type bonding MO formed by linear combination of the $3p_z$ AO on chlorine and the 1s AO on hydrogen (see Section 7.2.1.2 and Figure 7.15). The UPS spectrum of HCl, and the other hydrogen halides, confirms this simple picture.

Figure 8.10 shows the He I spectrum of HBr in which there are two band systems. The low ionization energy system shows a very short progression consistent with the removal of an electron from the doubly degenerate π_u lone pair $(4p_x, 4p_y)$ orbital on the bromine atom. The lone pair orbital picture and the expectation of only a short progression is further confirmed by the fact that the bond length r_e is known from high resolution electronic spectroscopy to be 1.4144 Å in the $X^1\Sigma^+$ ground state of HBr and to have a fairly similar value of 1.4484 Å in the $X^2\Pi$ ground state of HBr$^+$.

The splitting by 0.33 eV of the $X^2\Pi$ state into an inverted multiplet, with the $^2\Pi_{3/2}$ component, is relatively large due to the spin–orbit coupling caused by the high nuclear charge on the bromine atom and also the fact that the orbital from which the electron has been removed is close to this atom. The correspond-

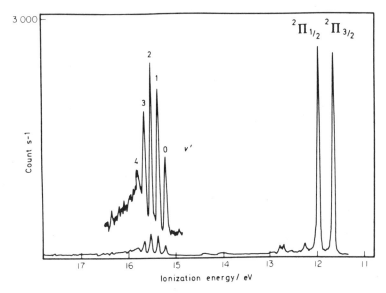

Figure 8.10 The He I UPS spectrum of HBr. Reproduced from Turner, D. W., Baker, C., Baker, A. D., and Brundle, C. R., *Molecular Photoelectron Spectroscopy*, p. 57, Wiley, London, 1970.)

ing splittings in the $X^2\Pi$ states of HCl and HI are 0.08 and 0.66 eV respectively, consistent with the magnitude of the nuclear charge on Cl and I.

Figure 8.10 shows that the next band system of HBr consists of a fairly long progression with maximum intensity at $v' = 2$. A long progression is to be expected following the removal of an electron from the strongly bonding σ MO to form the $A^2\Sigma^+$ state of HBr$^+$, in which r_e has the value of 1.6842 Å, an increase of about 0.27 Å from that of HBr. There is no spin–orbit interaction in $^2\Sigma^+$ states since they have no orbital angular momentum (see Section 7.2.2). Consequently, there is no doubling of bands in this system of the UPS spectrum.

8.1.3.2(d) Water The ground MO configuration of H_2O is (see Section 7.3.1.1 and Figure 7.31)

$$\ldots (2a_1)^2(1b_2)^2(3a_1)^2(1b_1)^2 \tag{8.8}$$

and the HOH angle in the $\tilde{X}^1A_1$ ground state is 104.5°. The He I UPS spectrum is shown in Figure 8.11.

Removal of an electron from the $1b_1$ orbital, which, as Figure 7.31 shows, does not favour any particular bond angle, to give the $\tilde{X}^2B_1$ ground state of H_2O^+ should not affect the angle very much. Both v_1, the symmetric stretching, and v_2, the angle bending, vibrations show only short progressions. The high resolution emission spectrum of H_2O^+ has shown that the angle is 110.5° in

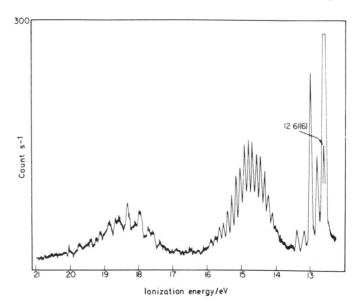

Figure 8.11 The He I UPS spectrum of H_2O. Reproduced from Turner, D. W., Baker, C., Baker, A. D., and Brundle, C. R., *Molecular Photoelectron Spectroscopy*, p. 113, Wiley, London, 1970.)

the $\tilde{X}^2B_1$ state, a rather larger change from the ground state of H_2O than might have been expected.

Removal of an electron from the $3a_1$ MO in H_2O should lead to a large increase in angle in the $\tilde{A}^2A_1$ state of H_2O^+ since, as Figure 7.31 shows, this orbital favours a linear molecule. The observation of a long progression in the bending vibration v_2, shown in Figure 8.11, is consistent with the increase of the angle to 180° (see Table 7.8).

The third band system, involving the removal of an electron from the $1b_2$ orbital, is vibrationally complex, consistent with the orbital being strongly bonding and favouring a linear molecule. Presumably both v_1 and v_2 are excited but the bands in this system are considerably broadened making analysis unreliable.

8.1.3.2(e) Benzene Figure 8.12 shows the He I UPS spectrum of benzene. This spectrum poses problems of interpretation which are typical for large molecules. These problems concern the difficulty of identifying the onset of a band system when there is overlapping with neighbouring systems and, generally, a lack of resolved vibrational structure. The recognition of the onset of a band system can be helped greatly by reliable MO calculations so that analysis of the spectrum of a large molecule is often a result of a combination of theory and experiment. In addition, study of the dependence on the angle of ejection

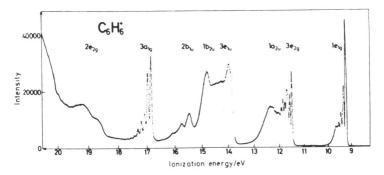

Figure 8.12 The He I UPS spectrum of benzene. (Reproduced from Karlsson, L., Mattsson, L., Jadrny, R., Bergmark, T., and Siegbahn, K., *Physica Scripta*, **14**, 230, 1976.)

with the kinetic energy of the photoelectrons, a topic which will not be discussed here, can aid identification of overlapping systems.

In Section 7.3.1.3 the π-electron MOs of benzene were obtained by the Hückel method using only the $2p_z$ AOs on the six carbon atoms. The ground configuration is

$$\ldots (1a_{2u})^2 \, (1e_{1g})^4 \tag{8.9}$$

The spectrum in Figure 8.12, together with a good MO calculation, shows that the band system with the lowest ionization energy is due to removal of an electron from the $1e_{1g}$ π orbital. However, the removal of an electron from the $1a_{2u}$ π orbital gives rise to the third, not the second, band system. The second, fourth, fifth, ... band systems are all due to removal of electrons from σ orbitals. This example illustrates the dangers of considering only π electrons occupying the higher energy MOs.

Vibrational structure is, at best, only partially resolved. The structure in the first and second systems is complex which could be due, in part, to there being a Jahn–Teller effect in each of them. Such an effect may arise when a molecule is in an orbitally degenerate E (or T) state. The Jahn–Teller effect involves a distortion of the molecule in order to destroy the degeneracy. The only relatively simple structure is observed in the seventh system in which an electron is removed from the $3a_{1g}$ orbital. There is a long progression in v_1, the a_{1g} ring-breathing vibration (see Figure 6.13f) consistent with the electron coming from a σ CC-bonding orbital.

8.1.3.3 X-ray photoelectron spectra of gases

We have seen in Section 8.1 how X-ray photons may eject an electron from the core orbitals of an atom, whether it is free or part of a molecule. So far, in all aspects of valence theory of molecules that we have considered, the core electrons have been assumed to be in orbitals which are unchanged from the AOs of the corresponding atoms. XPS demonstrates that this is almost, but not quite, true.

The point is illustrated in Figure 8.13 which shows the XPS spectrum of a 2:1 mixture of CO and CO_2 gases obtained with MgKα (1253.7 eV) source radiation. The ionization energy for removal of an electron from the 1s orbital on a carbon atom, referred to as the C 1s ionization energy, is 295.8 eV in CO and 297.8 eV in CO_2, these being quite comfortably resolved. The O 1s ionization energy is 541.1 eV in CO and 539.8 eV in CO_2, which are also resolved.

There are two features which stand out from these data. The first is that the ionization energy is much greater for O 1s than for C 1s due to the effect of the larger nuclear charge on oxygen. The second effect is that the ionization energy for a particular core orbital in a particular atom depends on the immediate environment in the molecule. This latter effect is known as the chemical shift. It is not to be confused with the chemical shift in n.m.r. spectroscopy which is the shift of the signal due to the nuclear spin of, for example, a proton due to neighbouring groups which may shield the proton more or less strongly from the applied magnetic field.

The XPS chemical shift, $\Delta E_{n\ell}$, for an atomic core orbital with principal and

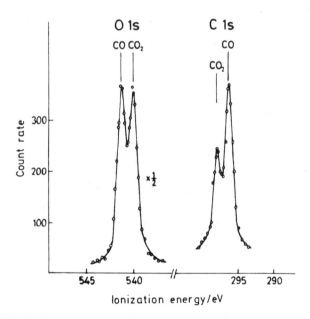

Figure 8.13 The MgKα oxygen 1s and carbon 1s XPS spectra of a 2:1 mixture of CO and CO_2 gases. (Reproduced, with permission, from Allan, C. J., and Siegbahn, K. (November 1971), *Publication No. UUIP-754*, p. 48, Uppsala University Institute of Physics.)

orbital angular momentum quantum numbers n and ℓ, respectively, in a molecule M is given by

$$\Delta E_{n\ell} = [E_{n\ell}(M^+) - E_{n\ell}(M)] - [E_{n\ell}(A^+) - E_{n\ell}(A)] \qquad (8.10)$$

The shift is measured relative to the ionization energy, $[E_{n\ell}(A^+) - E_{n\ell}(A)]$, of the corresponding orbital of the free atom A. In the approximation of Koopmans' theorem (equation 8.5),

$$\Delta E_{n\ell} \simeq -\varepsilon_{n\ell}(M) + \varepsilon_{n\ell}(A) \qquad (8.11)$$

where $\varepsilon_{n\ell}$ is the calculated orbital energy.

The chemical shift is related to the part of the electron density contributed by the valence electrons. It is a natural extension, therefore, to try to relate changes of chemical shift due to neighbouring atoms to the electronegativities of those atoms. A good illustration of this is provided by the XPS carbon $1s$ spectrum of ethyltrifluoroacetate, $CF_3COOCH_2CH_3$, in Figure 8.14, obtained with AlKα ionizing radiation which was narrowed with a monochromator.

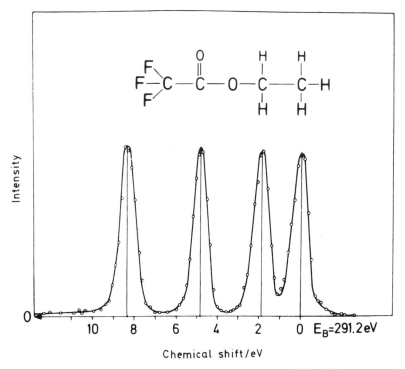

Figure 8.14 The monochromatized AlKα carbon $1s$ XPS spectrum of ethyltrifluoroacetate showing the chemical shifts relative to an ionization energy of 291.2 eV. (Reproduced, with permission, from Gelius, U., Basilier, E., Svensson, S., Bergmark, T., and Siegbahn, K., *J. Electron Spectrosc.*, **2**, 405, 1974.)

This spectrum shows that, at least in this molecule, a very simple electronegativity argument leads to the correct C $1s$ assignments. The carbon atom of the CF_3 group is adjacent to three strongly electronegative fluorine atoms which leads to a depletion of the electron density on the carbon atom and the fact that, consequently, it holds on to the $1s$ electron more strongly, resulting in the largest chemical shift. The carbon atom of the $C=O$ group has two neighbouring, fairly electronegative, oxygen atoms which have a similar but smaller effect, giving the second largest chemical shift. The carbon atom of the CH_2 group has one neighbouring oxygen atom and the third largest chemical shift, but that of the CH_3 group has none, leading to the smallest chemical shift (lowest ionization energy).

Figure 8.15 shows the C $1s$ spectra of furan, pyrrole, and thiophene. Due to the decreasing electronegativity of the order $O > N > S$ the C $1s$ line is shifted to low ionization energy from furan to pyrrole to thiophene. In addition the C

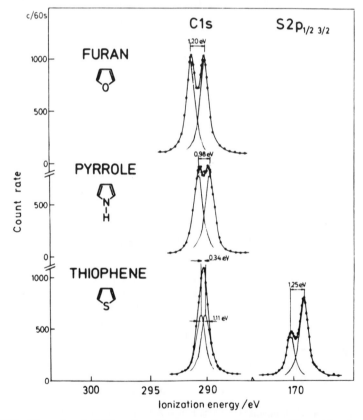

Figure 8.15 The carbon $1s$ XPS spectra of furan, pyrrole and thiophene. The sulphur $2p$ spectrum of thiophene is also shown. (Reproduced with permission from Gelius, U., Allan, C. J., Johansson, G., Siegbahn, H., Allison, D. A., and Siegbahn, K., *Physica Scripta*, **3**, 237, 1971.)

1s line is split into two components due to the two kinds of carbon atom environment. The carbon atom in position 2 is responsible for the C 1s component of higher ionization energy because it is nearer to the heteroatom than that in position 3. The splitting decreases with electronegativity and can be observed for thiophene only after deconvolution (breaking down composite lines into the contributing lines).

Also shown in Figure 8.15 is the line due to removal of a 2p electron from the sulphur atom in thiophene. Spin–orbit coupling is sufficient to split the resulting core term, as it is called, into $^2P_{3/2}$ and $^2P_{1/2}$ states, the multiplet being inverted.

Figure 8.16 shows the B 1s spectrum of the B_5H_9 molecule. The boron atoms are situated at the corners of a square pyramid. There are four B–H–B bridging hydrogen atoms and there is also a terminal hydrogen attached to each boron. The four equivalent boron atoms at the base of the pyramid give rise to the intense component at higher ionization energy than the component due to the one at the apex.

The example of B_5H_9 serves to show how the chemical shift may be used as an aid to determining the structure of a molecule and, in particular, in deciding between alternative structures. There are many examples in the literature of this kind of application which is reminiscent of the way in which the chemical shift in n.m.r. spectroscopy may be employed. However there is one important difference in using the two kinds of chemical shift. In XPS there are no interactions affecting closely spaced lines in the spectrum, however close they

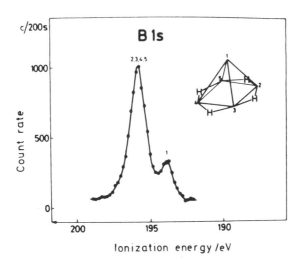

Figure 8.16 The Mg$K\alpha$ boron 1s XPS spectrum of B_5H_9. (Reproduced, with permission, from Allison, D. A., Johansson, G., Allan, C. J., Gelius, U., Siegbahn, H., Allison, J., and Siegbahn, K., *J. Electron Spectrosc.*, **1**, 269, 1972–73.)

may be. Figure 8.15 illustrates this for the C $1s$ lines of thiophene. In n.m.r. spectroscopy the spectrum becomes more complex, due to spin–spin interactions, when chemical shifts are similar.

We have seen in Section 8.1.1.4 that the best resolution, about 0.2 eV or 1600 cm^{-1}, of which XPS is capable, employing an X-ray monochromator for the source radiation, is very much worse than that of UPS. Nevertheless, it is possible, in favourable circumstances, to observe some vibrational structure in XPS spectra. Figure 8.17 shows such structure in the C $1s$ spectrum of methane. The removal of a C $1s$ electron causes the other electrons to be more tightly bound, resulting in a reduction of C–H bond lengths and a general shrinkage of the molecule. The shrinkage is in the direction of v_1, the symmetric C–H stretching vibration. The resulting progression in v_1, barely resolved in Figure 8.17, can be understood in terms of the Franck–Condon principle (Section 7.2.5.3). The wavenumber of the vibration increases from 2917cm^{-1} in CH$_4$ to about 3500 cm^{-1} in this state of CH$_4^+$, consistent with the shortening of the C–H bonds.

8.1.3.4 X-ray photoelectron spectra of solids

Both UPS and XPS of solids are useful techniques. So far as studies of adsorption by surfaces are concerned we would expect UPS, involving only

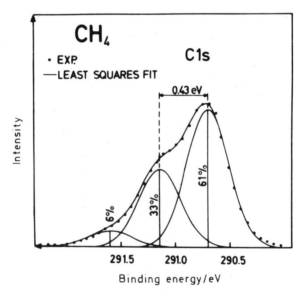

Figure 8.17 A short, barely resolved vibrational progression in the v_1 vibration of CH$_4^+$ in the carbon $1s$ XPS spectrum of methane obtained with a monochromatized X-ray source. (Reproduced, with permission, from Gelius, U., Svensson, S., Siegbahn, H., Basilier, E., Faxålv, Å., and Siegbahn, K., *Chem. Phys. Lett.*, **28**, 1, 1974.)

valence orbitals, to be more sensitive. For example, if we wish to determine whether nitrogen molecules are adsorbed onto an iron surface with the axis of the molecule perpendicular or parallel to the surface it would seem that the valence orbitals would be most affected. This is generally the case but, because UPS spectra of solids are considerably broadened, it is the XPS spectra which are usually the most informative.

Figure 8.18 shows an XPS spectrum of gold foil with mercury adsorbed onto the surface. Both the gold and mercury doublets result from the removal of a $4f$ electron leaving $^2F_{5/2}$ and $^2F_{7/2}$ core states for which $L = 3$, $S = \frac{1}{2}$, and $J = \frac{5}{2}$ or $\frac{7}{2}$. Less than 0.1 per cent of a monolayer of mercury on a gold surface can be detected in this way.

In Figure 8.19 is shown the XPS spectrum of Cu, Pd, and a 60 per cent Cu and 40 per cent Pd alloy (having a face-centred cubic lattice). In the Cu spectrum one of the peaks due to the removal of a $2p$ core electron, the one resulting from the creation of a $^2P_{3/2}$ core state, is shown (the one resulting from the $^2P_{1/2}$ state is outside the range of the figure).

The fact that a metal is a conductor of electricity is due to the valence electrons of all the metal atoms being able to move freely through the metal. They can be regarded as being in orbitals which are delocalized throughout the sample. The available electrons can then be fed into these orbitals, in order of increasing energy, until they are used up. The energy of the outermost filled orbital is known as the Fermi energy E_F. The spectrum of Cu in Figure 8.19 shows the characteristically broad valence band signal and the Fermi edge, at energy E_F, is indicated. The spectrum of Pd shows similar general features but with the formation of both the $^2D_{3/2}$ and $^2D_{5/2}$ core states, resulting from removal of a $3d$ electron, being indicated. The valence band is rather different from that of Cu with a much more pronounced Fermi edge.

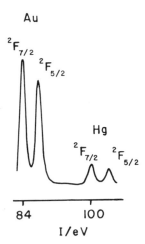

Figure 8.18 XPS spectrum of gold foil with adsorbed mercury. (Reproduced, with permission, from Brundle, C. R., Roberts, M. W., Latham, D., and Yates, K., *J. Electron Spectrosc.*, **3**, 241, 1974.)

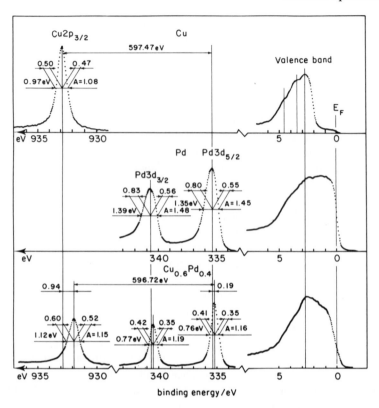

Figure 8.19 XPS spectrum, showing core and valence electron ionization energies, of Cu, Pd, and a 60% Cu and 40% Pd alloy (face-centred cubic lattice). The 'binding energy' is the ionization energy relative to the Fermi energy of Cu. (Reproduced, with permission, from Siegbahn, K., *J. Electron Spectrosc.*, **5**, 3, 1974.)

The spectrum of the alloy in Figure 8.19 shows pronounced differences. The shape of the Fermi edge is different from that of Cu or Pd and proves to be sensitive to the constitution of the alloy. The peak due to formation of the $^2P_{3/2}$ core state of Cu is shifted by 0.94 eV in the alloy and broadened slightly. The two Pd peaks are also shifted, but only slightly, and are narrowed to almost 50 per cent of their width in Pd itself.

Figure 8.20 shows an example of how XPS may be used to follow the take-up of an adsorbed layer and its behaviour at higher temperatures. The system is that of nitric oxide (NO) being adsorbed by an iron surface. In the region of nitrogen 1s ionization, curves 2, 3, and 4 show the effect of increasing surface coverage at a temperature of 85 K. At low coverage (curve 2) the NO is mostly dissociated to nitrogen and oxygen atoms, the peak at about 397 eV being due to removal of a 1s electron from an adsorbed nitrogen atom. The much weaker

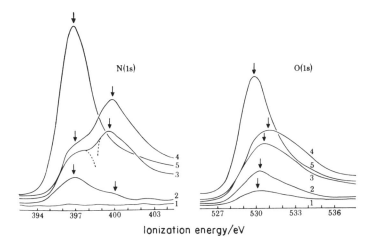

Figure 8.20 Nitrogen $1s$ and oxygen $1s$ XPS spectra of nitric oxide (NO) adsorbed on an iron surface. 1. Fe surface at 85 K. 2. Exposed at 85 K to NO at 2.65×10^{-5} Pa for 80 s. 3. As for 2 but exposed for 200 s. 4. As for 2 but exposed for 480 s. 5. After warming to 280 K. (Reproduced, with permission, from Kishi, K., and Roberts, M. W., *Proc. R. Soc. Lond.*, **A352**, 289, 1976.)

peak at about 400 eV is the nitrogen $1s$ peak of undissociated NO. As the coverage increases (curves 3 and 4) the proportion of undissociated NO increases. Curve 5 shows that warming up the sample to 280 K results in almost complete dissociation.

The corresponding oxygen $1s$ curves in Figure 8.20 support these conclusions. The peak at about 529 eV is due to oxygen atoms but a peak at about 530.5 eV, due to undissociated NO, is not as well resolved as for the nitrogen $1s$ peaks.

8.2 Auger Electron and X-ray Fluorescence Spectroscopy

Figure 8.1(c) illustrates schematically the kind of process occurring in Auger electron spectroscopy (AES). The process occurs in two stages. In the first, a high energy photon ejects an electron from a core orbital of an atom A:

$$A + h\nu \rightarrow A^+ + e \qquad (8.12)$$

and, in the second, an electron falls down from a higher energy orbital to fill the vacancy created thereby releasing sufficient energy to eject a second photoelectron, a so-called Auger electron, from one of the higher energy orbitals:

$$A^+ \rightarrow A^{2+} + e \qquad (8.13)$$

Although it is conceptually useful to think of two successive processes following the initial ionization to A^+, the electron transfer and the generation of the Auger electron occur simultaneously.

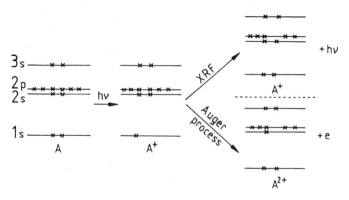

Figure 8.21 The competitive processes of X-ray fluorescence and Auger electron emission.

In Figure 8.1(c) the higher energy orbitals are indicated as being valence orbitals but, in most applications of AES, they are core orbitals. For this reason the technique is not usually concerned with atoms in the first row of the periodic table.

Figure 8.21 shows schematically a set of $1s$, $2s$, $2p$, and $3s$ core orbitals of an atom lower down the periodic table. The absorption of an X-ray photon produces a vacancy in, say, the $1s$ orbital to give A^+ and a resulting photo-electron which is of no further interest. The figure then shows that subsequent relaxation of A^+ may be by either of two processes. X-ray fluorescence (XRF) involves an electron dropping down from, say, the $2p$ orbital to fill the $1s$ vacancy with the consequent emission of an X-ray photon. XRF is an important analytical technique in its own right (see Section 8.2.2).

Competing with XRF is the alternative relaxation process in Figure 8.21 in which the energy released when an electron falls into the $1s$ orbital is used up in ejecting an Auger electron from, say, the $2s$ orbital with any excess energy available being in the form of kinetic energy of the Auger electron. The quantum yield for XRF, as opposed to the production of Auger electrons, is generally smaller for lighter elements when the initial vacancy is created in the $1s$ orbital.

There is an important difference between the two techniques in that photons, produced by XRF, can pass through a relatively large thickness of a solid sample, typically 40 000 Å, whereas electrons can penetrate only about 20 Å. This means that AES is more useful in the study of solid surfaces, while XRF gives information referring more to the bulk of a solid or liquid.

8.2.1 Auger Electron Spectroscopy

8.2.1.1 Experimental method

If monochromatic X-rays are used as the ionizing radiation the experimental technique is very similar to that for XPS (Section 8.1.1) except that it is the kinetic energy of the Auger electrons which is to be measured. Alternatively, a monochromatic electron beam may be used to eject an electron. The energy E of an electron in such a beam is given by

$$E = (2eVm_ec^2)^{1/2} \qquad (8.14)$$

when an electron of charge e and mass m_e is accelerated by a potential V. Since the sampling depth of a solid surface depends only on the escape depth of the Auger electron (*ca* 20 Å) it makes no difference whether they are generated at the greater depth possible with an X-ray source or at a smaller depth with an electron beam.

Various types of energy analysers have been used for the Auger electrons but one of the most successsful is that shown in Figure 8.22. There are two coaxial cylindrical plates, the inner one earthed and the outer having a variable negative voltage applied to it. Some Auger electrons emanating from the target (sample) pass through spaces in the inner plate and, when a voltage V is applied to the outer plate, only electrons whose kinetic energies correspond to that value of V will pass through the second set of spaces and be focused onto the detector. The Auger electron spectrum is then the number of electrons reaching the detector as a function of V. With this type of analyser Auger electrons are collected over a 360° angle.

8.2.1.2 *Processes in Auger electron ejection*

A common example of an Auger process involves the ejection of a photo-electron, as shown in Figure 8.23, from the K shell, i.e. a $1s$ electron, with energy E_P which is not considered further. Following the ejection of a K electron it is common for an electron from the L shell, specifically fom the L_I (or $2s$) orbital, to fill the vacancy releasing an amount of energy $E_K - E_{L_I}$. This energy is then used up in ejecting an Auger electron from the L_{II} orbital, any excess energy being converted into kinetic energy E_A of the electron.

The distinction between L_{II} and L_{III}, M_{II} and M_{III}, etc., is an important one. These labels are commonly used by those studying inner shell electron processes.

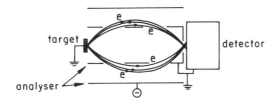

Figure 8.22 Cylindrical analyser used in an Auger spectrometer.

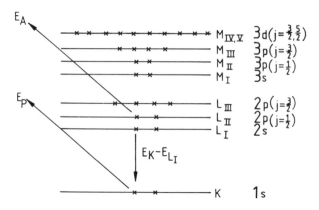

Figure 8.23 Illustration of a KL_1L_{II} Auger process.

If an electron is removed from, for example, a $2p$ orbital the ion is left in a $^2P_{1/2}$ or $^2P_{2/3}$ core state, ignoring valence shell electrons. These states result from the $2p^5$ core configuration of the ion (see Section 7.1.2.3) and are of different energies, so we can visualize the electron as having come from an orbital of different energy, namely a $2p_{1/2}$ or a $2p_{3/2}$ orbital where $j = \ell + s$, $\ell + s - 1, \ldots |\ell - s| = \frac{3}{2}, \frac{1}{2}$ from equation (7.7). Since $m_j = j, j - 1, \ldots, -j$ the $2p_{1/2}$ orbital is twofold degenerate and the $2p_{3/2}$ orbital fourfold degenerate so that they can take two or four electrons respectively, as in Figure 8.23. The distinction between the M_I, M_{II}, and M_{III} labels is similar.

The Auger electron kinetic energy E_A is obtained from

$$E_K - E_{L_1} = E'_{L_{II}} + E_A \tag{8.15}$$

where E_K and E_{L_1} are binding energies[†] for the neutral atom but $E'_{L_{II}}$ is the binding energy of an electron in L_{II} of the ion with a single positive charge due to a vacancy in L_1.

The Auger process described in this example is designated $K-L_1L_{II}$ or, alternatively, KL_1L_{II} or, simply, KLL. Unfortunately there is little standardization of the way in which the type of Auger process is indicated.

The lefthand side of equation (8.15) involves the difference between two electron binding energies, $E_K - E_{L_1}$. Each of these energies changes with the chemical (or physical) environment of the atom concerned but the changes in E_K and E_{L_1} are very similar so that the environmental effect on $E_K - E_{L_1}$ is small. It follows that the environmental effect on $E'_{L_{II}} + E_A$, the righthand side of equation (8.15), is also small. Therefore the effect on E_A is appreciable as it

---

[†] The binding energy is the negative of the orbital energy.

must be similar to that on $E'_{L_{II}}$. There is, then, a chemical shift effect in AES rather like that in XPS.

8.2.1.3 Examples of Auger electron spectra

The chemical shift in AES is illustrated for $S_2O_3^{2-}$, the thiosulphate ion, in solid $Na_2S_2O_3$ in Figure 8.24. The ion has a tetrahedral shape and, therefore, two

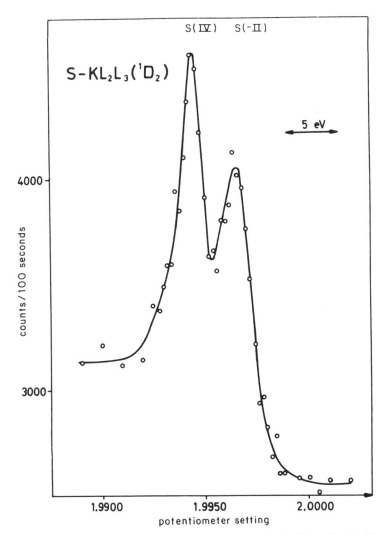

Figure 8.24 $KL_{II,III}L_{II,III}$ ($^1D_2$) Auger spectrum of sulphur in $Na_2S_2O_3$. (Reproduced, with permission, from Fahlmann, A., Hamrin, K., Nordberg, R., Nordling, C., and Siegbahn, K., *Physics Letters*, **20**, 156, 1966.)

types of sulphur atom, one in a $+6$ and the other in a -2 oxidation state, giving rise to two peaks in the spectrum. The Auger process involved is $KL_{II,III}L_{II,III}$, the difference in chemical shift between L_{II} and L_{III} not being resolved. The core state of sulphur produced by the processes in equations (8.12) and (8.13) is derived from the core configuration $1s^2 2s^2 2p^4 3s^2 \ldots$. We have seen

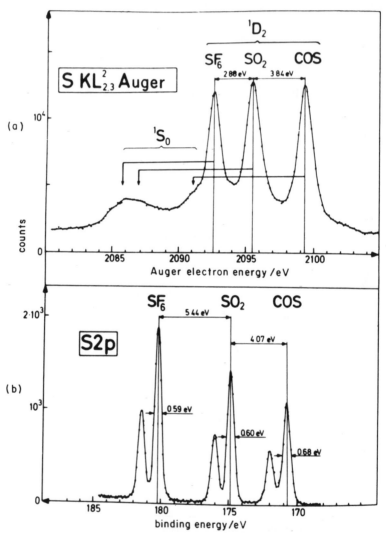

Figure 8.25 (a) The $KL_{II,III}L_{II,III}$ ($^1D_2$ and $^1S_0$) Auger spectrum of sulphur in a gaseous mixture of SF_6, SO_2, and OCS, compared with (b) the S $2p$ X-ray photoelectron spectrum of a mixture of the same gases. (Reproduced, with permission, from Aslund, L., Kelfve, P., Siegbahn, H., Goscinski, O., Fellner-Feldegg, H., Hamrin, K., Blomster, B., and Siegbahn, K., *Chem. Phys. Letters*, **40**, 353, 1976.)

in Section 7.1.2.3(b) that three terms, ^1S, ^3P, and ^1D result from two equivalent p electrons, or vacancies, but only the ^1D term, giving a $^1D_2$ state, is involved here.

Figure 8.25 shows the $KL_{II,III}L_{II,III}$ Auger spectrum of a gaseous mixture of SF_6, SO_2 and OCS, all clearly resolved. The three intense peaks are due to sulphur in a $^1D_2$ core state, but there are three weak peaks due to a $^1S_0$ core state also. The S $2p$ X-ray photoelectron spectrum of a mixture of the same gases is shown for comparison, each of the three doublets being due to sulphur in a $^2P_{1/2}$ or $^2P_{3/2}$ core state.

In Figure 8.26 is shown the $KL_{II,III}L_{II,III}$ Auger spectrum of sodium in crystalline NaCl. Once again, the formation of the $^1D_2$ and, weakly, the $^1S_0$ core states can be observed. Also shown are peaks resulting from additional processes in which the initial photoelectron with energy E_P provides the energy for removing a second photoelectron—in this case from the 2s orbital to give the configuration $1s^1 2s^1 2p^6$.... This is referred to as an electron shakeoff process. Following this an electron may fall down from $2p$ to fill the $1s$ vacancy, resulting in $^1P_1$ and $^3P_2$, $^3P_1$, $^3P_0$ core states (the latter three are unresolved), or an electron may fall down from $2s$ to $1s$ resulting in a $^1S_0$ core state.

The $KL_{II,III}L_{II,III}$ Auger spectrum of magnesium, and how it changes on conversion at the surface to magnesium oxide, is shown in Figure 8.27. Two peaks due to $^1D_2$ and $^1S_0$ core states resulting from the $1s^2 2s^2 2p^4$... configuration can be seen in Mg and in MgO with a chemical shift of about 5 eV. The inset in Figure 8.27 shows that, in the XPS spectrum, the chemical shift is much less. There are also several satellites in the Auger spectrum including those due to shakeoff processes resulting in $1s^2 2s^2 2p^5$... and $1s^2 2s^0 2p^6$... configurations, very similar to those for sodium in Figure 8.26.

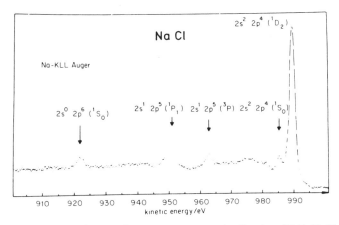

Figure 8.26 The $KL_{II,III}L_{II,III}$ ($^1D_2$ and $^1S_0$) Auger spectrum of sodium in solid NaCl. (Reproduced, with permission, from Siegbahn, K. (June 1976), *Publication No. UUIP-940*, p. 81, Uppsala University Institute of Physics.)

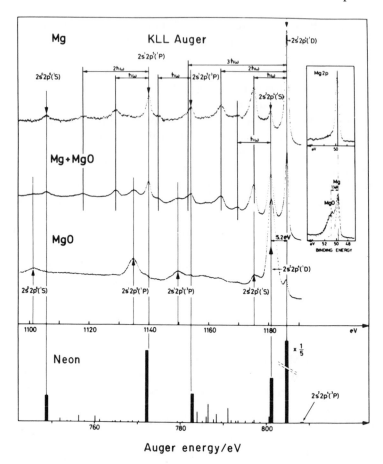

Figure 8.27 The $KL_{II,III}L_{II,III}$ Auger spectrum of magnesium showing how it changes as the surface is oxidized and how it compares with that of neon. (Reproduced, with permission, from Siegbahn, K., *J. Electron Spectrosc.*, **5**, 3, 1974.)

8.2.2 X-ray Fluorescence Spectroscopy

8.2.2.1 Experimental method

In XRF, as in AES, the ejection of the core electron from the atom A to produce the ion A^+, as illustrated in Figure 8.21, may be by an electron beam of appropriate energy or by X-rays. Much of the early work in XRF employed an electron beam but nowadays an X-ray source is used almost exclusively.

The usual source of X-rays is the Coolidge tube. Electrons are generated from a heated tungsten filament (the cathode) and accelerated towards the anode which consists of a target of such metals as W, Mo, Cr, Cu, Ag, Ni, Co, or Fe, depending on the range of wavelengths required. The Coolidge tube source can cover wavelengths up to 10 Å beyond which the intensity becomes too low. This restricts XRF investigations to atoms with atomic number Z greater than 12 (magnesium) but, since XRF spectra of light atoms are often broadened and are not very useful for analysis, this is not a serious limitation.

Figure 8.28 shows how the X-rays fall on the solid or liquid sample which then emits X-ray fluorescence in the region 0.2–20 Å. The fluorescence is dispersed by a flat crystal, often of lithium fluoride, which acts as a diffraction grating (rather like the quartz crystal in the X-ray monochromator in Figure 8.3). The fluorescence may be detected by a scintillation counter, a semiconductor detector or a gas flow proportional detector in which the X-rays ionize a gas such as argon and the resulting ions are counted.

In scanning the wavelength range of the fluorescence the crystal must be smoothly rotated to vary the angle θ and the detector must also be rotated, but at twice the angular speed since it is at an angle of 2θ to the direction of the X-ray fluorescence beam.

An alternative type of spectrometer is the energy dispersive spectrometer which dispenses with a crystal dispersion element. Instead, a type of detector is used which receives the undispersed X-ray fluorescence and outputs a series of pulses of different voltages which correspond to the different wavelengths (energies) that it has received. These energies are then separated with a multichannel analyser.

An energy dispersive spectrometer is cheaper and faster for multielement analytical purposes but has poorer detection limits and resolution.

8.2.2.2 Processes in X-ray fluorescence

Figure 8.29 shows two of the more common processes involved in XRF. Comparison with Figure 8.23 illustrating an Auger electron process shows that the same system of labelling energy levels is used in AES and XRF.

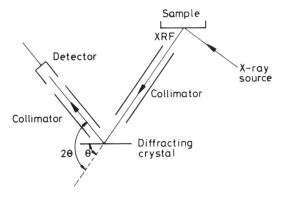

Figure 8.28 Wavelength dispersive X-ray spectrometer.

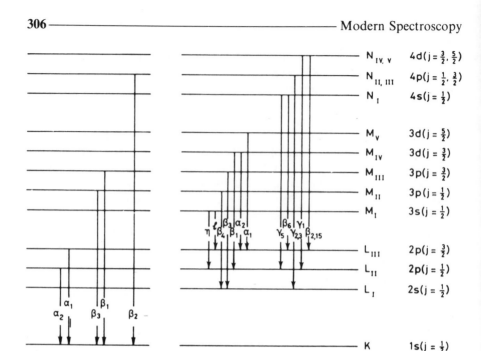

Figure 8.29 X-ray fluorescence transitions forming (a) a K emission spectrum and (b) an L emission spectrum. The energy levels are not drawn to scale.

In Figure 8.29(a) a $1s$ (K shell) electron has been ejected from the atom and fluorescence, forming a so-called K emission spectrum, is caused by an electron falling from another shell and filling the vacancy in the K shell. Similarly, Figure 8.29(b) shows transitions forming the L emission spectrum resulting from filling a vacancy created in the L shell.

The quantum yield for XRF decreases as the nuclear charge increases and also from K to L emission.

The selection rules which apply to XRF transitions are

$$\Delta n \geqslant 1; \quad \Delta \ell = \pm 1; \quad \Delta j = 0, \pm 1$$

where the quantum numbers refer to the electron which is filling the vacancy. There are important differences from the selection rules which apply to AES. For example, the $2s$–$1s$ (L_I–K) transition in the $KL_I L_{II}$ Auger process illustrated in Figure 8.23 is forbidden as an XRF transition since $\Delta \ell = 0$. This apparent contradiction occurs because the selection rules for an Auger process concern only the initial state of the atom and the final state of the doubly charged ion, regardless of the intermediate processes. The selection rules for the overall Auger process are

$$\Delta L = 0; \quad \Delta S = 0; \quad \Delta J = 0$$

where the quantum numbers refer to the atom or ion as a whole.

We have seen in Section 8.2.1.2 that there is a chemical shift effect in AES which is important in distinguishing atoms in different environments. In XRF there is only a very small chemical shift effect which is too small to be generally useful. The reason for the effect being small is that the XRF energy is the difference between two energy levels, such as those with energies E_K and $E_{L_{11}}$, and the energy shifts due to the environmental effects on each of them are very similar in magnitude and sign. As a result, XRF is a technique which is much more useful for very accurate quantitative detection of elements than for structural studies.

8.2.2.3 Examples of X-ray fluorescence spectra

The example of a K emission XRF spectrum of a solid sample of tin, shown in Figure 8.30, shows four prominent transitions. The method of labelling the transitions is, unfortunately, non-systematic but those in the lower energy group are labelled α and the higher energy group β. The subscripts $1, 2, \ldots$ are generally, but not always, used to indicate the order of decreasing intensity so α_1 and β_1 are the strongest transitions in their groups.

The splitting between the α_1 and α_2 transitions is due to the spin–orbit coupling (coupling of the electron spin and orbital angular momenta; see

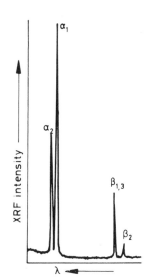

Figure 8.30 K emission spectrum of tin. The α_1 and β_2 lines are at 0.491 and 0.426 Å, respectively. (Reproduced, with permission, from Jenkins, R., *An Introduction to X-ray Spectrometry*, p. 22, Heyden, London, 1976.)

Section 7.1.2.3) which causes the splitting between the L_{II} and L_{III} levels. This coupling increases with nuclear charge Z and, with $Z = 50$ for tin, the resulting splitting of α_1 and α_2 is appreciable. The wavelengths of α_1 and α_2 are 0.491 and 0.495 Å, respectively, which are easily resolved. These correspond to an L_{II}–L_{III} energy separation of 206 eV. The transition α_1 is twice as intense as α_2 because of the degeneracy factor of $2j + 1$ associated with the upper levels of the transitions (see Section 8.1.3.1).

Spin–orbit coupling decreases as the orbital angular momentum quantum number ℓ increases. This is illustrated by the fact that the β_1 and β_3 transitions, split by only about 70 eV, are not resolved.

For elements with lower nuclear charges than tin, the α_1–α_2 splitting is smaller because of the reduced spin–orbit coupling. For example for calcium, with $Z = 20$, the splitting is only about 3 eV and usually unresolved.

There are two further effects on K emission XRF which become more important with decreasing nuclear charge. One is the appearance of weak satellite transitions, to lower wavelengths of the main transitions, occurring in the small proportion of doubly ionized atoms which may be produced by the initial X-ray bombardment. The other is a tendency for some transitions to be broadened into bands, rather than the usual sharp lines, due to the effects of the participation of the atomic orbitals concerned in molecular orbitals. This is observed in, for example, the K emission ($2s$–$1s$ and $2p$–$1s$) of oxygen.

In the XRF spectrum of tin, as in those of other elements, transitions such as $3d$–$1s$ and $4d$–$1s$, which are forbidden by the selection rules, may be observed very weakly due to perturbations by neighbouring atoms.

Figure 8.29(b) shows that an L emission XRF spectrum is much more complex than a K emission spectrum. This is illustrated by the L spectrum of gold in Figure 8.31. Apart from those labelled ℓ and η, the transitions fall into three groups labelled α, β and γ, the most intense within each group being α_1, β_1 and γ_1.

A feature of L spectra of atoms with $Z < 40$ is the absence of the β_2 transition which is intense in atoms such as gold with higher atomic numbers.

In atoms in which electrons in M or N shells take part to some extent in molecular orbital formation some transitions in the L spectrum may be broadened. Similarly in an M emission spectrum, in which the initial vacancy has been created in the M shell, there is a greater tendency towards broadening due to molecular orbital involvement.

8.3 Extended X-ray Absorption Fine Structure

As early as the 1930s X-ray absorption experiments were being carried out using a continuum source of X-rays (the *bremsstrahlung* mentioned in Section 8.1.1.1), a dispersive spectrometer, and a film detector. When the sample absorbs

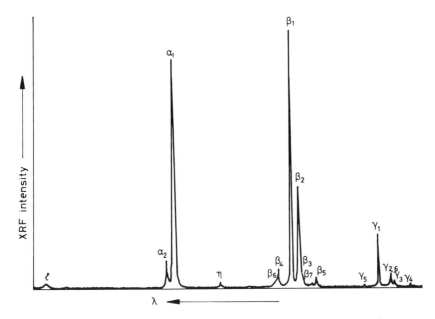

Figure 8.31 *L* emission spectrum of gold. The α_1 and γ_1 lines are at 1.277 and 0.927 Å, respectively. (Reproduced, with permission, from Jenkins, R., An Introduction to X-ray Spectrometry, p. 28, Heyden, London, 1976.)

X-rays the absorption coefficient a can be obtained from the Beer–Lambert law

$$a\ell = \log_{10}\left(\frac{I_0}{I}\right) \tag{8.16}$$

This is similar to equation (2.16) in which I_0 and I are the intensities of the incident and transmitted radiation respectively and ℓ is the pathlength of the absorbing sample.

In these experiments it was found that each absorption line is asymmetric showing a long tail extending to high photon energies and also some sharp structure to low energy of the sharp low energy absorption edge, as shown in Figure 8.32. The pre-edge structure of, say, a K-edge, corresponding to removal

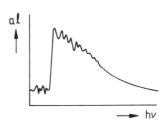

Figure 8.32 An X-ray absorption edge.

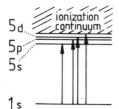

$5d$
$5p$
$5s$

$1s$

Figure 8.33 Formation of an absorption edge and pre-edge structure in X-ray absorption.

of a K-shell ($1s$) electron, is due to the electron being promoted to unoccupied orbitals, such as $5p$ or $5d$, rather than being removed altogether. The edge itself represents the onset of the ionization continuum, as illustrated in Figure 8.33 for the removal of a $1s$ electron. It is clear that X-ray absorption is closely related to the process involved in XPS (Section 8.1) except that, in the latter, monochromatic X-rays are used.

It was observed, in some cases of X-ray absorption, that there was some fine structure in the form of intensity variations superimposed on the otherwise smoothly declining intensity of the absorption tail, as shown in Figure 8.32. This structure is known as extended X-ray absorption fine structure (EXAFS).

Work on EXAFS then progressed very little until the advent of the synchrotron radiation source (storage ring) described in Section 8.1.1.1. This type of source produces X-ray radiation of the order of 10^5 to 10^6 times as intense as that of a conventional source and is continuously tunable. These properties led to the establishment of EXAFS as an important structural tool for solid materials.

Figure 8.34 illustrates the main features of the experimental method. The continuum radiation from the synchrotron storage ring is focused by a concave toroidal mirror, at low incidence angle, and falls on a double-crystal monochromator (see also Section 8.1.1.1), the crystals acting as diffraction gratings. The monochromator can be tuned smoothly through the photon energy range required. The intensity I_0 of the beam incident on the sample is measured by the ion chamber 1, in which the ions produced are counted, while the intensity I emerging from the sample is measured by the ion chamber 2.

If the sample is a monatomic gas, such as krypton, there is no EXAFS

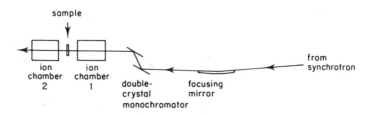

Figure 8.34 Experimental method for EXAFS.

(a)   (b)

Figure 8.35 Interference between scattered photoelectrons in (a) a gaseous diatomic molecule and (b) a crystalline material.

superimposed on the high energy side of the absorption peak but, for a diatomic gas such as Br_2, EXAFS is observed. The cause of the EXAFS is interference between the de Broglie wave representing the photoelectron (see Section 1.1) travelling outwards from one bromine atom and the same wave scattered by the other bromine atom, as illustrated in Figure 8.35(a). Clearly the interference pattern forming the EXAFS depends on the internuclear distance, as well as other parameters, and is reminiscent of electron diffraction in solids, mentioned in Section 1.1, and in gases.

The EXAFS technique is used primarily for investigations of disordered materials and amorphous solids. Figure 8.35(b) shows how interference occurs between the wave associated with a photoelectron generated on atom A and the waves scattered by nearest neighbour atoms B in a crystalline material.

In the theory of EXAFS it is usual to consider the wave vector k of the wave associated with the photoelectron rather than the wavelength λ. They are related by

$$k = \frac{2\pi}{\lambda} \tag{8.17}$$

and k is related to the photoelectron energy $(hv - I)$ of equation (8.3) by

$$k = \left[\frac{2m_e(hv - I)}{\hbar^2}\right]^{1/2} \tag{8.18}$$

where m_e is the electron mass.

In an EXAFS experiment the measurable quantity is the absorption coefficient a of equation (8.16). If a_0 is the absorption coefficient in the absence of EXAFS, deduced from the steeply falling background shown in Figure 8.32, then $\chi(k)$, the fractional change of a due to EXAFS, is given by

$$\chi(k) = \frac{a_0 - a}{a_0} \tag{8.19}$$

This is related to interatomic distances R_j between scattering atoms, and to other parameters, by

$$\chi(k) = \sum_j - \frac{N_j |f_j(k, \pi)|}{kR_j^2} e^{-2\sigma_j^2 k^2} e^{-2R_j/\lambda} \sin[2kR_j + \delta_j(k)] \qquad (8.20)$$

where N_j is the number of scattering atoms at a distance R_j from the one in which the photoelectron is generated, $|f_j(k, \pi)|$ is the amplitude of back-scattering by atom j, $\exp(-2\sigma_j^2 k^2)$ is known as the Debye–Waller function and allows for vibrational motion in a molecule, or disorder in a crystal, with root mean square amplitude σ_j, $\exp(-2R_j/\lambda)$ allows for the loss of photoelectrons due to inelastic scattering, and the $\sin[\ldots]$ term represents the wave nature of the photoelectron. When an electron wave is scattered there is likely to be a phase change. Figure 8.36(a) illustrates scattering with no phase change while Figure 8.36(b) shows a phase change of 180°, i.e. on scattering $\sin x \rightarrow \sin(x + \pi)$ where π is the phase angle. In equation (8.20) the phase angle is $\delta_j(k)$ for atom j.

$\chi(k)$ in equation (8.20) is the experimentally observed absorption, like that in Figure 8.32, after subtraction of the smoothly declining background. What is left is a sum of sine waves of which we require the wavelengths which can be related to R_j, provided the phase factor $\delta_j(k)$ is known. This process of obtaining wavelengths from a superposition of waves is that of Fourier transformation encountered in the discussion of interferometers in Section 3.3.3.

Figure 8.37 shows the example of germanium which has the same crystal structure as diamond. Figure 8.37(a) shows the absorption edge with the EXAFS superimposed. In Figure 8.37(b) the smooth background has been subtracted to give $\chi(k)$ which has then been multiplied by k, because of the k^{-1} factor in equation (8.20), and then by k^2, because the back-scattering amplitude $|f_j(k, \pi)|$ is proportional to k^{-2}, and plotted against k. The result consists of a sum of sine waves which, when Fourier transformed, gives the result in Figure 8.37(c). This shows a major peak at about 2.2 Å, corresponding to the four nearest neighbours in the diamond-type (tetrahedral) structure and other peaks corresponding to more distant neighbours. Then a 'filter' is applied to the curve in Figure 8.37(c) so that only the part between the dotted lines is retained—this

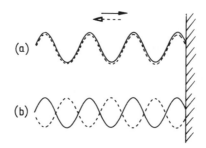

Figure 8.36 Scattering of an electron wave with (a) no phase change and (b) a π phase change.

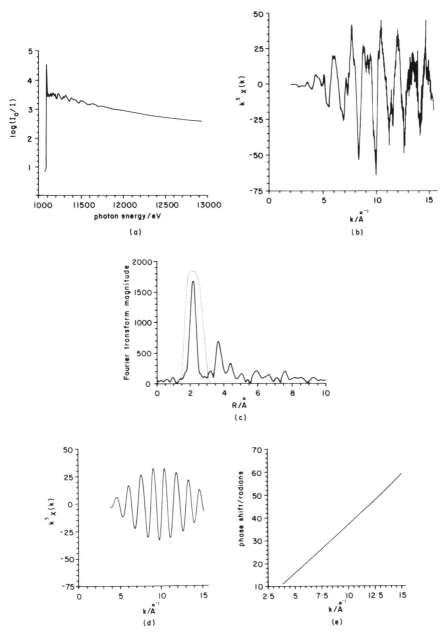

Figure 8.37 (a) X-ray absorption by powdered germanium. (b) The background has been subtracted and the remaining signal multiplied by k^3. (c) Fourier transform of (b). (d) Fourier transform of (b) back to k-space after filter window has been applied to (c). (e) Change of phase shift with k. (Reproduced, with permission, from Eisenberger, P., and Kincaid, B. M., *Science*, **200**, 1441, 1978. Copyright 1978 by The AAAS.)

is referred to as Fourier filtering. This part is then Fourier transformed back again to k-space giving the result in Figure 8.37(d). The phase shift $\delta(k)$ is obtained at the same time and Figure 8.37(e) shows how it varies with k.

All of the data processing is achieved by digitization of the raw data, like that in Figure 8.37(a), followed by treatment by computer.

In Figure 8.38 is shown an example of the determination from EXAFS of the Mo–S and Mo–N distances in crystalline tris(2-aminobenzenethiolate) Mo(VI), an octahedral compound with three bidentate ligands. Figure 8.38(a) shows that the Mo EXAFS experimental curve of $k^3\chi(k)$ versus k, after filtering, cannot be reproduced if the molybdenum atom is assumed to be coordinated to sulphur atoms at one distance R_j (dashed curve). There are clearly two waves of different wavelengths superimposed, indicating either sulphur atoms in two different positions with characteristic R_j's or, more likely, two kinds of near-neighbour atoms. The dashed curve in Figure 8.38(b) has been calculated assuming contributions from S and N atoms with an Mo–S distance of 2.42 Å and an Mo–N distance of 2.00 Å with typical uncertainties of 0.01 Å. There is now good agreement with the observed.

Figure 8.39 shows some results of EXAFS following absorption by iron atoms in proteins with three prototype iron–sulphur active sites. In the example in Figure 8.39(a) application of a 0.9 to 3.5 Å filter window before Fourier retransformation shows a single wave resulting from four identical Fe–S distances in a rubredoxin. In the examples of plant and bacterial ferrodoxins in Figure 8.39(b) and (c) there are two types of waves indicating two types of Fe–S linkage consistent with the structures shown.

The example in Figure 8.40 illustrates the use of EXAFS in indicating the degree of order in a sample of a catalyst. In this figure is shown the $k^3\chi(k)$

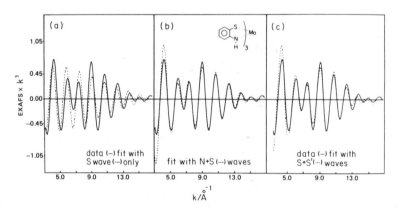

Figure 8.38 Curve fitting of Mo EXAFS for $Mo(SC_6H_4NH)_3$ taking into account (a) sulphur and (b) sulphur and nitrogen atoms as near neighbours. (Reproduced, with permission, from Winnick, H., and Doniach, S. (Eds) *Synchrotron Radiation Research*, p. 436, Plenum, New York, 1980.)

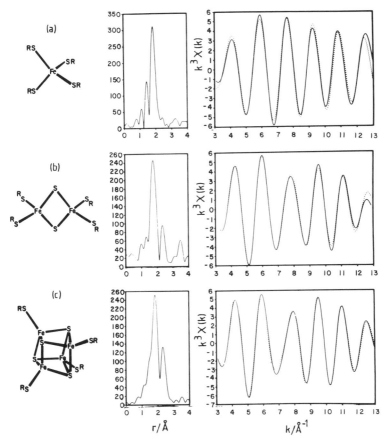

Figure 8.39 Fourier transformed Fe EXAFS and retransformation, after applying a 0.9–3.5 Å filter window, of (a) a rubredoxin, (b) a plant ferrodoxin, and (c) a bacterial ferrodoxin, whose structures are also shown. (Reproduced, with permission, from Teo, B. K., and Joy, D. C. (Eds), *EXAFS Spectroscopy*, p. 15, Plenum, New York, 1981.)

EXAFS signal from osmium atoms in the pure metal to be compared with that of a 1% osmium catalyst in small clusters dispersed on a silica surface. The main peak in the Fourier transformed spectrum is due to Os–Os nearest neighbours and is much more intense (*ca* 3.2) in the pure metal than in the catalyst (*ca* 0.8) indicating many more Os–Os bonds. When the filter window is applied, followed by retransformation, the main difference in the catalyst is that the single wave is damped much more rapidly than in the pure metal. The damping is due to the $\exp(-2\sigma_j^2 k^2)$ factor in equation (8.20). The quantity σ_j is a measure of the disorder in the material which is clearly greater in the catalyst resulting in more rapid damping. Analysis of the data has shown that (a) $N_1 = 8.3 \pm 2.0$

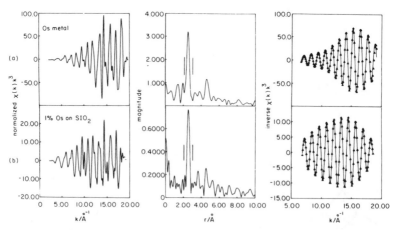

Figure 8.40 The $k^3\chi(k)$ EXAFS signal, Fourier transformed and then retransformed after application of the filter window indicated, in (a) osmium metal and (b) a 1% osmium catalyst supported on silica. (Reproduced, with permission, from Winnick, H., and Doniach, S. (Eds), *Synchrotron Radiation Research*, p. 413, Plenum, New York, 1980.)

for the catalyst compared with 12 for pure osmium, which has a face-centred cubic structure, and (b) $\Delta\sigma_1^2 = \sigma_1^2$ (catalyst) $- \sigma_1^2$ (metal) $= 0.0022 \pm 0.0002$, showing the great disorder in the catalyst.

The EXAFS method can be used for different elements in the same material or for different absorption edges, such as $1s$, $2s$, or $2p$, in the same element.

Auger electrons, produced in the way described in Section 8.2.1, may also produce scattering interference in the same way as ordinary photoelectrons. The technique is useful for studying surfaces and is known as surface EXAFS or SEXAFS. An example of its use is an investigation of Br_2 adsorbed onto carbon in the form of 50% carbon crystallites and 50% graphite. The fact that the X-ray radiation from a storage ring is plane polarized was used to show that, for a 20% Br_2 monolayer coverage, the Br–Br bond is randomly orientated relative to the surface whereas, for 60 to 90 per cent coverage, the alignment is parallel to the surface.

Questions

1. What would you expect to be the most important features of the He I UPS spectrum of mercury vapour?
2. Which low-lying states of NO^+ would you expect to feature in the He I UPS spectrum of NO? Indicate whether a long or short vibrational progression would be anticipated in each case.
3. Make measurements of band positions in the UPS spectrum in Figure 8.7 as accurately as you can and use these to determine ω_e and $\omega_e x_e$ (equations 7.82 and 6.18) for the ground electronic state of H_2^+.

4. For XPS spectra of a mixture of acetone and carbon dioxide gases explain, giving reasons, what you would expect to observe regarding the relative binding enegies in the O $1s$ and C $1s$ spectra.
5. Write down the configurations and derive the core states arising from Auger processes of the *KLM* type in krypton.
6. EXAFS spectra of platinum metal, having a face-centred cubic crystal structure, have been obtained at 300 and 673 K. Explain what qualitative differences you might expect. How many nearest-neighbour atoms are there in this structure? Illustrate your answer with a diagram.

Bibliography

Burhop, E. H. S. (1952). *The Auger Effect and Other Radiationless Transitions*, Cambridge University Press, London.
Carlson, T. A. (1975). *Photoelectron and Auger Spectroscopy*, Plenum, New York.
Eland, J. H. D. (1974). *Photoelectron Spectroscopy*, Butterworths, London.
Rabalais, J. W. (1977). *Principles of Ultraviolet Photoelectron Spectroscopy*, Wiley, New York.
Roberts, M. W. and McKee, C. S. (1978). *Chemistry of the Metal-Gas Interface*, Oxford University Press, Oxford.
Siegbahn, K., Nordling, C., Fahlman, A., Nordberg, R., Hamerin, K., Hedman, J., Johansson, G., Bergmark, T., Karlsson, S.-E., Lindgren, I., and Lindberg, B. (1967). *Electron Spectroscopy for Chemical Analysis—Atomic, Molecular, and Solid State Structure Studies by Means of Electron Spectroscopy*, Almqvist and Wiksells, Uppsala.
Teo, B. K., and Joy, D. C. (Eds) (1981). *EXAFS Spectroscopy*, Plenum, New York.
Winnick, H., and Doniach, S. (Eds) (1980) *Synchrotron Radiation Research*, Plenum, New York.

CHAPTER 9

Lasers and Laser Spectroscopy

9.1 General Discussion of Lasers

9.1.1 General Features and Properties

The word 'laser' is an acronym derived from light amplification by the stimulated emission of radiation. If the light concerned is in the microwave region then the alternative acronym 'maser' is often used. Although the first such device to be constructed was the ammonia maser in 1954 it is the lasers made subsequently which operate in the infrared, visible, or ultraviolet regions of the spectrum which have made a greater impact.

In Section 2.2 we have seen that emission of radiation by a source may be by a spontaneous (equation 2.4) or by an induced, or stimulated, process

$$M^* + hc\tilde{v} \rightarrow M + 2hc\tilde{v} \tag{9.1}$$

as in equation (2.5).

Laser radiation is emitted entirely by the process of stimulated emission, unlike the more conventional sources of radiation discussed in Chapter 3 which emit through a spontaneous process.

For induced emission from the upper energy level n of the two-level system in Figure 2.2(a) to dominate absorption (equation 2.3) there must be a population inversion between the two levels, that is $N_n > N_m$ where N_i refers to the population of state i. To disturb the normal Boltzmann population distribution (equation 2.11), in which $N_n < N_m$, requires an input of energy. The process by which such a population inversion is brought about is known as pumping. A system, which may be gaseous, solid, or liquid and in which a population inversion has been created is referred to as an active medium. According to equation (9.1) the active medium is capable of acting as an amplifier of radiation falling on it. The equation shows that, for every photon entering the active medium, two photons are emitted from it.

To make an oscillator from an amplifier requires, in the language of electronics, positive feedback. In lasers this is provided by the active medium being between two mirrors, both of them highly reflecting but one rather less so in order to allow some of the stimulated radiation to leak out and form the laser beam. The region bounded by the mirrors is called the laser cavity. Various mirror systems are used but that shown in Figure 9.1, consisting of two plane mirrors a distance d apart, is one of the simplest. The mirror separation must be an integral number of half-wavelengths, $n\lambda/2$, apart, necessitating extremely accurate alignment. The resonant frequency v of the cavity is then given by

$$v = \frac{nc}{2d} \tag{9.2}$$

The reflecting surfaces of the mirrors are specially coated, with alternate layers of high and low dielectric materials such as TiO_2 and SiO, to give almost total reflection at the specific laser wavelength. The usual aluminium, silver, or gold coatings are not sufficiently highly reflecting. One of the mirrors is coated so as to allow 1 to 10 per cent of the radiation to emerge as the laser beam. Photons of energy $hc\tilde{v}$ are generated initially in the cavity through spontaneous emission. Those that strike the cavity mirrors at 90° are retained within the cavity causing the photon flux to reach a level which is sufficiently high to cause stimulated emission to occur and the active medium is said to lase.

Laser radiation has four very remarkable properties:

1. *Directionality.* The laser beam emerging from the output mirror of the cavity is highly parallel, which is a consequence of the strict requirements for the alignment of the cavity mirrors. Divergence of the beam is typically a few milliradians.
2. *Monochromaticity.* If the energy levels n and m in Figure 2.2(a) are sharp, as they are in a gaseous active medium, the Planck relation of equation (2.2) limits the wavelength range of the radiation. However, whatever the nature of the active medium, the fact that the laser cavity is resonant only for the frequencies given by equation (9.2) limits the wavelength range.
3. *Brightness.* This is defined as the power emitted per unit area of the output mirror per unit solid angle and is extremely high compared with a conventional source. The reason for this is that, although the power may be only modest, as in, for example, a low power 0.5 mW helium–neon gas laser, the solid angle over which it is distributed is very small.
4. *Coherence.* Conventional sources of radiation are incoherent, which means that the electromagnetic waves associated with any two photons of the same wavelength are, in general, out-of-phase. The coherence of laser radiation is

Figure 9.1   Laser cavity with two plane mirrors.

both temporal and spatial, the coherence lasting for a relatively long time and extending over a relatively large distance. Coherence of laser radiation is responsible for its use as a source of intense local heating, as in metal cutting and welding, and in holography.

9.1.2 Methods of Obtaining Population Inversion

Equation (9.1) may give the impression that in the stimulated emission process we are getting something for nothing—putting in one quantum of energy and getting out two. The process does involve an amplification of the radiation (hence the 'light amplification' which appears in the acronym) but energy has to be put into the system to excite M to M* so that, in the overall process of M being excited and then undergoing stimulated emission, there is no energy gain. Not only is there no energy gain but the efficiency of the overall process is very low. For example, a nitrogen gas laser has an efficiency of less than 0.1 per cent and a semiconductor (diode) laser, one of the best in this respect, has an efficiency of about 30 per cent.

Before we look at the various methods of pumping we shall consider the types of energy level scheme encountered in lasing materials.

So far we have thought of the stimulated emission occurring in a lasing material as being in a simple two-level system like that in Figure 9.2(a). In fact, a laser operating through such a two-level system is very unusual—the excimer laser discussed in Section 9.2.8 is such an example. The reason for this is that, under equilibrium conditions, level 2 will have a much lower population than level 1 (see equation 2.11). If level 2 is a high-lying vibrational or an electronic energy level the population will be negligibly small. Pumping with energy $E_2 - E_1$ results, initially, in net absorption which continues until the populations are equal, a condition known as saturation and encountered in Section

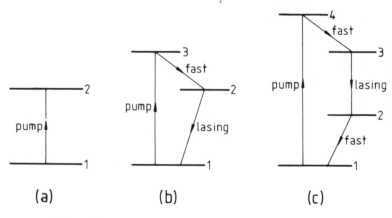

Figure 9.2 (a) Two-, (b) three-, and (c) four-level lasing systems.

2.3.4.2. At this point further pumping results in absorption and induced emission occurring at the same rate so that population inversion cannot normally be achieved.

Commonly a three- or four-level system, illustrated in Figure 9.2(b) and (c), is necessary for population inversion to be obtained between two of the levels.

In the three-level system of Figure 9.2(b) population inversion between levels 2 and 1 is achieved by pumping the 3–1 transition. The 3–2 process must be efficient and fast in order to build up the population of level 2 while that of level 1 is depleted. Lasing occurs in the 2–1 transition.

The four-level system in Figure 9.2(c) is even more efficient in the creation of a population inversion, in this case between levels 3 and 2. The reason for the greater efficiency is that, not only is level 3 populated through the fast 4–3 process, but the population of level 2 is rapidly depleted by the fast 2–1 process.

In more complex systems there may be more levels between 4 and 3 and between 2 and 1, all involved in fast processes to lower levels, but they are still referred to as four-level systems.

Population inversion is not only difficult to achieve but also to maintain. Indeed, for many laser systems there is no method of pumping which will maintain a population inversion continuously. For such systems inversion can be brought about only by means of a pumping source which delivers short, high energy pulses. The result is a pulsed laser as opposed to a continuous wave, or CW, laser which operates continuously.

Methods of pumping, irrespective of the type of level system and of whether lasing is to be pulsed or CW, fall into two general categories—optical and electrical pumping.

Optical pumping involves the transfer of energy to the system from a high intensity light source and is used particularly for solid and liquid lasers. A very high photon flux can be obtained over a short period of time from inert gas flashlamps of the type used in flash photolysis described in Section 3.5.4. The result is a pulsed laser, the repetition rate being that of the pumping source. CW optical pumping may be achieved in some lasers by a continuously acting tungsten–iodine, krypton, or high pressure mercury arc lamp.

Electrical pumping is used for gas and semiconductor lasers. In a gas laser this involves an electrical discharge in the gas which may be induced by microwave radiation outside the gas cell or by a high voltage across electrodes inside the cell. The pumping is achieved through collisions between the gaseous atoms or molecules and electrons produced with high translational energy in the discharge.

In some gas lasers it is preferable to use a mixture of the lasing gas M and a second gas N, where N serves only to be excited to N* by collisions with electrons and to transfer this energy to M by further collisions:

$$M + N^* \rightarrow M^* + N \tag{9.3}$$

Ideally N* is a long-lived, metastable state with an energy similar to that of the level of M* which is being pumped.

9.1.3 Laser Cavity Modes

The cavity of a laser may resonate in various ways during the process of generation of radiation. The cavity, which we can regard as a rectangular box with a square cross-section, has modes of oscillation, referred to as cavity modes, which are of two types, transverse and axial (or longitudinal) modes. These are, respectively, normal to and along the direction of propagation of the laser radiation.

The transverse modes are labelled $\text{TEM}_{m\ell}$ where TEM stands for transverse electric and magnetic (field): m and ℓ are integers which refer to the number of vertical and horizontal nodal planes of the oscillation respectively. Usually it is preferable to use only the TEM_{00} mode which produces a laser beam with a gaussian intensity distribution (see Figure 2.5) normal to the direction of propagation.

The variety of possible axial modes is generally of greater consequence. The various frequencies possible are given by equation (9.2) so that the separation Δv of the axial modes is given by

$$\Delta v = \frac{c}{2d} \tag{9.4}$$

For a cavity length of, for example, 50 cm the axial mode separation is 300 MHz (0.01 cm^{-1}).

In practice the laser can operate only when n, in equation (9.2), takes values such that the corresponding resonant frequency v lies within the line width of the transition between the two energy levels involved. If the active medium is a gas this line width may be the Doppler line width (see Section 2.3.2). Figure 9.3 shows a case where there are twelve axial modes within the Doppler profile. The number of modes in the actual laser beam depends on how much radiation is allowed to leak out of the cavity. In the example in Figure 9.3 the output level has been adjusted so that the so-called threshold condition allows six axial

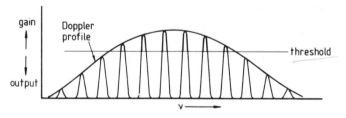

Figure 9.3 Doppler limited laser line with twelve axial modes within the line width.

modes in the beam. The gain, or the degree of amplification, achieved in the laser is a measure of the intensity.

It is clear from Figure 9.3 that the laser line width for a single axial mode is much less than the Doppler line width. Normally a laser will operate in multimode fashion but for many purposes, e.g. in high resolution spectroscopy using a laser source, it is desirable that the laser be made to operate in a single mode. One possible way of achieving this is clear from equation (9.4). By making the cavity length d sufficiently short that only one axial mode lies within the Doppler profile single-mode operation results. This method is applicable mostly to infrared lasers where the Doppler line width, of the order of 100 MHz, is relatively small. Whatever method is used for single-mode operation there is a considerable loss of laser power compared to multimode operation.

9.1.4   Q-switching

The quality factor Q of a laser cavity is defined as

$$Q = \frac{v}{\Delta v} \tag{9.5}$$

where Δv is the laser line width. Q can be regarded as the 'resolving power' of the cavity, as defined in equation (3.3) for the dispersing element of a spectrometer. Single-mode operation results in a smaller Δv and, therefore, a higher Q than multimode operation. Q is related to the energy E_c stored in the cavity and the energy E_t allowed to leak out in time t by

$$Q = \frac{2\pi v E_c t}{E_t} \tag{9.6}$$

Q-switching is an operation whereby the Q of a laser cavity is reduced for a short period of time by preventing the radiation from being reflected backwards and forwards between the cavity mirrors. During this time the population of the upper of the two levels involved in laser action builds up to a much higher value than it would if Q remained high. Then Q is allowed to increase rapidly and the cycle is repeated, resulting in very short laser pulses. Since the pulse duration Δt is related to the pulse power P_p and energy E_p by

$$P_p = \frac{E_p}{\Delta t} \tag{9.7}$$

a much shorter pulse leads to a large increase in power. The resulting pulse is referred to as a giant pulse.

Various methods of Q-switching are used. A rotating mirror in the laser cavity is simple in principle but a Pockels cell is used more often.

A Pockels cell is made from an electro-optic material which becomes

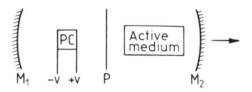

Figure 9.4   Use of Pockels cell (PC) in a laser cavity to produce Q-switching.

birefringent (doubly refracting) when a voltage is applied across it. The result is that, if plane polarized radiation passes through the material, it emerges, in general, elliptically polarized. Such materials are pure crystalline ammonium dihydrogen phosphate $((NH_4)H_2PO_4)$ or ADP, potassium dihydrogen phosphate (KH_2PO_4) or KDP, and potassium dideuterium phosphate (KD_2PO_4) or KD*P. All may be used in a laser operating in the visible region.

Figure 9.4 shows how a Pockels cell may be used in a laser cavity. The radiation in the cavity is plane-polarized by the polarizer P and passes to the Pockels cell where it becomes circularly polarized when the voltage is applied. On reflection at the mirror M_1 the direction of circular polarization is reversed. On passing through the Pockels cell a second time it is plane-polarized but at 90° to the original plane and, therefore, is not transmitted by the polarizer. It is only when the voltage is switched off that a Q-switched giant pulse emerges from the output mirror M_2. The timing of the voltage-switching determines the power and length of the output pulses.

9.1.5   Mode Locking

Although Q-switching produces shortened pulses, typically 10 to 200 ns long, if we require pulses in the picosecond (10^{-12} s) range the technique of mode locking may be used. This technique is applicable only to multimode operation of a laser and involves exciting many axial cavity modes but with the correct amplitude and phase relationship. The amplitudes and phases of the various modes are normally quite random.

Each axial mode has its own characteristic pattern of nodal planes and the frequency separation Δv between modes is given by equation (9.4). If the radiation in the cavity can be modulated at a frequency of $c/2d$ then the modes of the cavity are locked both in amplitude and phase since t_r, the time for the radiation to make one round-trip of the cavity (a distance $2d$), is given by

$$t_r = \frac{2d}{c} \tag{9.8}$$

The result is that, for the case of a cavity operated with, say, seven modes, the output is like that in Figure 9.5 when the cavity is mode locked. Only the

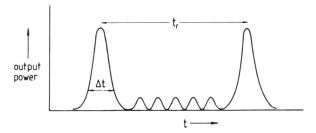

Figure 9.5 Suppression of five out of seven axial cavity modes by mode locking.

modes which have a node at one end of the cavity are output from the laser and all others are suppressed.

The width Δt of the pulse at half-height is given by

$$\Delta t = \frac{2\pi}{(2N + 1)\Delta v} \tag{9.9}$$

where $(2N + 1)$ is the number of axial modes excited and Δv is their frequency separation.

One method of mode locking a visible laser is by placing an acoustic modulator in the cavity and driving it at a frequency of $c/2d$.

An important consequence of shortening a laser pulse is that the line width is increased as a result of the uncertainty principle as stated in equation (1.16). When the width of the pulse is very small there is difficulty in measuring the energy precisely because of the rather small number of wavelengths in the pulse. For example, for a pulse width of 40 ps there is a frequency spread of the laser, given approximately by $(2\pi\Delta t)^{-1}$, of about 4.0 GHz (0.13 cm^{-1}). Pulse lengths of < 100 fs (1 femtosecond $= 10^{-15}$ s) have been achieved in this way.

9.1.6 Harmonic Generation

In the context of discussion of the Raman effect, equation (5.43) relates the oscillating electric field E of the incident radiation, the induced electric dipole μ, and the polarizability α by

$$\mu = \alpha E \tag{9.10}$$

In fact, this equation is only approximate and μ should really be expressed as a power series in E

$$\mu = \mu^{(1)} + \mu^{(2)} + \mu^{(3)} + \cdots \tag{9.11}$$
$$= \alpha E + \tfrac{1}{2}\beta E.E + \tfrac{1}{6}\gamma E.E.E + \cdots$$

where β is known as the hyperpolarizability and γ the second hyperpolarizability. Any effects due to the second (or higher) terms in the series are referred to as non-linear effects because they arise from terms which are non-linear in E. These effects are usually small but the very high power, and therefore E, characteristic of laser radiation causes them to be important.

The magnitude of the oscillating electric field is given by

$$E = A \sin 2\pi vt \qquad (9.12)$$

where A is the amplitude and v the frequency. Since

$$E^2 = A^2(\sin 2\pi vt)^2 = \tfrac{1}{2}A^2(1 - \cos 2\pi 2vt) \qquad (9.13)$$

the radiation scattered by the sample contains, due to the $\mu^{(2)}$ term, some radiation with *twice* the frequency (or *half* the wavelength) of the incident radiation. The phenomenon is called frequency doubling or second harmonic generation. In general, higher order terms in equation (9.11) can result in third, fourth, etc., harmonic generation.

There are several pure crystalline materials which may be used for frequency doubling. Examples are ADP, KDP, and KD*P, mentioned in Section 9.1.4, potassium pentaborate (KB_5O_8) or KPB, β-barium borate (BaB_2O_4) or BBO, and lithium niobate (Li_3NbO_4), each material being suitable for incident radiation of only a limited wavelength range in the visible region. The importance of these materials is that a laser operating in the visible region, of which there are a relatively large number, can be made to operate in the near-ultraviolet, where there are relatively few.

The efficiency of frequency doubling is quite low, often only a few per cent but it may be as high as 20 to 30 per cent.

9.2 Examples of Lasers

9.2.1 The Ruby and Alexandrite Lasers

After the ammonia maser, operating in the microwave region and constructed by Townes *et al.* in 1954, the next major step forward was the ruby laser, operating in the red region of the spectrum and demonstrated in 1960 by Maiman. This is a solid state laser employing a ruby crystal consisting of aluminium oxide, Al_2O_3, containing 0.5 per cent by weight of Cr_2O_3 giving it a pale pink colour.

The lasing constituent is the Cr^{3+} ion which is in such low concentration that it can be regarded as a free ion. The ground configuration of Cr^{3+} (see Table 7.1) is $KL3s^23p^63d^3$ which gives rise to eight terms (see Table 7.2) of which ^4F is, according to Hund's rules (see Section 7.1.2.3b), the lowest lying (ground) term. Of the others ^2G is the lowest excited term. Each Cr^{3+} ion is in a crystal field (see Section 7.3.1.4a) of approximately octahedral symmetry. In the octahedral point group O_h, the ^4F ground term gives $^4A_2$, $^4T_1$, and $^4T_2$ states while the ^2G excited term gives $^2A_1$, ^2E, $^2T_1$, and $^2T_2$ states. Of these, $^4A_2$ is the ground state and $^4T_1$, $^4T_2$, ^2E, and $^2T_2$ are relatively low-lying excited states: all are shown in Figure 9.6(a).

The $^4T_1$ and $^4T_2$ states are broadened due to slight variations in the crystal field. The $^2T_2$ and ^2E states are sharper but the ^2E state is split into two components, 29 cm^{-1} apart, because of the slight distortion of the octahedral field. Population inversion and consequent laser action occurs between the ^2E and $^4A_2$ states. This is achieved by optical pumping into the $^4T_2$ or $^4T_1$ states with 510 to 600 or 360 to 450 nm radiation respectively. The ruby laser is seen to be a three-level laser illustrated in Figure 9.2(b). The broadness of the $^4T_2$ and $^4T_1$ states contributes to the efficiency of pumping which is achieved with a flashlamp of the type described in Section 3.5.4. This may be in the form of a helix around the ruby crystal as shown in Figure 9.6(b), the whole being contained in a reflector. The mirror material is deposited directly on to the ends of the crystal which may be as large as 2 cm in diameter and 20 cm in length.

The transitions labelled R$_1$ and R$_2$ in Figure 9.6(a) are at 694.3 and 693.4 nm respectively, but laser action involves principally R$_1$.

The laser normally operates in the pulsed mode because of the necessity of the dissipation of a large amount of heat between pulses.

The efficiency of a ruby laser is less than 0.1 per cent, typically low for a three-level laser.

Alexandrite, like ruby, contains Cr^{3+} ions but they are substituted in the lattice of chrysoberyl, $BeAl_2O_4$. The chromium ions occupy two symmetrically non-equivalent positions which would otherwise be occupied by aluminium ions. In this environment the $^4A_2$ ground state of Cr^{3+} is broadened, compared to that in ruby, by coupling to vibrations of the crystal lattice.

Lasing occurs at 680.4 nm in alexandrite, the transition involved being analogous to R_1 in ruby (see Figure 9.6a) and not involving any vibrational excitation in the $^4A_2$ state. However, of much greater importance in alexandrite

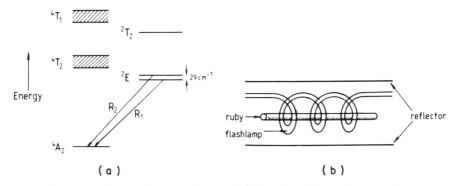

Figure 9.6 (a) Low-lying energy levels of Cr^{3+} in ruby. (b) Design for a ruby laser.

is laser action between the $^4T_2$ and $^4A_2$ states. Because both the $^4T_2$ and $^4A_2$ states are broadened by vibrational coupling the alexandrite laser is sometimes referred to as a vibronic laser. As a result of this broadening of both the electronic states involved in the $^4T_2$–$^4A_2$ transition the laser can be tuned over a wide wavelength range, 720–800 nm. This tunability gives the alexandrite laser a great advantage over the ruby laser which is limited to just two wavelengths.

A further advantage is the higher efficiency of the alexandrite laser due to its being a four-level laser. In the illustration in Figure 9.2(c), level 4 is a vibronic level and 3 the zero-point level of the $^4T_2$ state. Level 2 is a vibronic level of the $^4A_2$ state and 1 the zero-point level. Because of the excited nature of level 2 it is almost depopulated at room temperature so that a population inversion between levels 3 and 2 is relatively easy to achieve. In fact level 2 is a continuous band of vibronic levels covering a wide energy range resulting in the wide wavelength range over which the laser is tunable.

Pumping is with a flashlamp, as in the case of the ruby laser, and a pulse energy of the order 1 J may be achieved. Frequency doubling (second harmonic generation) can provide tunable radiation in the 360–400 nm region.

9.2.2 The Titanium–Sapphire Laser

Despite the fact that the first laser to be produced (the ruby laser, Section 9.2.1) has the remarkable property of having all its power concentrated into one or two wavelengths, a property possessed by most lasers, it was soon realized that the inability to change these wavelengths appreciably, that is to tune the laser, is a serious drawback which limits the range of possible applications.

Historically, the first type of laser to be tunable over an appreciable wavelength range was the dye laser, to be described in Section 9.2.10. The alexandrite laser (Section 9.2.1), a tunable solid state laser, was first demonstrated in 1978 and then, in 1982, the titanium–sapphire laser. This is also a solid state laser but tunable over a larger wavelength range, 670–1100 nm, than the alexandrite laser which has a range of 720–800 nm.

The lasing medium in the titanium–sapphire laser is crystalline sapphire (Al_2O_3) with about 0.1 per cent by weight of Ti_2O_3. The titanium is present as Ti^{3+} and it is between energy levels of this ion that lasing occurs.

The ground configuration of Ti^{3+} (see Table 7.1) is $KL3s^23p^63d^1$. The crystal field experienced by the ion splits the $3d$ orbital into a triply degenerate lower energy t_2 orbital and a doubly degenerate higher energy e orbital (see Figure 7.38). If the electron is in the lower orbital a $^2T_2$ ground state results and, if it is in the upper orbital, a ^2E excited state results. These states are about 19 000 cm^{-1} apart but each is split into further components and is also coupled to the vibrations of the crystal lattice. In a similar way to that in alexandrite (Section 9.2.1) population inversion can be created between these two sets of levels resulting in a four-level vibronic laser with a tunable range of 670–1100 nm.

A further advantage, compared to the alexandrite laser, apart from a wider tuning range, is that it can operate in the CW as well as in the pulsed mode. In the CW mode the Ti^{3+}–sapphire laser may be pumped by a CW argon ion laser (see Section 9.2.5) and is capable of producing an output power of 5 W. In the pulsed mode pumping is usually achieved by a pulsed Nd^{3+}:YAG laser (see Section 9.2.3) and a pulse energy of 100 mJ may be achieved.

9.2.3 The Neodymium–YAG Laser

Laser action can be induced in Nd^{3+} ions embedded in a suitable solid matrix. Several matrices, including some special glasses, are suitable but one of the most frequently used is yttrium aluminium garnet ($Y_3Al_5O_{12}$) which is referred to as YAG.

The neodymium atom has the ground configuration ... $4d^{10}4f^45s^25p^66s^2$ and a $^5I_4$ ground state (see Table 7.1). The ground configuration of Nd^{3+} is ... $4d^{10}4f^35s^25p^6$ and, of the terms arising from it, ^4I and ^4F are important in the laser. For the ^4I term $L = 6$ and $S = \frac{3}{2}$ giving $J = \frac{15}{2}, \frac{13}{2}, \frac{11}{2}, \frac{9}{2}$ in the Russell–Saunders approximation (see Section 7.1.2.3). The multiplet is normal, i.e. the

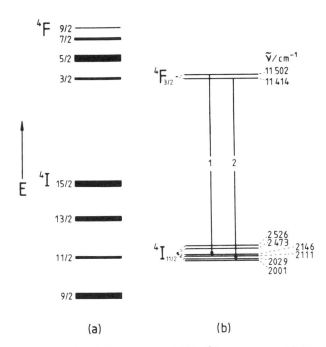

(a)　　　　　　　　　　(b)

Figure 9.7 Energy levels in (a) free Nd^{3+} and (b) Nd^{3+} split by crystal field interactions.

lowest value of J has the lowest energy, as in Figure 9.7(a). Also shown is the normal multiplet arising from the ^4F term.

Laser action involves mainly the $^4F_{3/2} - {}^4I_{11/2}$ transition at about 1.06 μm. Since $^4I_{11/2}$ is not the ground state, the laser operates on a four-level system (see Figure 9.2c) and consequently is much more efficient than the ruby laser.

In free Nd^{3+} the $^4F_{3/2} - {}^4I_{11/2}$ transition is doubly forbidden, violating the $\Delta L = 0, \pm 1$ and $\Delta J = 0, \pm 1$ selection rules (see Section 7.1.6). In the YAG crystal the $^4I_{11/2}$ state of Nd^{3+} is split by crystal field interactions into six and the $^4F_{3/2}$ state into two components, as shown in Figure 9.7(b). There are eight transitions, grouped around 1.06 μm, between the components but only the two marked in the figure are important. At room temperature transition 1 at 1.0648 μm is dominant but, at 77 K, transition 2 at 1.0612 μm is dominant.

A krypton arc lamp may be used for CW pumping or a flashlamp for much higher power, pulsed operation.

The Nd^{3+}:YAG rod is a few centimetres long and contains 0.5 to 2.0 per cent by weight of Nd^{3+}. In pulsed operation the peak power of each pulse is sufficiently high for generation of second, third, or fourth harmonics at 533, 355, and 266 nm respectively, using suitable crystals.

9.2.4 The Diode or Semiconductor Laser

A diode, or semiconductor, laser operates in the near-infrared and just into the visible region of the spectrum. Like the ruby and Nd^{3+}:YAG lasers it is a solid state laser but the mechanism involved is quite different.

Figure 9.8(a) shows how the conduction band[†] C and the empty valence band V are not separated in a conductor whereas Figure 9.8(c) shows that they are well separated in an insulator. The situation in a semiconductor, shown in Figure 9.8(b), is that the band gap, between the conduction and valence bands, is sufficiently small that promotion of electrons into the conduction band is possible by heating the material. For a semiconductor the Fermi energy E_F, such that at $T = 0$ K all levels with $E < E_F$ are filled, lies between the bands as shown.

Semiconductors may also be made from a material which is normally an insulator by introducing an impurity, a process known as doping. Figure 9.9 shows two ways in which an impurity may promote semiconducting properties. In Figure 9.9(a) the dopant has one more valence electron per atom than the host and contributes a band of filled impurity levels I close to the conduction band of the host. This characterizes an n-type semiconductor. An example is silicon ($KL3s^23p^2$) doped with phosphorus ($KL3^23p^3$) which reduces the band gap to about 0.05 eV. Since kT at room temperature is about 0.025 eV, the

---

[†] Bands in the solid state can be regarded as grossly delocalized orbitals extending throughout the sample.

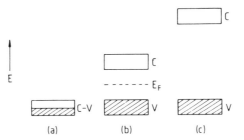

Figure 9.8 Conduction band, C, and valence band, V, in (a) a conductor, (b) a semiconductor, and (c) an insulator.

phosphorus converts silicon from a high temperature semiconductor into a room temperature semiconductor.

Alternatively, as in Figure 9.9(b), a dopant with one valence electron fewer than the host contributes an impurity band I which is empty but more accessible to electrons from the valence band. An example of such a p-type semiconductor is silicon doped with aluminium ($KL3s^2 3p^1$) in which the band gap is about 0.08 eV.

A semiconductor laser takes advantage of the properties of a junction between a p-type and an n-type semiconductor made from the same host material. Such an n–p combination is called a semiconductor diode. Doping concentrations are quite high and, as a result, the conduction and valence band energies of the host are shifted in the two semiconductors, as shown in Figure 9.10(a). Bands are filled up to the Fermi level with energy E_F.

If a voltage is applied to the junction with the negative and positive terminals attached to the n and p regions respectively, electrons flow from the n to the p region and positive holes in the opposite direction. The levels are also displaced, as shown in Figure 9.10(b), and the Fermi energies $E_F'(n)$ and $E_F''(p)$ are now unequal, resulting in a population inversion in the region of the junction and leading to laser action. The semiconductor laser is, unusually, an example of a two-level system, but the population inversion is not obtained by pumping: we saw in Section 9.1.2 that this could not be done.

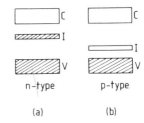

Figure 9.9 Impurity levels I in (a) an n-type and (b) a p-type semiconductor.

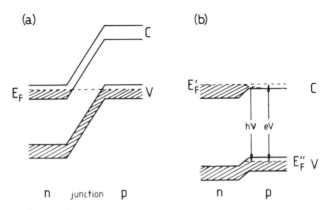

Figure 9.10 (a) The Fermi level in the region of a p-n junction. (b) The result of applying a voltage across the junction.

A typical semiconductor laser, shown in Figure 9.11, is small, only a few millimetres long and with an effective thickness of about 2 μm.

A variety of materials are used depending on the region in which the laser is required to operate. For example, a range of lead alloy semiconductors such as $Pb_{1-x}Sn_xSe$ and $PbS_{1-x}Se_x$ covers the range 2.8 to 30 μm. Semiconductor lasers can be tuned but the tuning range of a particular laser is small so that a whole series of them is necessary to cover an appreciable wavelength range. Gross tuning of the wavelength is achieved by surrounding the laser with a refrigeration unit to control and vary the temperature.

The two ends of the laser diode in Figure 9.11 are polished to increase internal reflection. As a consequence of the cavity geometry the laser beam is, unlike that of most lasers, highly divergent.

Semiconductor lasers are some of the most efficient of all lasers with an efficiency of about 30 per cent.

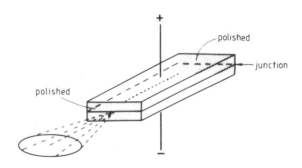

Figure 9.11 A semiconductor, or diode, laser.

9.2.5 The Helium–Neon Laser

The helium–neon laser is a CW gas laser which is simple and reliable to operate and, if the laser is of relatively low power, quite inexpensive.

Laser action takes place between excited levels of the neon atoms, in a four-level scheme, while the helium atoms serve only to mop up energy from the pump source and transfer it to neon atoms on collision. The energy level scheme is shown in Figure 9.12.

Pumping is electrical, a discharge being created in a helium–neon gas mixture by applying either a high voltage through internal electrodes, or by applying microwave radiation externally. Helium atoms are excited, on collisions with electrons present in the discharge, to various excited states. Of these the 2^3S_1 and 2^1S_0 states are metastable, and therefore long-lived, because transitions to the 1^1S_0 ground state are forbidden (see Section 7.1.5).

The ground configuration of Ne is $1s^2 2s^2 2p^6$, giving a $^1S_0$ state. The excited configurations give rise to states to which the Russell–Saunders approximation does not apply. Nevertheless, any ... $2p^5 ns^1$ or ... $2p^5 np^1$ configuration, with $n > 2$, gives rise to four or ten states respectively, as would be the case in the Russell–Saunders approximation (see Section 7.1.2.3) and as is indicated in boxes in Figure 9.12. We shall not be concerned with the approximation which is appropriate for describing these states.

The states arising from the ... $2p^5 5s^1$ configuration of Ne have very nearly the same energy as that of the 2^1S_0 state of He so that collisional energy transfer results in efficient population of these Ne states. Similarly the states arising

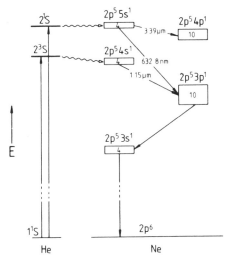

Figure 9.12 Energy levels of the He and Ne atoms relevant to the helium–neon laser. The number of states arising from each Ne configuration is given in a 'box'.

from the ... $2p^54s^1$ configuration of Ne lie just below the 2^3S_1 state of He and are also populated by collisions. All the ... $2p^5ns^1$ states have lifetimes of the order of 100 ns compared with 10 ns for the ... $2p^5np^1$ states. These conditions are ideal for four-level lasing in Ne with population inversion between the ... $2p^5ns^1$ and ... $2p^5np^1$ states.

The first laser lines to be discovered in the He–Ne system were a group of five in the infrared close to 1.15 μm and involving ... $2p^54s^1 - $ ... $2p^53p^1$ transitions, the strongest being at 1.1523 μm. Similarly, the ... $2p^55s^1 - $ ... $2p^53p^1$ transitions give laser lines in the red region, the one at 632.8 nm being the strongest.

Infrared laser lines involving ... $2p^55s^1 - $ ... $2p^54p^1$ transitions in the 3.39 μm region are not particularly useful. However, they do cause some problems in a 632.8 nm laser because they deplete the populations of the ... $2p^55s^1$ states and decrease the 632.8 nm intensity. The 3.39 μm transitions are suppressed by using multilayer cavity mirrors designed specifically for the 632.8 nm wavelength or by placing a prism in the cavity orientated so as to deflect the infrared radiation out of the cavity.

Decay from the ... $2p^53p^1$ states to the ... $2p^53s^1$ states is rapid but the ... $2p^53s^1$ states are relatively long-lived. Their populations tend to build up and this increases the probability of the ... $2p^53p^1 \rightarrow$ ... $2p^53s^1$ radiation being reabsorbed, a process known as radiation trapping, thereby increasing the ... $2p^53p^1$ population and decreasing the laser efficiency at 632.8 nm (and 1.15 μm). Depopulation of the ... $2p^53s^1$ states occurs on collisions with the walls of the discharge tube. For this reason narrow tubes, only a few millimetres in diameter, are used.

Figure 9.13 illustrates the construction of a helium–neon laser. In this example the discharge in the gas mixture (typically ten parts of He to one of Ne at a total pressure of about 1 Torr) is excited by internal electrodes. On the ends of the discharge tube are Brewster angle windows to prevent excessive light loss from multiple transmissions. If a window is at 90° to the optic axis of the laser a certain percentage is lost every time the radiation passes through. On the other hand, if the window is oriented at Brewster's angle ϕ, as shown, some is lost by reflection on the first transmission but no more is lost on subsequent transmissions. As the figure also shows, for unpolarized radiation

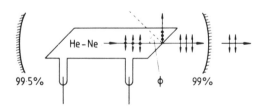

Figure 9.13 A helium–neon laser.

incident from inside the cavity, the transmitted and reflected radiation are plane polarized, the planes being at $90°$ to each other. The laser beam is, therefore, plane polarized.

Brewster's angle is given by

$$\tan \phi = n \tag{9.14}$$

where n is the refractive index of the window material. Since n varies with wavelength, so does ϕ but, for glass in the visible region, $\phi \simeq 57°$ and varies little with wavelength.

9.2.6 The Argon Ion and Krypton Ion Lasers

Laser action occurs in the noble gas ions Ne^+, Ar^+, Kr^+, and Xe^+ but that in Ar^+ and Kr^+ produces the most useful lasers.

These ion lasers are very inefficient, partly because energy is required first to ionize the atom and then to produce the population inversion. This inefficiency leads to a serious problem of heat dissipation which is partly solved by using a plasma tube, in which a low voltage high current discharge is created in the Ar or Kr gas, made from beryllium oxide, BeO, which is an efficient heat conductor. Water cooling of the tube is also necessary.

Most Ar^+ and Kr^+ lasers are CW. A gas pressure of about 0.5 Torr is used in a plasma tube of 2 to 3 mm bore. Powers of up to 40 W distributed among various laser wavelengths can be obtained.

The spectroscopy of ion lasers is generally less well understood than that of neutral atom lasers because of the lack of detailed knowledge of ion energy level schemes. Indeed, ion lasers were first produced accidentally and attempts to assign the transitions came later.

The ground configuration of Ar^+ is $KL3s^23p^5$, giving an inverted $^2P_{3/2}$, $^2P_{1/2}$ multiplet. The excited states involved in laser action involve promotion of an electron from the $3p$ orbital into excited $4s,5s,4p,5p,3d,4d,\ldots$ orbitals. Similarly, excited states of Kr^+ involved arise from promotion of an electron from the $4p$ orbital. In Ar^+ the $KL3s^23p^4$ configuration gives rise to $^1S,^3P,^1D$ terms (see Section 7.1.2.3). Most laser transitions involve the core in one of the ^3P states and the promoted electron in the $4p$ orbital.

The Ar^+ laser produces about ten lines in the 454 to 529 nm region, the most intense being at 488.0 and 514.5 nm. The Kr^+ laser produces about nine lines in the 476 to 800 nm region, with the 647.1 nm line being the most intense. Quite commonly a laser contains a mixture of argon and krypton gases and is capable of producing a fairly wide range of wavelengths.

9.2.7 The Nitrogen (N_2) Laser

The molecular orbital configuration of N_2 has been described in Section 7.2.1.1. The ground configuration of equation (7.57) can be abbreviated to

Table 9.1 Configurations and bond lengths of N_2.

State	MO configuration	$r_e/\text{Å}$
$X^1\Sigma_g^+$	$\ldots(\sigma_u^*2s)^2(\pi_u2p)^4(\sigma_g2p)^2$	1.0977
$A^3\Sigma_u^+$	$\ldots(\sigma_u^*2s)^2(\pi_u2p)^3(\sigma_g2p)^2(\pi_g^*2p)^1$	1.2866
$B^3\Pi_g$	$\ldots(\sigma_u^*2s)^2(\pi_u2p)^4(\sigma_g2p)^1(\pi_g^*2p)^1$	1.2126
$C^3\Pi_u$	$\ldots(\sigma_u^*2s)^1(\pi_u2p)^4(\sigma_g2p)^2(\pi_g^*2p)^1$	1.1487

$\ldots(\sigma_u^*2s)^2(\pi_u2P)^4(\sigma_g2p)^2$ and gives rise to the $X^1\Sigma_g^+$ ground state. When an electron is promoted to a higher energy orbital, singlet and triplet states result. We shall be concerned here with only the triplet states and, in particular, the $A^3\Sigma_u^+$, $B^3\Pi_g$, and $C^3\Pi_u$ states[†]. The orbital configurations and values of r_e, the equilibrium internuclear distance, for these states are given in Table 9.1.

In a high voltage discharge through nitrogen gas there is a deep pink glow due mainly to two electronic band systems in emission. The B–A system, or so-called first positive system because it was thought initially to be due to N_2^+, stretches from the red to the green region while the C–B system, or so-called second positive system, stretches from the blue into the near-ultraviolet.

Laser action has been obtained in a few transitions in both these systems but the C–B laser action has proved more important because it resulted in the first ultraviolet laser. It is only this system that we shall consider here.

The values of the equilibrium internuclear distance r_e for the various states in Table 9.1 indicate that the minimum of the potential for the C state lies almost vertically above that of the X state, as in Figure 7.20(b), whereas those of states B and A are shifted to high r. The result is that the electron–molecule collisional cross-section for the transition from $v'' = 0$ in the X state to $v' = 0$ in the C state is greater than that for analogous transitions in the A–X and B–X systems. A population inversion is created between the $v = 0$ level of the C state and the $v = 0$ level of the B state. Lasing has been observed in the 0–0 transition, as well as in the 0–1 transition, of the C–B system. However, the laser action is self-terminating because the lifetime of the lower state B (10 µs) is *longer* than that of the upper state C (40 ns). This does not render laser action impossible but necessitates pulsing of the input energy with a pulse length shorter than the lifetime of the C state.

A design for a nitrogen laser is shown in Figure 9.14. A pulsed high voltage of about 20 kV, triggered by a spark gap or a thyratron, is applied transversely across the cavity. A single mirror is used to double the output. Laser pulses of about 10 ns length are typical. Peak power can be as much as 1 MW. The

---

[†] The reader is reminded that the labels A, B, C rather than a, b, c for triplet states of N_2 do not follow the usual convention.

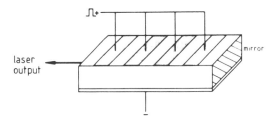

Figure 9.14   Nitrogen laser cavity.

maximum repetition rate is about 100 Hz with longitudinally flowing gas. Much higher repetition rates are possible for transverse flow.

The operating wavelength of a nitrogen laser is 337 nm for the 0–0 transition of the C–B system.

9.2.8   The Excimer and Exciplex Lasers

An excimer is a dimer which is stable only in an excited electronic state but dissociates readily in the ground state. Examples of these are the noble gas dimers such as He_2 discussed in Section 7.2.5.6. This molecule has a repulsive $X^1\Sigma_g^+$ ground state but a bound $A^1\Sigma_u^+$ excited state, as illustrated in Figure 7.24(a).

Such a situation suggests the possibility of creating a population inversion and laser action between two such states, since any molecules in the repulsive ground state have an extremely short lifetime, typically a few picoseconds. A laser operating by this mechanism is a two-level laser but population of the upper state is not, of course, caused by pumping ground state molecules. Molecules in the upper state are created in a discharge by collisions between two atoms, one or both of which may be in an excited state. The efficiency of such lasers is high, about 20 per cent.

An Xe_2 excimer laser has been made to operate in this way but of much greater importance are the noble gas halide lasers. These halides also have repulsive ground states and bound excited states and are examples of exciplexes. An exciplex is a complex consisting, in a diatomic molecule, of two *different* atoms, which is stable in an excited electronic state but dissociates readily in the ground state. In spite of this clear distinction between an excimer and an exciplex it is now common for all such lasers to be called excimer lasers.

Excimer lasers employing NeF, ArF, KrF, XeF, ArCl, KrCl, XeCl, ArBr, KrBr, XeBr, KrI, and XeI as the active medium have been made.

The method of excitation was, in the early days, by an electron beam but now a transverse electrical discharge, like that for the nitrogen laser in Figure 9.14, is used. Indeed such an excimer laser can be converted to a nitrogen laser by changing the gas.

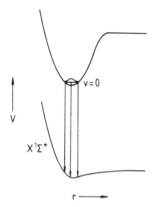

$v = 0$

$X\,^1\Sigma^+$

$r \longrightarrow$

Figure 9.15 Potential curves for a very weakly bound ground state and a strongly bound excited state of a noble gas halide.

In an excimer laser the mixture of inert gas, halogen gas, and helium, used as a buffer, is pumped around a closed system consisting of a reservoir and the cavity.

The examples of ArF (193 nm), KrF (248 nm), XeF (351 nm), KrCl (222 nm), XeCl (308 nm), and XeBr (282 nm) indicate the range of wavelengths from excimer lasers. Because the ground states of these molecules are not totally repulsive but very weakly bound, there is a very shallow minimum in the potential curve, as illustrated in Figure 9.15. In the case of XeF the potential energy minimum is relatively deep, about 1150 cm^{-1}, and supports a few vibrational levels. As a result the laser may be tuned over several transitions.

The excimer laser radiation is pulsed with a typical maximum rate of about 200 Hz. Peak power of up to 5 MW is high compared to that of a nitrogen laser.

9.2.9 The Carbon Dioxide Laser

The CO_2 laser is a near-infrared gas laser capable of very high power and with an efficiency of about 20 per cent.

CO_2 has three normal modes of vibration, v_1, the symmetric stretch, v_2, the bending vibration, and v_3, the antisymmetric stretch with symmetry species σ_g^+, π_u, and σ_u^+, and fundamental vibration wavenumbers of 1354, 673, and 2396 cm^{-1} respectively. Figure 9.16 shows some of the vibrational levels, the nomenclature of which is explained on page 87, which are involved in the laser action. This occurs principally in the $3_0^1 2_2^0$ transition, at about 10.6 μm, but may also be induced in the $3_0^1 1_1^0$ transition, at about 9.6 μm.

Population of the 3^1 level is partly by electron–molecule collisions and partly by energy transfer from nitrogen molecules in the $v = 1$ level, this being metastable due to the fact that the transition to $v = 0$ is forbidden (see Section 6.1.1). Energy transfer from nitrogen is particularly efficient because the $v = 1$ level is only 18 cm^{-1} below the 3^1 level of CO_2 (Figure 9.16). Because of

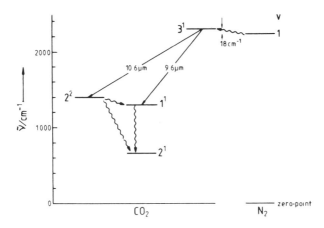

Figure 9.16 Vibrational levels of N_2 and CO_2 relevant to the CO_2 laser.

near-degeneracies of higher vibrational levels of nitrogen and the v_3 stack of CO_2, transfer to levels such as 3^2, 3^3, ... also occurs. Transitions down the v_3 stack are fast until the 3^1 level is reached.

Decay of the 1^1 and 2^2 lower levels[†] of the laser transitions are rapid down to the 2^1 level; this is depopulated mostly by collisions with helium atoms in the CO_2:N_2:He gas mixture which is used.

Lifetimes of upper and lower states are governed by collisions and that of the upper is always longer than that of the lower in the gas mixtures used.

The energy input into a CO_2 laser is in the form of an electrical discharge through the mixture of gases. The cavity may be sealed, in which case a little water vapour must be added in order to convert back to CO_2 any CO which is formed. More commonly longitudinal or, preferably, transverse gas flow through the cavity is used. The CO_2 laser can operate in a CW or pulsed mode with power up to 1 kW possible in the CW mode.

Each of the lasing vibrational transitions has associated rotational fine structure, discussed for linear molecules in Section 6.2.3.1. The $3^1_0 1^0_1$ transition is $\Sigma_u^+ - \Sigma_g^+$ with associated P and R branches, for which $\Delta J = -1$ and $+1$ respectively, similar to the 3^1_0 band of HCN in Figure 6.23. The $3^1_0 2^0_2$ band is, again, $\Sigma_u^+ - \Sigma_g^+$ with a P and R branch.

Unless the cavity is tuned to a particular wavelength the vibration–rotation transition with the highest gain is the P-branch transition involving the rotational level which has the highest population in the 3^1 state. This is $P(22)$, with $J'' = 22$ and $J' = 21$, at normal laser temperatures. The reason why this

---

[†] The assignments of these levels have recently been reversed. The new assignments are used here.

P-branch line is so dominant is that thermal redistribution of rotational level populations is faster than the population depletion due to emission.

The cavity may be tuned to a particular transition by a prism or, preferably, by replacing one of the mirrors (not the output mirror) at one end of the cavity by a diffaction grating.

9.2.10 The Dye Lasers

Laser action in some dye solutions was first discovered by Lankard and Sorokin in 1966. This led to the first laser which was continuously tunable over an appreciable wavelength range. Dye lasers are also unusual in that the active medium is a liquid.

One characteristic property of dyes is their colour, due to absorption, from the ground electronic state S_0 to the first excited singlet state S_1, lying in the visible region. Also typical of a dye is a high absorbing power characterized by a value of the oscillator strength f (see equation 2.18) close to 1, and also a value of the fluorescence quantum yield ϕ_F (see equation 7.135) close to 1.

Figure 9.17 illustrates these features in the case of the dye rhodamine B. The maximum of the typically broad $S_1 - S_0$ absorption occurs at about 548 nm with a very high value of 80 000 L mol^{-1} cm^{-1} for ε_{max}, the maximum value of the molar absorption coefficient (equation 2.16). The fluorescence curve shows, as usual, an approximate mirror image relationship to the absorption

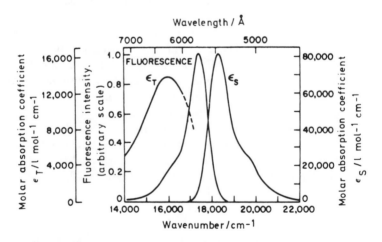

Figure 9.17 Absorption and fluorescence spectra of rhodamine B in methanol (5×10^{-5} mol L^{-1}). The curve marked ε_T is for the $T_2 - T_1$ absorption (process 8 in Figure 9.18) and that marked ε_S for process 1. (Reproduced, with permission, from Dienes, A., and Shank, C. V., Chapter 4 in *Creation and Detection of the Excited State* (Ed. W. R. Ware), Vol. 2, p. 154, Marcel Dekker, New York, 1972.)

curve. It has the additional property, important for all laser dyes, that the fluorescence and absorption maxima do not coincide: if they did, a large proportion of the fluorescence would be reabsorbed.

Figure 9.18 shows a typical energy level diagram of a dye molecule including the lowest electronic states S_0, S_1, and S_2 in the singlet manifold and T_1 and T_2 in the triplet manifold. Associated with each of these states are vibrational and rotational sub-levels broadened to such an extent by collisions in the liquid that they form a continuum. As a result the absorption spectrum, such as that in Figure 9.17, is typical of a liquid phase spectrum showing almost no structure within the band system.

Depending on the method of pumping, the population of S_1 may be achieved by $S_1 - S_0$ or $S_2 - S_0$ absorption processes, labelled 1 and 2 in Figure 9.18, or both. Following either process collisional relaxation to the lower vibrational levels of S_1 is rapid by process 3 or 4: e.g. the vibrational–rotational relaxation of process 3 takes of the order of 10 ps. Following relaxation the distribution among the levels of S_1 is that corresponding to thermal equilibrium, i.e. there is a Boltzmann population (equation 2.11).

The state S_1 may decay by radiative (r) or non-radiative (nr) processes, labelled 5 and 7 respectively in Figure 9.18. Process 5 is the fluorescence which forms the laser radiation and the figure shows it terminating in a vibrationally excited level of S_0. The fact that it does so is vital to the dye being usable as an active medium and is a consequence of the Franck–Condon principle (see Section 7.2.5.3).

The shape of the broad absorption curve in Figure 9.17 is typical of that of any dye suitable for a laser. It shows an absorption maximum to low wave-

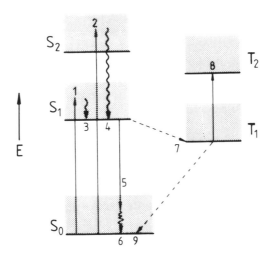

Figure 9.18 Energy level scheme for a dye molecule showing processes important in laser action.

length of the 0_0^0 band position, which is close to the absorption–fluorescence crossing point. The shape of the absorption curve results from a change of shape of the molecule, from S_0 to S_1, in the direction of one or more normal coordinates, so that the most probable transition in absorption is to a vibrationally excited level of S_1. Similarly, in emission from the zero-point level of S_1, the most probable transition is to a vibrationally excited level of S_0. The fluorescence lifetime τ_r for spontaneous emission from S_1 is typically of the order of 1 ns while the relaxation process 6, like process 3, takes only about 10 ps. The result is that, following processes 1 and 3, there is a population inversion between the zero-point level of S_1 and vibrationally excited levels of S_0 to which emission may occur, provided that these levels are sufficiently highly excited to have negligible thermal population.

The population of S_1 may also be reduced by absorption of the fluorescence taking the molecule from S_1 into S_2, if the wavelengths of the two processes correspond, as well as by non-radiative transitions to either S_0 (internal conversion) or T_1 (intersystem crossing), as discussed in Section 7.3.6. In dye molecules it is the $S_1 - T_1$ process, labelled 7 in Figure 9.18, which is the most important. This is a spin-forbidden process with a lifetime τ_{nr} of the order of 100 ns. The lifetime τ of the state S_1 is related to τ_{nr} and the radiative lifetime τ_r by

$$\frac{1}{\tau} = \frac{1}{\tau_r} + \frac{1}{\tau_{nr}} \tag{9.15}$$

Since τ_r is of the order of 1 ns, fluorescence is the dominant decay process for S_1.

The lifetime τ_T of the state T_1 is long because the $T_1 - S_0$ transition, process 9 in Figure 9.18, is spin-forbidden. Depending on the molecule and on the conditions, particularly the amount of dissolved oxygen, it may be anywhere in the range 100 ns to 1 ms. If $\tau_T > \tau_{nr}$ the result is that the concentration of molecules in T_1 can build up to a high level. It happens in many dye molecules that the intense, spin-allowed, $T_2 - T_1$ absorption, process 8 in Figure 9.18, overlaps with, and therefore can be excited by, the $S_1 - S_0$ emission, thereby decreasing the efficiency of the laser considerably. Figure 9.17 shows how important this process is in rhodamine B.

In order to prevent this occurring a pulsed method of pumping is used with a repetition rate low enough to allow time for $T_1 - S_0$ relaxation. For CW operation either τ_T must be sufficiently short or another dye has to be used whose $T_2 - T_1$ absorption does not overlap with the fluorescence.

There are many dyes available, each of which can be used over a 20 to 30 nm range and which, together, cover a wavelength range from about 365 nm in the ultraviolet to about 930 nm in the near-infrared. Dye concentrations are low, typically in the range 10^{-2} to 10^{-4} mol L^{-1}.

Taking into account the possibility of frequency doubling (Section 9.1.6) dye lasers can provide tunable radiation throughout the range 220 to 930 nm but with varying levels of intensity and degrees of difficulty. The tunability and the extensive wavelength range make dye lasers probably the most generally useful of all visible or ultraviolet lasers.

A pulsed dye laser may be pumped with a flashlamp surrounding the cell through which the dye is flowing. With this method of excitation pulses from the dye laser about 1 μs long and with an energy of the order of 100 mJ can be obtained. Repetition rates are typically low—up to about 30 Hz.

More commonly a pulsed dye laser is pumped with a nitrogen, excimer, or Nd^{3+}:YAG laser. The nitrogen laser, operating at 337 nm, and a xenon fluoride excimer laser, operating at 351 nm, both excite the dye initially into a singlet excited state higher in energy than S_1. The Nd^{3+}:YAG laser is either frequency doubled to operate at 532 nm or frequency tripled to operate at 355 nm depending on the dye which is being pumped. However, because of the low efficiency of frequency tripling, it is more usual to mix the frequency doubled dye radiation, of wavelength λ_D, with the Nd^{3+}:YAG laser fundamental, of wavelength λ_F ($= 1.0648$ μm), in a non-linear crystal such as KDP (Section 9.1.6) to give a wavelength λ where

$$\frac{1}{\lambda} = \frac{1}{\lambda_0} + \frac{1}{\lambda_F}$$

(9.15a)

Pulse rates of about 50 Hz are usual.

CW dye lasers are usually pumped with an argon ion laser, up to about 1 W of continuous dye laser power being produced compared to about 1 MW peak power which may be produced in a pulsed dye laser.

In both CW and pulsed lasers the dye solution must be kept moving to prevent overheating and decomposition. In a pulsed laser the dye is continuously flowed through the containing cell. Alternatively magnetic stirring may be adequate for low repetition rates and relatively low power. In a CW laser the dye solution is usually in the form of a jet flowing rapidly across the laser cavity.

9.2.11 Laser Materials in General

In choosing the examples of lasers discussed in Sections 9.2.1 to 9.2.10 many have been left out. These include the CO, H_2O, HCN, colour centre, and chemical lasers, all operating in the infrared region, and the green copper vapour laser. The examples that we have looked at in some detail serve to show how disparate and arbitrary the materials seem to be. For example, the fact that Ne atoms lase in a helium–neon laser does not mean that Ar, Kr, and Xe will lase also—they do not. Nor is it the case that because CO_2 lases the chemically similar CS_2 will lase also.

The potential for laser activity is not anything we can demand of any atom or molecule. We should regard it as accidental that among the extremely complex sets of energy levels associated with a few atoms or molecules there happens to be one (or more) pairs between which it is possible to produce a population inversion and thereby create a laser.

9.3 Uses of Lasers in Spectroscopy

From 1960 onwards, the increasing availability of intense, monochromatic laser sources provided a tremendous impetus to a wide range of spectroscopic investigations.

The most immediately obvious application of early, essentially non-tunable, lasers was to all types of Raman spectroscopy in the gas, liquid, or solid phase. The experimental techniques, employing laser radiation, were described in Section 5.3.1. Examples of the quality of spectra which can be obtained in the gas phase are to be found in Figure 5.17, which shows the pure rotational Raman spectrum of $^{15}N_2$, and Figure 6.9 which shows the vibration-rotation Raman spectrum of the $v = 1$–0 transition in CO. Both of these spectra were obtained with an argon ion laser.

Laser radiation is very much more intense than that from, for example, a mercury arc which was commonly used as a Raman source before 1960. As a result of the greater source intensity, much weaker Raman scattering can now be observed. In addition, the narrower line width of laser radiation increases the resolution obtainable.

In addition to carrying out conventional Raman experiments with laser sources new kinds of Raman experiments became possible using Q-switched, giant pulse lasers to investigate effects which arise from the non-linear relationship between the induced electric dipole and the oscillating electric field (equation 9.11). These are grouped under the general heading of non-linear Raman effects.

For branches of spectroscopy other than Raman spectroscopy most laser sources may appear to have a great disadvantage, that of non-tunability. In regions of the spectrum, particularly the infrared, where tunable lasers are not readily available ways have been devised for tuning, i.e. shifting, the atomic or molecular energy levels concerned until the transition being studied moves into coincidence with the laser radiation. This may be achieved by applying an electric field to the sample and the technique is called laser Stark spectroscopy. The corresponding technique using a magnetic field is that of laser magnetic resonance (or laser Zeeman) spectroscopy.

A useful way of changing the wavelength of some lasers, e.g. the CO_2 infrared laser, is to use isotopically substituted material in which the wavelengths of laser transitions are appreciably altered.

In regions of the spectrum where a tunable laser is available it may be possible to use it to obtain an absorption spectrum in the same way as a tunable klystron or backward wave oscillator is used in microwave or millimetre wave spectroscopy (see Section 3.4.1). Absorbance (equation 2.16) is measured as a function of frequency or wavenumber. This technique can be used with a diode laser to produce an infrared absorption spectrum. When electronic transitions are being studied, greater sensitivity is usually achieved by monitoring secondary processes which follow, and are directly related to, the absorption which has occurred. Such processes include fluorescence, dissociation, or predissociation and, following the absorption of one or more additional photons, ionization. The spectrum resulting from monitoring these processes usually resembles the absorption spectrum very closely.

It may be apparent to the reader at this stage that, when lasers are used as spectroscopic sources, we can no longer think in terms of generally applicable experimental methods. A wide variety of ingenious techniques have been devised using laser sources and it will be possible to describe only a few of them here.

9.3.1 Hyper Raman Spectroscopy

We have seen in equation (9.11) how the dipole moment induced in a material by radiation falling on it contains a small contribution which is proportional to the square of the oscillating electric field E of the radiation. This field can be sufficiently large, when using a Q-switched laser focused on the sample, that hyper Raman scattering, involving the hyperpolarizability β introduced in equation (9.11), is sufficiently intense to be detected.

Hyper Raman scattering is at a wavenumber $2\tilde{v}_0 \pm \tilde{v}_{HR}$, where $\tilde{v}_0$ is the wavenumber of the exciting radiation and $-\tilde{v}_{HR}$ and $+\tilde{v}_{HR}$ are the Stokes and anti-Stokes hyper Raman displacements respectively. The hyper Raman scattering is well separated from the Raman scattering, which is centred on $\tilde{v}_0$, but is extremely weak, even with a Q-switched laser.

Scattering of wavenumber $2\tilde{v}_0$ is called hyper Rayleigh scattering, by analogy with Raleigh scattering of wavenumber $\tilde{v}_0$ (see Section 5.3.2). However, whereas Rayleigh scattering *always* occurs, hyper Rayleigh scattering occurs only if the scattering material does not have a centre of inversion (see Section 4.1.3). Frequency doubled radiation, discussed in Section 9.1.6, consists of hyper Rayleigh scattering from a pure crystal. Consequently one of the necessary properties of the crystals used, such as ADP and KDP, is that the unit cell does not have a centre of inversion.

The selection rules for molecular vibrations involved in hyper Raman scattering are summarized by

$$\Gamma(\psi'_v) \times \Gamma(\beta_{ijk}) \times \Gamma(\psi''_v) = A(\text{or} \supset A) \qquad (9.16)$$

analogous to equations (6.64) and (6.65) for Raman scattering, where ψ'_v and ψ''_v are the upper and lower state vibrational wave functions respectively, i, j, and k can be x, y, or z, and A is the totally symmetric species of the point group to which the molecule belongs. If, as is usually the case, the lower vibrational state is the zero-point level $\Gamma(\psi_v'') = A$ and equation (9.16) becomes

$$\Gamma(\psi_v') = \Gamma(\beta_{ijk}) \tag{9.17}$$

The hyperpolarizability is a tensor with eighteen elements β_{ijk}. We shall not go further into their symmetry properties but important results of equation (9.17) include:

1. Vibrations allowed in the infrared are also allowed in the hyper Raman effect.
2. In a molecule with a centre of inversion all hyper Raman active vibrations are u vibrations, antisymmetric to inversion.
3. Some vibrations which are both Raman and infrared inactive may be allowed in the hyper Raman effect. Indeed, the occasional appearance of such vibrations in Raman spectra in a condensed phase has sometimes been attributed to an effect involving the hyperpolarizability.

Figure 9.19 shows the hyper Raman spectrum of gaseous ethane, C_2H_6, which belongs to the D_{3d} point group (see Figure 4.11(i) and Table A.28 in the Appendix). Ethane has a centre of inversion and therefore there is no hyper Rayleigh scattering at $2\tilde{v}_0$. In the hyper Raman spectrum a_{1u}, a_{2u}, and e_u

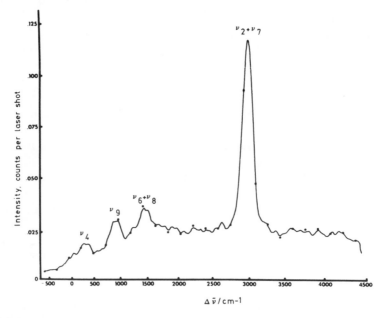

Figure 9.19 The hyper Raman spectrum of ethane. (Reproduced, with permission, from Verdick, J. F., Peterson, S. H., Savage, C. M., and Maker, P. D., *Chem. Phys. Letters*, **7**, 219, 1970.)

vibrations are allowed. None of these is allowed in the Raman spectrum and only the a_{2u} and e_u vibrations are allowed in the infrared. The intense scattering at about 3000 cm^{-1} from $2\tilde{\nu}_0$ is a combination of two bands: one is the 2_0^1 (ν_2) and the other the 7_0^1 (ν_7) band where ν_2 and ν_7 are a_{2u} and e_u CH-stretching vibrations respectively. The scattering at $\Delta\tilde{\nu} \simeq 1400$ cm^{-1} is again due to two coincident bands, 6_0^1 and 8_0^1, where ν_6 and ν_8 are a_{2u} and e_u CH$_3$–deformation vibrations respectively. The 9_0^1 band is at $\Delta\tilde{\nu} \simeq 900$ cm^{-1} and ν_9 is an e_u bending vibration of the whole molecule. The 4_0^1 band, at $\Delta\tilde{\nu} \simeq 300$ cm^{-1} is the most interesting as ν_4 is the a_{1u} torsional vibration about the C–C bond (see Section 6.2.4.4c) which is forbidden in the infrared and Raman spectra.

9.3.2 Stimulated Raman Spectroscopy

Stimulated Raman spectroscopy is experimentally different from normal Raman spectroscopy in that the scattering is observed in the *forward* direction emerging from the sample in the same direction as that of the emerging exciting radiation, or at a very small angle to it.

Figure 9.20(a) shows how stimulated Raman scattering can be observed by focusing radiation from a Q-switched ruby laser with a lens L into a cell C containing, for example, liquid benzene. The forward scattering, within an angle of about 10°, is collected by the detector D. If the detector is a photographic colour film, broad concentric coloured rings ranging from dark red in the centre to green on the outside are observed, as Figure 9.20(b) indicates. The wave-

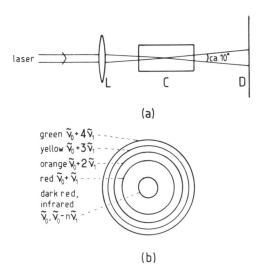

(a)

(b)

Figure 9.20 (a) Stimulated Raman scattering experiment. (b) Concentric rings observed, in the forward direction, from liquid benzene.

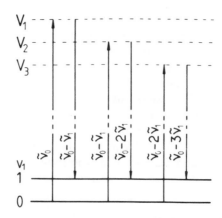

Figure 9.21 Transitions in the stimulated Raman effect in benzene.

numbers corresponding to these rings range from $\tilde{v}_0$ (and $\tilde{v}_0 - n\tilde{v}_1$) in the centre to $\tilde{v}_0 + 4\tilde{v}_1$ on the outside. v_1 is the ring-breathing vibration of benzene (see Figure 6.13f) and the series of $\tilde{v}_0 + n\tilde{v}_1$ rings, with $n = 0-4$, shows a *constant* separation of $\tilde{v}_1$, which is the $v_1 = 1$–0 interval of 992 cm^{-1}.

The reason why the spacings are equal, and not the 1–0, 2–1, 3–2,... anharmonic intervals, is explained in Figure 9.21. The laser radiation of wavenumber $\tilde{v}_0$ takes benzene molecules into the virtual state V_1 from which they may drop down to the $v_1 = 1$ level. The resulting Stokes scattering is, as mentioned above, extremely intense in the forward direction with about 50 per cent of the incident radiation scattered at a wavenumber of $\tilde{v}_0 - \tilde{v}_1$. This radiation is sufficiently intense to take other molecules into the virtual state V_2 resulting in intense scattering at $\tilde{v}_0 - 2\tilde{v}_1$, and so on.

In the stimulated Raman effect it is only the vibration which gives the most intense Raman scattering which is involved: this is the case for $\tilde{v}_1$, in benzene.

The high efficiency of conversion of the laser radiation $\tilde{v}_0$ into Stokes radiation allows the effect to be used for shifting to higher wavelengths the radiation from a pulsed laser which is otherwise non-tunable. High pressure hydrogen gas, having a $v = 1$–0 interval of 4160 cm^{-1}, is often used in such a Raman shifting device.

9.3.3 Coherent Anti-Stokes Raman Scattering Spectroscopy

Coherent anti-Stokes Raman scattering, or CARS as it is usually known, depends on the general phenomenon of wave mixing, as occurs, for example, in a frequency doubling crystal (see Section 9.1.6). In that case three-wave mixing occurs involving two incident waves of wavenumber $\tilde{v}$ and the outgoing wave of wavenumber $2\tilde{v}$.

In CARS, radiation from two lasers of wavenumbers $\tilde{\nu}_1$ and $\tilde{\nu}_2$, where $\tilde{\nu}_1 > \tilde{\nu}_2$, fall on the sample. As a result of *four*-wave mixing, radiation of wavenumber $\tilde{\nu}_3$ is produced where

$$\tilde{\nu}_3 = 2\tilde{\nu}_1 - \tilde{\nu}_2 = \tilde{\nu}_1 + (\tilde{\nu}_1 - \tilde{\nu}_2) \qquad (9.18)$$

The wave mixing is much more efficient when $(\tilde{\nu}_1 - \tilde{\nu}_2) = \tilde{\nu}_i$, where $\tilde{\nu}_i$ is the wavenumber of a Raman-active vibrational or rotational transition of the sample.

The scattered radiation $\tilde{\nu}_3$ is to high wavenumber of $\tilde{\nu}_1$, i.e. on the anti-Stokes side, and is coherent, unlike spontaneous Raman scattering: hence the name CARS. As a consequence of the coherence of the scattering and the very high conversion efficiency to $\tilde{\nu}_3$, the CARS radiation forms a collimated, laser-like beam.

The selection rules for CARS are precisely the same as for spontaneous Raman scattering but CARS has the advantage of vastly increased intensity.

Figure 9.22 illustrates how a CARS experiment might be carried out. In order to vary $(\tilde{\nu}_1 - \nu_2)$ in equation (9.18) one laser wavenumber, $\tilde{\nu}_1$, is fixed and $\tilde{\nu}_2$ is varied. Here $\tilde{\nu}_1$ is frequency-doubled Nd^{3+}:YAG laser radiation at 532 nm while the $\tilde{\nu}_2$ radiation is that of a dye laser which is pumped by the same Nd^{3+}:YAG laser. The two laser beams are focused with a lens L into the sample cell C making a small angle 2α with each other. The collimated CARS radiation emerges at an angle 3α to the optic axis, is spatially filtered from $\tilde{\nu}_1$ and $\tilde{\nu}_2$ by a filter F in the form of a pinhole, and passes to a detector D. The sample may be solid, liquid, or gaseous.

In equation (9.18) we have treated $\tilde{\nu}_1$ and $\tilde{\nu}_2$ differently by involving two photons of $\tilde{\nu}_1$ and only one of $\tilde{\nu}_2$. However four-wave mixing involving one photon of $\tilde{\nu}_1$ and two of $\tilde{\nu}_2$ to produce $\tilde{\nu}_4$, represented by

$$\tilde{\nu}_4 = 2\tilde{\nu}_2 - \tilde{\nu}_1 = \tilde{\nu}_2 - (\tilde{\nu}_1 - \tilde{\nu}_2) \qquad (9.19)$$

is equally probable. In this case the radiation $\tilde{\nu}_4$ is to low wavenumber of $\tilde{\nu}_2$, i.e. on the Stokes side. This radiation is referred to as coherent Stokes

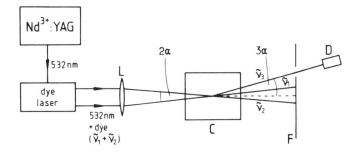

Figure 9.22 Experimental arrangement for CARS.

Raman scattering or CSRS. From the symmetry of equations (9.18) and (9.19) there seems to be no reason to favour CARS or CSRS but, since $(2\tilde{v}_2 - \tilde{v}_1)$ is to low wavenumber of both $\tilde{v}_1$ and $\tilde{v}_2$, there is a tendency in CSRS for the region of $\tilde{v}_4$ to be overlapped by fluorescence from the sample. For this reason the CARS technique is used more frequently.

9.3.4 Laser Stark (or Laser Electric Resonance) Spectroscopy

We have seen in Section 5.2.3 that an electric field splits the rotational levels of a diatomic, linear, or symmetric rotor molecule—the Stark effect. Such splitting occurs for rotational levels associated with all vibrational levels so that a gas phase vibrational spectrum will show corresponding splitting of the rotational fine structure. Using a fixed wavenumber infrared laser a smooth variation of an electric field applied to the sample will bring various transitions into coincidence with the laser wavenumber. This type of spectroscopy is usually called laser Stark spectroscopy but is sometimes referred to as laser electric resonance, a name which parallels laser magnetic resonance, the name given to a corresponding experiment using a magnetic field.

Early laser Stark spectra were obtained with the absorption cell outside the laser cavity but there are advantages in placing it inside the cavity, an arrangement shown in Figure 9.23. The laser cavity is bounded by the mirror M and the grating G, used for selecting wavelengths in a multiple-line laser such as CO_2 or CO. The sample compartment is divided from the laser compartment by a window W (all the windows are at Brewster's angle—see equation 9.14). The Stark electrodes S are only a few millimetres apart in order to produce a large field between them, of the order of $50\ kV\ cm^{-1}$. Some of the laser radiation leaks out to a detector D.

Figure 9.24 shows part of the laser Stark spectrum of the bent triatomic molecule FNO obained with a CO infrared laser operating at $1837.430\ cm^{-1}$. All the transitions shown are Stark components of the $^qP_7(8)$ rotational line of the 1_0^1 vibrational transition, where v_1 is the N–F stretching vibration. The rotational symbolism is that for a symmetric rotor (which FNO nearly is) for which q implies that $\Delta K = 0$, P that $\Delta J = -1$ and the numbers indicate that

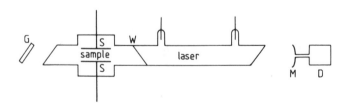

Figure 9.23   Laser Stark spectroscopy with the sample inside the cavity.

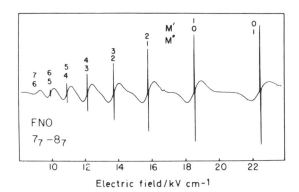

Figure 9.24 Laser Stark spectrum of FNO showing Lamb dips in the components of the $^qP_7(8)$ line of the 1_0^1 vibrational transition. (Reproduced, with permission, from Allegrini, M., Johns, J. W. C., and McKellar, A. R. W., *J. Molec. Spectrosc.*, **73**, 168, 1978.)

$K'' = 7$ and $J'' = 8$ (see Section 6.2.3.2). In an electric field each J level is split into $(J + 1)$ components (see Section 5.2.3), each specified by its value of $|M_J|$. The selection rule when the radiation is polarized perpendicular to the field (as here) is $\Delta M_J = \pm 1$. Eight of the resulting Stark components are shown.

As well as resulting in rotational constants for the two vibrational states involved, such a spectrum also yields the dipole moment in each state.

An important feature of the spectrum in Figure 9.24 is the unusual shape of the lines. The gross ∿-shape of each is due to modulation of the electric field followed by phase-sensitive detection. Figure 9.25 shows the effect on a line limited to the Doppler width and observed by sweeping the potential V between the plates while keeping the laser wavenumber fixed. Modulation of V is sinusoidal with small amplitude. On the 'up-slope' of the line in Figure 9.25(a) a small decrease in the modulated V produces a small decrease in signal and a small increase in V produces a small increase in signal: in other words the modulation and the signal are in-phase. Similarly, on the 'down-slope' of the line they are out-of-phase.

Figure 9.25(b) shows the effect of using a phase-sensitive detector. A positive signal at the detector results when the modulation and the ordinary signal are in-phase, a negative signal when they are out-of-phase, and a zero signal corresponds to the maximum intensity of the line. The result is the first derivative of the signal in Figure 9.25(a).

A further feature of the spectrum in Figure 9.24 is the sharp spike at the centre of each ∿-shaped transition. The reason for this is that saturation of the transition has occurred. This was discussed in Section 2.3.4.2 in the context of Lamb dips in microwave and millimetre wave spectroscopy and referred to the situation in which the two energy levels involved, m(lower) and n(upper),

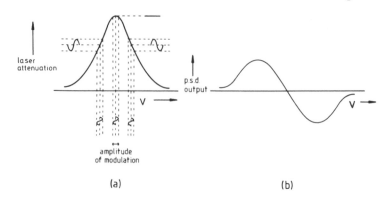

Figure 9.25 (a) A Doppler-limited line. (b) The effect of modulation and phase-sensitive detection.

are close together. Under these circumstances saturation occurs when the populations N_m and N_n are nearly equal. If a reflecting mirror is placed at one end of the absorption cell, a Lamb dip may be observed in the absorption line profile, as shown in Figure 2.5.

In other regions of the spectrum, such as the infrared, visible, and ultraviolet regions, the levels m and n are further apart but it turns out that the effects of saturation may be observed when N_n is high, but considerably less than N_m.

For the sample inside the laser cavity, as in Figure 9.23, saturation may well occur producing a line shape like that in Figure 9.26(a) showing a Lamb dip. Modulation and phase-sensitive detection give the signal as the first derivative, shown in Figure 9.26(b). It is these first derivative Lamb dips which are seen in Figure 9.24. Clearly the accuracy of measurement of the line centre is increased considerably when such Lamb dips are observed.

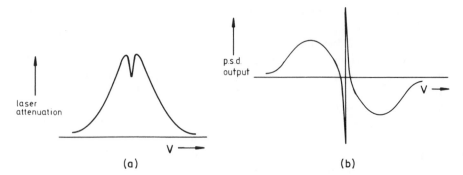

Figure 9.26 (a) Doppler line shape with a Lamb dip. (b) As in (a) but with modulation and phase-sensitive detection.

9.3.5 Two-photon and Multiphoton Absorption

In the discussion in Section 9.1.6 of harmonic generation of laser radiation we have seen how the high photon density produced by focusing a laser beam into certain crystalline materials may result in doubling, tripling, etc., of the laser frequency. Similarly, if a laser beam of wavenumber $\tilde{v}_L$, is focused into a cell containing a material which is known to absorb at a wavenumber $2\tilde{v}_L$, in an ordinary one-photon process, the laser radiation may be absorbed in a two-photon process provided it is allowed by the relevant selection rules.

The similarity between a two-photon absorption and a Raman scattering process is even closer. Figure 9.27(a) shows that a Raman transition between states 1 and 2 is really a two-photon process. The first photon is absorbed at a wavenumber $\tilde{v}_a$ to take the molecule from state 1 to the virtual state V and the second photon is emitting at a wavenumber $\tilde{v}_b$.

In a two-photon absorption process the first photon takes the molecule from the initial state 1 to a virtual state V and the second takes it from V to 2. As in Raman spectroscopy, the state V is not an eigenstate of the molecule. The two photons absorbed may be of equal or unequal energies, as shown in Figure 9.27(b) and (c). It is possible that more than two photons may be absorbed in going from state 1 to 2. Figure 9.27(d) illustrates three-photon absorption.

Two-photon absorption has been observed in the microwave region with an intense klystron source but in the infrared, visible, and ultraviolet regions laser sources are necessary.

Because Raman scattering is also a two-photon process the selection rules for two-photon absorption are the same as for vibrational Raman transitions. For example, for a two-photon electronic transition to be allowed between a lower state ψ_e'' and an upper state ψ_e',

$$\Gamma(\psi_e') \times \Gamma(S_{ij}) \times \Gamma(\psi_e'') = A \ (\text{or} \supset A) \tag{9.20}$$

where the S_{ij} are elements of the two-photon tensor S which is similar to

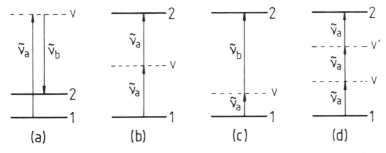

Figure 9.27 Multiphoton processes shown are (a) Raman scattering, (b) absorption of two identical photons, (c) absorption of two different photons, (d) absorption of three identical photons.

the polarizability tensor α in equation (5.42) in that

$$\Gamma(S_{ij}) = \Gamma(\alpha_{ij}) \tag{9.21}$$

Then equation (9.20) is seen to be analogous to equations (6.64) and (6.65) for vibrational Raman transitions.

Because two-photon selection rules are different from one-photon (electric dipole) selection rules, two-photon transitions may allow access to states which otherwise could not be reached. We shall consider just one example in detail—a two-photon electronic absorption spectrum.

A two-photon, or any multiphoton, electronic absorption process may be monitored in various ways and Figure 9.28 illustrates two of them. If a laser, typically a tunable dye laser, is scanned through an absorption system then, if two photons match a transition to an excited electronic or vibronic state, fluorescence may be detected from that state, as in Figure 9.28(a). The intensity of the total, undispersed fluorescence as a function of laser wavenumber gives the two-photon fluorescence excitation spectrum. Figure 9.28(b) illustrates a second method of monitoring absorption. In the case shown two photons take the molecule into an eigenstate 2 and a third ionizes it. This process is known as a 2 + 1 multiphoton ionization process but other processes, such as 2 + 2 or 3 + 1, may also be observed. The number of ions, collected by plates with a negative potential, is counted as a function of laser wavenumber to produce the multiphoton ionization spectrum. Multiphoton ionization is advantageous in cases where the fluorescence quantum yield is too small for the method of two-photon fluorescence excitation to be used.

The example we consider is the two-photon fluorescence excitation spectrum of 1,4-difluorobenzene, shown in Figure 9.29 and belonging to the D_{2h} point group. The transition between the ground and first singlet excited state is $\tilde{A}^1B_{2u} - \tilde{X}^1A_g$. Table A.32 in the Appendix shows that $B_{2u} = \Gamma(T_y)$ and,

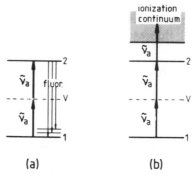

(a)   (b)

Figure 9.28 A two-photon (or more) absorption process may be monitored by (a) measuring total, undispersed fluorescence or (b) counting the ions produced by a further photon (or photons).

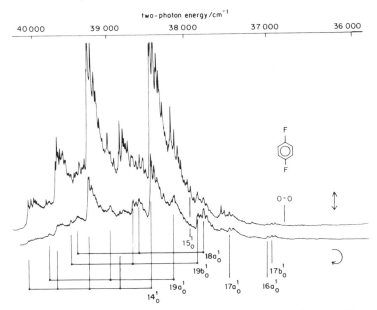

Figure 9.29 Two-photon fluorescence excitation spectrum of 1,4-difluorobenzene. The upper and lower traces are obtained with plane and circularly polarized radiation respectively, but the differences are not considered here. (Reproduced, with permission, from Robey, M. J., and Schlag, E. W., *Chem. Phys.*, **30**, 9, 1978.)

therefore, according to equation (7.122), the electronic transition is allowed as a one-photon process polarized along the y axis which is in-plane and perpendicular to the F—C ---- C—F line: the 0_0^0 band is shown in Figure 7.44(a). However, since Table A.32 in the Appendix also shows that $B_{2u} \neq \Gamma(\alpha_{ij})$ the transition is forbidden as a two-photon process. As in Raman spectroscopy $u \leftrightarrow g$ transitions are forbidden while $g \leftrightarrow g$ or $u \leftrightarrow u$ transitions are allowed for a molecule with a centre of inversion.

Nevertheless, 1,4-difluorobenzene has a rich two-photon fluorescence excitation spectrum shown in Figure 9.29. The position of the forbidden 0_0^0 (labelled 0–0) band is shown. All the vibronic transitions observed in the band system are induced by non-totally symmetric vibrations, rather like the one-photon case of benzene discussed in Section 7.3.4.2(b). The two-photon transition moment may become non-zero when certain vibrations are excited.

The general vibronic selection rule replacing that in equation (9.20) is

$$\Gamma(\psi'_{ev}) \times \Gamma(S_{ij}) \times \Gamma(\psi''_{ev}) = A \text{ (or } \supset A) \qquad (9.22)$$

If the lower state is the zero-point level of the ground electronic state, $\Gamma(\psi''_{ev}) = A$ and equations (9.22) and (9.21) reduce to

$$\Gamma(\psi'_{ev}) = \Gamma(S_{ij}) = \Gamma(\alpha_{ij}) \qquad (9.23)$$

or

$$\Gamma(\psi'_e) \times \Gamma(\psi'_v) = \Gamma(\alpha_{ij}) \qquad (9.24)$$

Figure 9.29 shows that the most important inducing vibration is $v_{14}{}^{\dagger}$, a b_{2u} vibration involving stretching and contracting of alternate C–C bonds in the ring. Using Table A.32 in the Appendix for the 14^1_0 transition, equation (9.24) becomes

$$\Gamma(\psi'_e) \times \Gamma(\psi'_v) = B_{2u} \times b_{2u} = A_g = \Gamma(\alpha_{xx}, \alpha_{yy}, \alpha_{zz}) \qquad (9.25)$$

and the transition is allowed. In Figure 9.29 it can be seen that other non-totally symmetric vibrations are more weakly active in vibronic coupling.

9.3.6 Multiphoton Dissociation and Laser Separation of Isotopes

In 1971 it was discovered that luminescence (fluorescence or phosphorescence) occurs in various molecular gases when a pulsed CO_2 laser is focused into the body of the gas. To observe this effect requires a pulsed laser in order to achieve the high power necessary (a peak power of ca. 0.5 MW was used) and also to be able to observe the luminescence when each pulse has died away. The gases used included CCl_2F_2, SiF_4 and NH_3, all of which have an infrared vibration–rotation absorption band in a region of the spectrum in which one of the CO_2 laser lines falls. In CCl_2F_2, SiF_4, and NH_3 the species responsible for the luminescence were identified as C_2, SiF, and NH_2 respectively.

The process of dissociation by the absorption of infrared photons clearly involves the simultaneous absorption of many photons—of the order of 30, depending on the dissociation energy and the photon energy—and is called multiphoton dissociation.

The theory of the process is not simple. Figure 9.30 illustrates the mechanism as being one in which a laser photon of wavenumber $\tilde{v}_L$ is resonant with a $v = 1$–0 transition in the molecule and subsequent photons are absorbed to take the molecule to successively higher vibrational levels—like climbing the rungs of a ladder. The figure shows that an effect of vibrational anharmonicity is that the laser radiation is resonant *only* with the $v = 1$–0 transition and the higher the vibrational energy level, the greater is the possibility of the laser being off-resonance—the rungs of the ladder are not equally separated. This difficulty can be overcome to some extent by taking into account the fact that each vibrational level has rotational levels associated with it and this 'rotational compensation' of vibrational anharmonicity may mean that resonance of the laser radiation with some vibration–rotation levels of the molecule may occur up to, say, $v = 3$. At higher vibrational energy than, say, that of the $v = 3$ level, a polyatomic molecule has a high density of vibration–rotation states, the

---

$\dagger$ Based on the Wilson numbering for benzene (see Bibliography for Chapter 6).

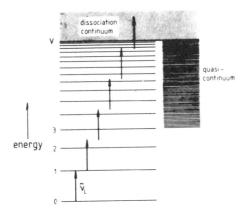

Figure 9.30 Vibrational energy level scheme for multiphoton dissociation.

density increasing with the number of vibrational modes and, therefore, with the number of atoms. This high density results in a quasi-continuum of states similar to that discussed in Section 7.3.6 as a possible cause of diffuseness in electronic spectra of polyatomic molecules. The quasi-continuum, contributed to by all vibrational modes except that for which discrete levels are drawn, is shown in Figure 9.30. It is this quasi-continuum, together with an effect which results in broadening of a transition by the high power of the laser, which provides the higher rungs of the multiphoton dissociation ladder.

The yield of dissociation products may be small but sensitive methods of detection can be used. One of these is laser-induced fluorescence, shown schematically in Figure 9.31, in which a second, probe, laser is used to excite fluorescence in one of the products of dissociation. The CO_2 and probe laser beams are at 90° to each other and the fluorescence is detected by a photo-multiplier at 90° to both beams. This technique has been used, for example, to

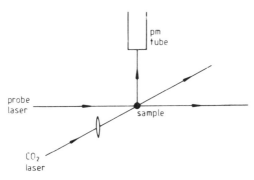

Figure 9.31 Detection of products of multiphoton dissociation by laser-induced fluorescence.

monitor the production of NH_2 from the dissociation of hydrazine (N_2H_4) or methylamine (CH_3NH_2). The probe laser was a tunable dye laser set at a wavelength of 598 nm corresponding to absorption in the 2_0^9 band, where v_2 is the bending vibration, of the $\tilde{A}^2A_1 - \tilde{X}^2B_1$ electronic system of NH_2 and total fluorescence from the 2^9 level was monitored.

The phenomenon of multiphoton dissociation finds a very important application in the separation of isotopes. For this purpose it is not only the high power of the laser which is important but the fact that it is highly monochromatic. This latter property makes it possible, in favourable circumstances, for the laser radiation to be absorbed selectively by a single isotopic molecular species. This species is then selectively dissociated resulting in isotopic enrichment in both the dissociation products and in the undissociated material.

One of the first applications of this technique was to the enrichment of ^{10}B and ^{11}B isotopes, present as 18.7 and 81.3 per cent respectively in natural abundance. Boron trichloride, BCl_3, dissociates when irradiated with a pulsed CO_2 laser in the 3_0^1 vibrational band at 958 cm^{-1} (v_3 is an e' vibration of the planar, D_{3h}, molecule). One of the products of dissociation was detected by reaction with O_2 to form BO which then produced chemiluminescence (emission of radiation as a result of energy gained by chemical reaction) in the visible region due to $A^2\Pi - X^2\Sigma^+$ fluorescence. Irradiation in the 3_0^1 band of $^{10}BCl_3$ or $^{11}BCl_3$ resulted in ^{10}BO or ^{11}BO chemiluminescence. The fluorescence of ^{10}BO is easily resolved from that of ^{11}BO.

Figure 9.32 illustrates the isotopic enrichment of SF_6 following irradiation

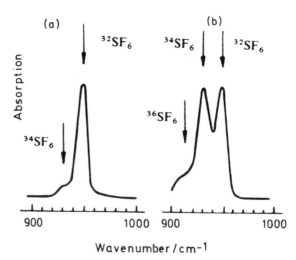

Figure 9.32 Isotopic enrichment of SF_6 by multiphoton dissociation following irradiation in the 3_0^1 vibrational band of $^{32}SF_6$. The absorption spectrum is shown (a) before and (b) after irradiation. (Reproduced, with permission, from Letokhov, V. S., *Nature, Lond.*, **277**, 605, 1979 Copyright © 1979 Macmillan Journals Limited.)

with a pulsed CO_2 laser in the 3_0^1 vibrational band, at 945 cm^{-1}, of $^{32}SF_6$, v_3 being a strongly infrared active t_{1u} bending vibration. The natural abundances of the isotopes of sulphur are ^{32}S (95.0 per cent), ^{34}S (4.24 per cent) ^{33}S (0.74 per cent), and ^{36}S (0.017 per cent). The figure shows that depletion of $^{32}SF_6$ has been achieved to such an extent that equal quantities of $^{34}SF_6$ and $^{32}SF_6$ remain.

9.3.7 Single Vibronic Level Fluorescence

It was mentioned in Section 7.2.5.2, in a discussion of progressions in electronic spectra of diatomic molecules, that in a gas phase emission spectrum a progression with, say, $v' = 2$, shown in Figure 7.18, could be observed on its own only under rather special conditions. These conditions are that the method of excitation takes the molecule from $v'' = 0$ to $v' = 2$ only and that no collisions occur between molecules or with the walls of the container during the lifetime of the emission. The emission, usually fluorescence, is then from the single vibronic level with $v' = 2$ and the technique is known as single vibronic level fluorescence (SVLF) spectroscopy or, alternatively, dispersed fluorescence (DF) spectroscopy. Dispersion of this fluorescence in a spectrometer produces a spectrum which gives information about ground state vibrational levels. It is a particularly powerful technique for the investigation of electronic spectra of large molecules where band congestion causes problems both in absorption and, under normal conditions, in emission.

From about 1970, before the availability of suitable lasers, Parmenter and others obtained SVLF spectra, particularly of benzene, using radiation from an intense high pressure xenon arc source (see Section 3.4.4) and passing it through a monochromator to select a narrow band (*ca* 20 cm^{-1} wide) of radiation to excite the sample within a particular absorption band.

Dye lasers, frequency doubled if necessary, provide ideal sources for such experiments. The radiation is very intense, the line width is small (≤ 1 cm^{-1}) and the wavenumber may be tuned to match any absorption band in the visible or near-ultraviolet region.

Figure 9.33 shows examples of SVLF spectra obtained by tuning a frequency doubled dye laser to the 0_0^0 absorption band of the $\tilde{A}^1 B_{3u} - \tilde{X}^1 A_g$ system of pyrazine (1,4-diazabenzene) and of perdeuteropyrazine. At a pressure of about 3 Torr the collision-free dispersed fluorescence is from only the zero-point level of the $\tilde{A}^1 B_{3u}$ state.

Pyrazine belongs to the D_{2h} point group and Table A.32 in the Appendix shows that, since $B_{3u} = \Gamma(T_x)$, the 0_0^0 band is polarized along the x axis, which is perpendicular to the molecular plane.

The spectra in Figure 9.33 show progressions in v_{6a}, v_{9a}, and v_{10a}.[†] Two

---

[†] Numbering based on the Wilson numbering for the analogous vibrations of benzene.

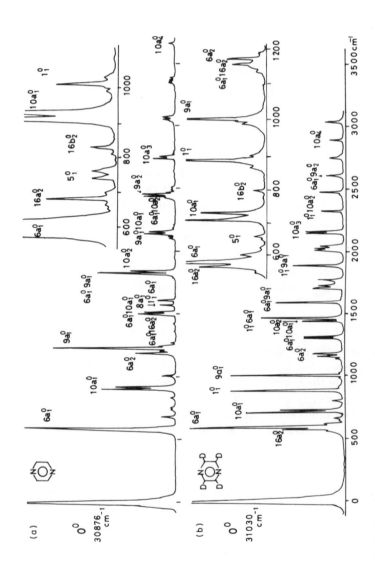

Figure 9.33 Single vibronic level fluorescence spectra obtained by collision-free emission from the zero-point level of the A^1B_{3u} state of (a) pyrazine, (b) perdeuteropyrazine. (Reproduced, with permission, from Udagawa, Y., Ito, M., and Suzuka, I., *Chem. Phys.*, **46**, 237, 1980.)

of them, v_{6a} and v_{9a}, are totally symmetric a_g vibrations so each band in each of these progressions is also polarized along the x axis. But v_{10a} is a b_{1g} vibration so that, for the $10a_1^0$ band, equation (7.127) gives

$$\Gamma(\psi'_e) \times \Gamma(\psi''_e) \times \Gamma(\psi''_v) = B_{3u} \times A_g \times b_{1g} = B_{2u} = \Gamma(T_y) \qquad (9.26)$$

and the band is polarized along the y axis which is in-plane and perpendicular to the N $\cdots$ N line. Perdeuteropyrazine is an oblate asymmetric rotor and, since the x axis is the c inertial axis and the y axis is the a inertial axis, the 0_0^0 band shows a type C and the $10a_1^0$ band a type A rotational contour due to the corresponding rotational selection rules operating (see Section 7.3.5).

When the laser line is tuned to match the 0_0^0 absorption band it does not overlie the whole rotational contour but only a part of it. The result is that only a limited number of rotational levels in the excited electronic state are populated. Under the low pressure conditions of the sample not only is there no vibrational relaxation (reversion towards a Boltzmann population distribution) but there is no rotational relaxation either. So what is observed in Figure 9.33, so far as rotational structure of the bands is concerned, are fragments of complete rotational contours. It is clear from a comparison of, say, the $6a_1^0$ and the $10a_1^0$ bands in the spectrum of perdeuteropyrazine shown that the fragmented type C contour is characterized by a single sharp peak while the fragmented type A contour shows two sharp peaks. Similar fragmented contours are observed in the pyrazine spectrum in Figure 9.33 in spite of a slight complication due to the y axis being the a inertial axis in the $\tilde{A}$ state but the b inertial axis in the $\tilde{X}$ state due to the different geometry in the two states.

The value of these characteristic fragmented contours in making assignments is apparent.

The $10a_1^0$ band, as well as many bands built on it, and also the $10a_3^0$ band obtain their intensity by the Herzberg–Teller vibronic coupling mechanism discussed in Section 7.3.4.2(b).

We have seen in Section 6.1.3.2 that, for diatomic molecules, vibrational energy levels, other than those with $v = 1$, in the ground electronic state are very often obtained not from vibrational spectroscopy but from electronic emission spectroscopy. In the emission process we rely on the Franck–Condon principle to allow access to high-lying vibrational levels in the ground electronic state. From these levels the potential energy curve may be constructed.

If we require similar information regarding the ground state potential energy surface in a polyatomic molecule the electronic emission spectrum may again provide valuable information and SVLF spectroscopy is a particularly powerful technique for providing it.

9.3.8 Spectroscopy of Molecules in Supersonic Jets

In Section 2.3.4.1 we have seen that pumping of atoms or molecules through a narrow slit, or small pinhole, whose width (or diameter) is about 20 μm at a pressure of a few torr on the high pressure side of the aperture produces an effusive beam. Removal of pressure broadening and a considerable reduction in Doppler broadening of spectral lines are useful properties of such a beam. However, because the slit width or pinhole diameter d is much less than the mean free path λ_0 for collisions between the atoms or molecules passing through it, i.e.

$$d \ll \lambda_0 \tag{9.27}$$

there are no collisions in or beyond the pinhole. As a result the Maxwellian velocity distribution among the particles remains the same as it was in the reservoir of the gas forming the beam.

In 1951, Kantrowitz and Grey suggested that if conditions are changed such that

$$d \gg \lambda_0 \tag{9.28}$$

where d may be typically 100 μm and the pressure several atmospheres, then the numerous collisions occurring in, and immediately beyond, the pinhole (or nozzle) will convert the random motions of the particles in the gas into mass flow in the direction of the resulting beam. Not only is the flow highly directional but the range of velocities is very much reduced. One result of this is that the translational temperature T_{tr} of the beam produced may be extremely low, even less than 1 K, because any particle travelling in the beam has a very small velocity relative to neighbouring particles and experiences collisions only rarely. Such cooling continues in the region beyond the nozzle where collisions are still occurring. In this region the type of flow of the gas is called hydro-dynamic whereas in the region beyond this, where collisions no longer occur, the flow is said to be molecular. In the region of molecular flow there is no further decrease of T_{tr} and there are decreases in pressure and Doppler line broadening similar to those produced in an effusive beam, in which the flow is always molecular.

Why are these beams, or jets, distinguished from effusive beams by their description as supersonic? In some ways this description is rather misleading, firstly because particles in an effusive beam may well be travelling at supersonic velocities and, secondly, because the name implies that something special happens when the particle velocities become supersonic whereas this is not the case. What 'supersonic' is meant to imply is that the particles may have *very* high Mach numbers (of the order of 100). The Mach number M is defined as

$$M = \frac{u}{a} \tag{9.29}$$

where u is the mass flow velocity and a is the local speed of sound, given by

$$a = \left(\frac{\gamma k T_{tr}}{m} \right)^{1/2} \tag{9.30}$$

where $\gamma = C_p/C_V = 5/3$ for a monatomic gas, such as helium or argon which are commonly used in supersonic jets, and m is the mass of the gas particles. Very high Mach numbers arise not so much because u is large (although it might be twice the speed of sound in air) but because a is so small as a result of the low value of T_{tr}.

If a particularly low background pressure is required in the chamber into which the beam is flowing the beam may be skimmed in the region of hydrodynamic flow. A skimmer is a collimator which is specially constructed in order to avoid shock waves travelling back into the gas and increasing T_{tr}. The gas which has been skimmed away may be pumped off in a separate vacuum chamber. Further collimation may be carried out in the region of molecular flow and a so-called supersonic beam results. When a skimmer is not used, a supersonic jet results: this may or may not be collimated. Unskimmed jets are commonly used for spectroscopic purposes. On the high pressure side of the nozzle molecules may be seeded into the jet of helium or argon and are also cooled by the many collisions that take place. However, in discussing temperature in molecules, we must distinguish between translational, rotational and vibrational temperatures. The translational temperature is the same as that of the helium or argon carrier gas and may be less than 1 K.

The rotational temperature is defined as the temperature which describes the Boltzmann population distribution among rotational levels. For example, for a diatomic molecule, this is the temperature in equation (5.15). Since collisions are not so efficient in producing rotational cooling as for translational cooling, rotational temperatures are rather higher but may be as low as 1 K.

The vibrational temperature, defined for a diatomic harmonic oscillator by the temperature in equation (5.22), is considerably higher because of the low efficiency of vibrational cooling. A vibrational temperature of about 100 K is typical although, in a polyatomic molecule, it depends very much on the nature of the vibration.

This vibrational cooling is sufficient to stabilize complexes which are weakly bound by van der Waals or hydrogen-bonding forces. The pure rotational spectra and structure of species such as $Ar \cdots H$–Cl,

$$
\begin{array}{ccc}
& O & H \\
& \parallel & / \\
Ar \cdots C & \text{and} & O \cdots H{-}O \quad, \\
& \parallel & H\diagup | \\
& O & H
\end{array}
$$

produced in supersonic jets have been investigated by Klemperer and others who were the first to use supersonic jets for spectroscopic purposes. They observed transitions between molecular rotational levels perturbed by an electric (Stark) field.

Electronic transitions in molecules in supersonic jets may be investigated by intersecting the jet with a tunable dye laser in the region of molecular flow and observing the total fluorescence intensity. As the laser is tuned across the absorption band system a fluorescence excitation spectrum results which strongly resembles the absorption spectrum. The spectrum may be of individual molecules, van der Waals complexes with the helium or argon carrier gas, hydrogen bonded or van der Waals dimers, or larger clusters.

Figure 9.34 shows an example of a fluorescence excitation spectrum of hydrogen bonded dimers of s-tetrazine (1,2,4,5-tetraazabenzene). The pressure of s-tetrazine seeded into helium carrier gas at 4 atm pressure was about 0.001 atm. Expansion was through a 100 μm diameter nozzle. A high resolution (0.005 cm^{-1}) dye laser crossed the supersonic jet 5 mm downstream from the nozzle.

The three bands in Figure 9.34 show resolved rotational structure and a

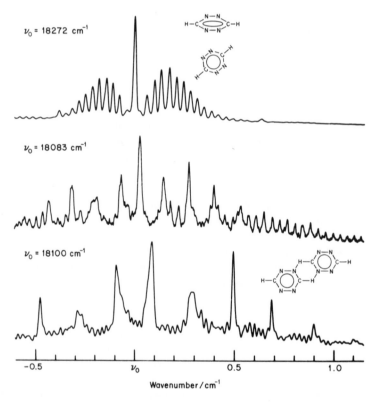

Figure 9.34 Rotational structure of the 0^0_0 bands in the fluorescence excitation spectra of s-tetrazine dimers at about 552 nm. Bottom: 0^0_0 band of planar dimer. Middle: 0^0_0 band of T-shaped dimer with transition in monomer unit in stem of T. Top: 0^0_0 band of T-shaped dimer with transition in monomer unit in top of T. (Reproduced, with permission, from Haynam, C. A., Brumbaugh, D. V., and Levy, D. H., *J. Chem. Phys.*, **79**, 1581, 1983.)

rotational temperature of about 1 K. Computer simulation has shown that they are all 0_0^0 bands of dimers. The bottom spectrum is the 0_0^0 band of the planar, doubly hydrogen bonded dimer illustrated. The electronic transition moment is polarized perpendicular to the ring in the $\tilde{A}^1B_{3u} - \tilde{X}^1A_g$, $\pi^* - n$ transition of the monomer and the rotational structure of the bottom spectrum is consistent only with it being perpendicular to the molecular plane in the dimer also, as expected.

The top two bands in Figure 9.34 show rotational structure consistent with their being 0_0^0 bands of an approximately T-shaped dimer in which one of the hydrogen atoms in the monomer unit in the stem of the T is attracted towards the π-electron density of the ring at the top of the T. The planes of the two rings are perpendicular to each other as shown but the T is not symmetrical. Instead of being 90°, the stem-top angles are about 50° and 130°. The electronic transition is localized in one of the rings. The top 0_0^0 band in Figure 9.34 is the result of the transition being in the ring at the top of the T while the middle 0_0^0 band results from the transition being in the ring forming the stem of the T.

It might be thought that the small number of molecules in a typical supersonic jet would seriously limit the sensitivity of observation of the spectra. However, the severe rotational cooling which may be produced results in a collapsing of the overall intensity of a band into many fewer rotational transitions. Vibrational cooling, which greatly increases the population of the zero-point level, concentrates the intensity in few vibrational transitions and these two effects tend to compensate for the small number of molecules.

Fluorescence spectra of fairly large molecules in a supersonic jet may be vibrationally simplified, due to depopulation of low-lying vibrational levels in the ground electronic state, thereby revealing vibrational structure which, in the corresponding gas phase spectrum, may be buried in a vibrationally congested spectrum. This vibrational structure may be extremely useful in obtaining information on the structure of the molecule in the ground or excited electronic state.

Figure 9.35 shows part of the $\tilde{A}^1B_{2u} - \tilde{X}^1A_g$ ($S_1 - S_0$) fluorescence excitation spectrum of 1,2,4,5-tetrafluorobenzene in a supersonic jet. This spectrum is dominated by a long progression in v_{11}, a b_{3u} vibration in which all the fluorine atoms move above, or all below, the plane of the benzene ring in a 'butterfly' type of motion. The vibronic selection rules restrict the transitions to those with Δv_{11} even. Transitions are observed up to $v'_{11} = 10$ with an intensity maximum at $v'_{11} = 2$. The Franck–Condon principle tells us that there is a geometry change from the ground to the excited electronic state. The ground state is planar and the equilibrium structure in the excited state is non-planar, butterfly-shaped. The C–F bonds are about 11° out-of-plane.

Since, in the excited state, the fluorine atoms may be above or below the plane of the benzene ring the potential function for v_{11} is W-shaped, like that

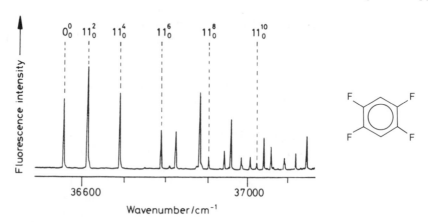

Figure 9.35 Part of the fluorescence excitation spectrum of 1,2,4,5-tetrafluorobenzene in a supersonic jet. (Reproduced, with permission, from Okuyama, K., Kakinuma, T., Fujii, M., Mikami, N. and Ito, M., *J. Phys. Chem.*, **90**, 3948, 1986.)

in Figure 6.38. Fitting the observed vibrational energy levels to the potential function in equation (6.93) gives the height of the barrier to planarity as 78 cm^{-1}.

By fixing the laser wavelength so that it corresponds to that of a particular vibronic transition, population of a single vibronic level may be achieved. Because of the collision-free conditions in the region of the jet where molecular flow is occurring, fluorescence from molecules in this region will be from the single vibronic level which was populated by absorption of the laser radiation. Resulting single vibronic level fluorescence (SVLF) or dispersed fluorescence (DF) spectra at low rotational and vibrational temperatures are analogous to the room temperature gas-phase spectra described in Section 9.3.7 but without the attendant congestion which makes the band in which absorption takes place likely to suffer overlap from nearby bands.

Figure 9.36 shows the SVLF spectrum of styrene ($C_6H_5CH=CH_2$) with excitation in the 0^0_0 band of the $\tilde{A}^1 A' - \tilde{X}^1 A'$ ($S_1 - S_0$) band system. There is a prominent progression in the vibration ν_{42} which is a torsional motion of the vinyl group about the C(1)–C(α) bond. The vibronic selection rules allow only transitions with $\Delta \nu_{42}$ even. Those with $\nu''_{42} = 0$, 2, 4, 6 and 10 are observed. More vibrational levels, with ν''_{42} even and odd, have been identified and fitted to a torsional potential function of the type in equation (6.96) giving

$$V(\phi)/\mathrm{cm}^{-1} = [1070(1 - \cos 2\phi) - 275(1 - \cos 4\phi) + 7(1 - \cos 6\phi)] \quad (9.31)$$

where $V_2 = 1070$ cm^{-1}, $V_4 = -275$ cm^{-1} and $V_6 = 7$ cm^{-1}; and $\phi = 0°$ corresponds to the planar configuration.

In styrene there are two competing effects so far as the possible planarity of the molecule is concerned. Conjugation between the double bond of the vinyl

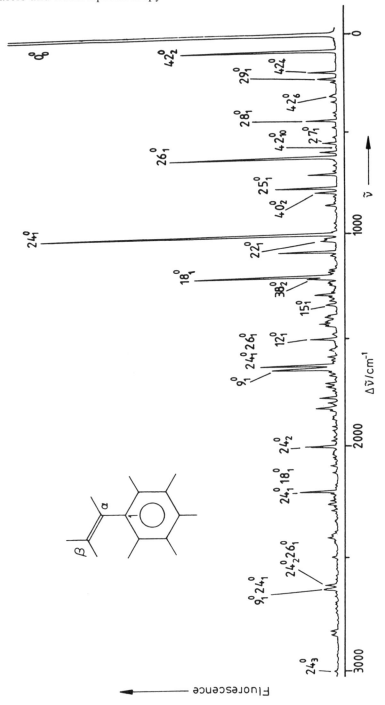

Figure 9.36 Single vibronic level fluorescence spectrum of styrene, in a supersonic jet, with excitation in the 0_0^0 band.

group and the benzene ring is favoured by planarity but through-space interaction between the double bond of the vinyl group and the nearest C–C bond of the benzene ring, together, perhaps, with some steric hindrance between hydrogen atoms of the vinyl group and the benzene ring, are reduced by the vinyl group being twisted out of plane. Conjugation turns out to be the dominating effect and the molecule is planar in the ground electronic state.

The potential in equation (9.31) repeats every π radians and $V_2 = 1070 \text{ cm}^{-1}$ is the height of the barrier corresponding to the energy required to twist the vinyl group so that it is perpendicular to the benzene ring. However, the large, negative V_4 term makes the potential very flat-bottomed. The vibrational energy levels are very closely-spaced: for example the $v''_{42} = 1{-}0$ interval is only 38 cm^{-1}, and it takes very little energy to twist the vinyl group appreciably out of plane. The molecule is planar, but only just.

Questions

1. A dye laser cavity is 10.34 cm long and is operating at 533.6 nm. How many half-wavelengths are there along the length of the cavity and by how much would the cavity length have to be changed to increase this number by one? What consequences does this result have for the tuning of such a cavity? How long does it take for the radiation to complete one round-trip of the cavity?

2. Show that the third term in equation (9.11) results in generation of radiation of frequency 3ν when radiation of frequency ν is incident on a crystal capable of harmonic generation.

3. Show that the 10.6 and 9.6 µm radiation from a CO_2 laser is due to transitions allowed by electric dipole selection rules.

4. Draw a diagram similar to that in Figure 9.21 to illustrate the stimulated Raman effect in H_2. High pressure H_2 is used to Raman shift radiation from a KrF laser. Calculate the two wavelengths of the shifted radiation which are closest to that of the KrF laser.

5. Write down the symmetry species of vibrations which are allowed in the CARS spectrum of CH_4.

6. In the two-photon spectrum in Figure 9.29 the vibrations ν_{18a}, ν_{17b}, ν_{16a}, and ν_{15} have symmetry species b_{1u}, b_{3u}, a_u and b_{2u} respectively. Show that the $18a_0^1$, $17b_0^1$, $16a_0^1$ and 15_0^1 transitions are allowed by symmetry.

Bibliography

Andrews, D. L. (1985). *Lasers in Chemistry*, Springer-Verlag, Berlin.
Beesley, M. J. (1976). *Lasers and Their Applications*, Taylor and Francis, London.
Demtröder, W. (1981). *Laser Spectroscopy*, Springer-Verlag, Berlin.
Lengyel, B. A. (1971). *Lasers*, Wiley–Interscience, New York.
Siegman, A. E. (1971). *An Introduction to Lasers and Masers*, McGraw-Hill, New York.
Svelto, O. (1976). *Principles of Lasers* (translated by D. C. Hanna), Heyden, London.

APPENDIX

Character tables

Index to tables

Table A.1

C_s	I	σ		
A'	1	1	T_x, T_y, R_z	$\alpha_{xx}, \alpha_{yy}, \alpha_{zz}, \alpha_{xy}$
A''	1	-1	T_z, R_x, R_y	α_{yz}, α_{xz}

Table A.2

C_i	I	i		
A_g	1	1	R_x, R_y, R_z	$\alpha_{xx}, \alpha_{yy}, \alpha_{zz}, \alpha_{xy}, \alpha_{xz}, \alpha_{yz}$
A_u	1	−1	T_x, T_y, T_z	

Table A.3

C_1	I	
A	1	All R, T, α

Table A.4

C_2	I	C_2		
A	1	1	T_z, R_z	$\alpha_{xx}, \alpha_{yy}, \alpha_{zz}, \alpha_{xy}$
B	1	−1	T_x, T_y, R_x, R_y	α_{yz}, α_{xz}

Table A.5

C_3	I	C_3	C_3^2		
A	1	1	1	T_z, R_z	$\alpha_{xx} + \alpha_{yy}, \alpha_{zz}$
E	$\left\{\begin{matrix}1\\1\end{matrix}\right.$	$\begin{matrix}\varepsilon\\\varepsilon^*\end{matrix}$	$\left.\begin{matrix}\varepsilon^*\\\varepsilon\end{matrix}\right\}$	$(T_x, T_y), (R_x, R_y)$	$(\alpha_{xx} - \alpha_{yy}, \alpha_{xy}), (\alpha_{yz}, \alpha_{xz})$

$\varepsilon = \exp(2\pi i/3)$, $\varepsilon^* = \exp(-2\pi i/3)$

Table A.6

C_4	I	C_4	C_2	C_4^3		
A	1	1	1	1	T_z, R_z	$\alpha_{xx} + \alpha_{yy}, \alpha_{zz}$
B	1	−1	1	−1		$\alpha_{xx} - \alpha_{yy}, \alpha_{xy}$
E	$\left\{\begin{matrix}1\\1\end{matrix}\right.$	$\begin{matrix}i\\-i\end{matrix}$	$\begin{matrix}-1\\-1\end{matrix}$	$\left.\begin{matrix}-i\\i\end{matrix}\right\}$	$(T_x, T_y), (R_x, R_y)$	$(\alpha_{yz}, \alpha_{xz})$

Table A.7

C_5	I	C_5	C_5^2	C_5^3	C_5^4		
A	1	1	1	1	1	T_z,R_z	$\alpha_{xx}+\alpha_{yy},\alpha_{zz}$
E_1	$\begin{cases}1\\1\end{cases}$	$\begin{matrix}\varepsilon\\\varepsilon^*\end{matrix}$	$\begin{matrix}\varepsilon^2\\\varepsilon^{2*}\end{matrix}$	$\begin{matrix}\varepsilon^{2*}\\\varepsilon^2\end{matrix}$	$\begin{matrix}\varepsilon^*\\\varepsilon\end{matrix}\Big\}$	$(T_x,T_y),(R_x,R_y)$	$(\alpha_{yz},\alpha_{xz})$
E_2	$\begin{cases}1\\1\end{cases}$	$\begin{matrix}\varepsilon^2\\\varepsilon^{2*}\end{matrix}$	$\begin{matrix}\varepsilon^*\\\varepsilon\end{matrix}$	$\begin{matrix}\varepsilon\\\varepsilon^*\end{matrix}$	$\begin{matrix}\varepsilon^{2*}\\\varepsilon^2\end{matrix}\Big\}$		$(\alpha_{xx}-\alpha_{yy},\alpha_{xy})$

$\varepsilon = \exp(2\pi i/5),\ \varepsilon^* = \exp(-2\pi i/5)$

Table A.8

C_6	I	C_6	C_3	C_3^2	C_6^5			
A	1	1	1	1	1	1	T_z,R_z	$\alpha_{xx}+\alpha_{yy},\alpha_{zz}$
B	1	-1	1	-1	1	-1		
E_1	$\begin{cases}1\\1\end{cases}$	$\begin{matrix}\varepsilon\\\varepsilon^*\end{matrix}$	$\begin{matrix}-\varepsilon^*\\-\varepsilon\end{matrix}$	$\begin{matrix}-1\\-1\end{matrix}$	$\begin{matrix}-\varepsilon\\-\varepsilon^*\end{matrix}$	$\begin{matrix}\varepsilon^*\\\varepsilon\end{matrix}\Big\}$	$(T_x,T_y),(R_x,R_y)$	$(\alpha_{xz},\alpha_{yz})$
E_2	$\begin{cases}1\\1\end{cases}$	$\begin{matrix}-\varepsilon^*\\-\varepsilon\end{matrix}$	$\begin{matrix}-\varepsilon\\-\varepsilon^*\end{matrix}$	$\begin{matrix}1\\1\end{matrix}$	$\begin{matrix}-\varepsilon^*\\-\varepsilon\end{matrix}$	$\begin{matrix}-\varepsilon\\-\varepsilon^*\end{matrix}\Big\}$		$(\alpha_{xx}-\alpha_{yy},\alpha_{xy})$

$\varepsilon = \exp(2\pi i/6),\ \varepsilon^* = \exp(-2\pi i/6)$

Table A.9

C_7	I	C_7	C_7^2	C_7^3	C_7^4	C_7^5	C_7^6		
A	1	1	1	1	1	1	1	T_z,R_z	$\alpha_{xx}+\alpha_{yy},\alpha_{zz}$
E_1	$\begin{cases}1\\1\end{cases}$	$\begin{matrix}\varepsilon\\\varepsilon^*\end{matrix}$	$\begin{matrix}\varepsilon^2\\\varepsilon^{2*}\end{matrix}$	$\begin{matrix}\varepsilon^3\\\varepsilon^{3*}\end{matrix}$	$\begin{matrix}\varepsilon^{3*}\\\varepsilon^3\end{matrix}$	$\begin{matrix}\varepsilon^{2*}\\\varepsilon^2\end{matrix}$	$\begin{matrix}\varepsilon^*\\\varepsilon\end{matrix}\Big\}$	$(T_x,T_y),(R_x,R_y)$	$(\alpha_{xz},\alpha_{yz})$
E_2	$\begin{cases}1\\1\end{cases}$	$\begin{matrix}\varepsilon^2\\\varepsilon^{2*}\end{matrix}$	$\begin{matrix}\varepsilon^{3*}\\\varepsilon^3\end{matrix}$	$\begin{matrix}\varepsilon^*\\\varepsilon\end{matrix}$	$\begin{matrix}\varepsilon\\\varepsilon^*\end{matrix}$	$\begin{matrix}\varepsilon^3\\\varepsilon^{3*}\end{matrix}$	$\begin{matrix}\varepsilon^{2*}\\\varepsilon^2\end{matrix}\Big\}$		$(\alpha_{xx}-\alpha_{yy},\alpha_{xy})$
E_3	$\begin{cases}1\\1\end{cases}$	$\begin{matrix}\varepsilon^3\\\varepsilon^{3*}\end{matrix}$	$\begin{matrix}\varepsilon^*\\\varepsilon\end{matrix}$	$\begin{matrix}\varepsilon^2\\\varepsilon^{2*}\end{matrix}$	$\begin{matrix}\varepsilon^{2*}\\\varepsilon^2\end{matrix}$	$\begin{matrix}\varepsilon\\\varepsilon^*\end{matrix}$	$\begin{matrix}\varepsilon^{3*}\\\varepsilon^3\end{matrix}\Big\}$		

$\varepsilon = \exp(2\pi i/7),\ \varepsilon^* = \exp(-2\pi i/7)$

Table A.10

C_8	I	C_8	C_4	C_8^3	C_2	C_8^5	C_4^3	C_8^7		
A	1	1	1	1	1	1	1	1	T_z, R_z	$\alpha_{xx} + \alpha_{yy}, \alpha_{zz}$
B	1	-1	1	-1	1	-1	1	-1		
E_1	$\begin{cases}1 \\ 1\end{cases}$	$\begin{matrix}\varepsilon \\ \varepsilon^*\end{matrix}$	$\begin{matrix}i \\ -i\end{matrix}$	$\begin{matrix}-\varepsilon^* \\ -\varepsilon\end{matrix}$	$\begin{matrix}-1 \\ -1\end{matrix}$	$\begin{matrix}-\varepsilon \\ -\varepsilon^*\end{matrix}$	$\begin{matrix}-i \\ i\end{matrix}$	$\begin{matrix}\varepsilon^* \\ \varepsilon\end{matrix}$	$(T_x, T_y), (R_x, R_y)$	$(\alpha_{xz}, \alpha_{yz})$
E_2	$\begin{cases}1 \\ 1\end{cases}$	$\begin{matrix}i \\ -i\end{matrix}$	$\begin{matrix}-1 \\ -1\end{matrix}$	$\begin{matrix}-i \\ i\end{matrix}$	$\begin{matrix}1 \\ 1\end{matrix}$	$\begin{matrix}i \\ -i\end{matrix}$	$\begin{matrix}-1 \\ -1\end{matrix}$	$\begin{matrix}-i \\ i\end{matrix}$		$(\alpha_{xx} - \alpha_{yy}, \alpha_{xy})$
E_3	$\begin{cases}1 \\ 1\end{cases}$	$\begin{matrix}-\varepsilon^* \\ -\varepsilon\end{matrix}$	$\begin{matrix}-i \\ i\end{matrix}$	$\begin{matrix}\varepsilon \\ \varepsilon^*\end{matrix}$	$\begin{matrix}-1 \\ -1\end{matrix}$	$\begin{matrix}\varepsilon^* \\ \varepsilon\end{matrix}$	$\begin{matrix}i \\ -i\end{matrix}$	$\begin{matrix}-\varepsilon \\ -\varepsilon^*\end{matrix}$		

$\varepsilon = \exp(2\pi i/8), \ \varepsilon^* = \exp(-2\pi i/8)$

Table A.11

C_{2v}	I	C_2	$\sigma_v(xz)$	$\sigma_v'(yz)$		
A_1	1	1	1	1	T_z	$\alpha_{xx}, \alpha_{yy}, \alpha_{zz}$
A_2	1	1	-1	-1	R_z	α_{xy}
B_1	1	-1	1	-1	T_x, R_y	α_{xz}
B_2	1	-1	-1	1	T_y, R_x	α_{yz}

Table A.12

C_{3v}	I	$2C_3$	$3\sigma_v$		
A_1	1	1	1	T_z	$\alpha_{xx} + \alpha_{yy}, \alpha_{zz}$
A_2	1	1	-1	R_z	
E	2	-1	0	$(T_x, T_y), (R_x, R_y)$	$(\alpha_{xx} - \alpha_{yy}, \alpha_{xy}), (\alpha_{xz}, \alpha_{yz})$

Table A.13

C_{4v}	I	$2C_4$	C_2	$2\sigma_v$	$2\sigma_d$		
A_1	1	1	1	1	1	T_z	$\alpha_{xx} + \alpha_{yy}, \alpha_{zz}$
A_2	1	1	1	-1	-1	R_z	
B_1	1	-1	1	1	-1		$\alpha_{xx} - \alpha_{yy}$
B_2	1	-1	1	-1	1		α_{xy}
E	2	0	-2	0	0	$(T_x, T_y), (R_x, R_y)$	$(\alpha_{xz}, \alpha_{yz})$

Table A.14

C_{5v}	I	$2C_5$	$2C_5^2$	$5\sigma_v$		
A_1	1	1	1	1	T_z	$\alpha_{xx}+\alpha_{yy},\alpha_{zz}$
A_2	1	1	1	-1	R_z	
E_1	2	$2\cos 72°$	$2\cos 144°$	0	$(T_x,T_y),(R_x,R_y)$	$(\alpha_{xz},\alpha_{yz})$
E_2	2	$2\cos 144°$	$2\cos 72°$	0		$(\alpha_{xx}-\alpha_{yy},\alpha_{xy})$

Table A.15

C_{6v}	I	$2C_6$	$2C_3$	C_2	$3\sigma_v$	$3\sigma_d$		
A_1	1	1	1	1	1	1	T_z	$\alpha_{xx}+\alpha_{yy},\alpha_{zz}$
A_2	1	1	1	1	-1	-1	R_z	
B_1	1	-1	1	-1	1	-1		
B_2	1	-1	1	-1	-1	1		
E_1	2	1	-1	-2	0	0	$(T_x,T_y),(R_x,R_y)$	$(\alpha_{xz},\alpha_{yz})$
E_2	2	-1	-1	2	0	0		$(\alpha_{xx}-\alpha_{yy},\alpha_{xy})$

Table A.16

$C_{\infty v}$	I	$2C_\infty^\phi$	$\cdots$	$\infty\sigma_v$		
$A_1\equiv\Sigma^+$	1	1	$\cdots$	1	T_z	$\alpha_{xx}+\alpha_{yy},\alpha_{zz}$
$A_2\equiv\Sigma^-$	1	1	$\cdots$	-1	R_z	
$E_1\equiv\Pi$	2	$2\cos\phi$	$\cdots$	0	$(T_x,T_y),(R_x,R_y)$	$(\alpha_{xz},\alpha_{yz})$
$E_2\equiv\Delta$	2	$2\cos 2\phi$	$\cdots$	0		$(\alpha_{xx}-\alpha_{yy},\alpha_{xy})$
$E_3\equiv\Phi$	2	$2\cos 3\phi$	$\cdots$	0		
$\vdots$	$\vdots$	$\vdots$	$\cdots$	$\vdots$		

Table A.17

D_2	I	$C_2(z)$	$C_2(y)$	$C_2(x)$		
A	1	1	1	1		$\alpha_{xx},\alpha_{yy},\alpha_{zz}$
B_1	1	1	-1	-1	T_z,R_z	α_{xy}
B_2	1	-1	1	-1	T_y,R_y	α_{xz}
B_3	1	-1	-1	1	T_x,R_x	α_{yx}

Table A.18

D_3	I	$2C_3$	$3C_2$		
A_1	1	1	1		$\alpha_{xx} + \alpha_{yy}, \alpha_{zz}$
A_2	1	1	-1	T_z, R_z	
E	2	-1	0	$(T_x, T_y), (R_x, R_y)$	$(\alpha_{xx} - \alpha_{yy}, \alpha_{xy}), (\alpha_{xz}, \alpha_{yz})$

Table A.19

D_4	I	$2C_4$	$C_2(=C_4^2)$	$2C_2'$	$2C_2''$		
A_1	1	1	1	1	1		$\alpha_{xx} + \alpha_{yy}, \alpha_{zz}$
A_2	1	1	1	-1	-1	T_z, R_z	
B_1	1	-1	1	1	-1		$\alpha_{xx} - \alpha_{yy}$
B_2	1	-1	1	-1	1		α_{xy}
E	2	0	-2	0	0	$(T_x, T_y), (R_x, R_y)$	$(\alpha_{xz}, \alpha_{yz})$

Table A.20

D_5	I	$2C_5$	$2C_5^2$	$5C_2$		
A_1	1	1	1	1		$\alpha_{xx} + \alpha_{yy}, \alpha_{zz}$
A_2	1	1	1	-1	T_z, R_z	
E_1	2	$2\cos 72°$	$2\cos 144°$	0	$(T_x, T_y), (R_x, R_y)$	$(\alpha_{xz}, \alpha_{yz})$
E_2	2	$2\cos 144°$	$2\cos 72°$	0		$(\alpha_{xx} - \alpha_{yy}, \alpha_{xy})$

Table A.21

D_6	I	$2C_6$	$2C_3$	C_2	$3C_2'$	$3C_2''$		
A_1	1	1	1	1	1	1		$\alpha_{xx}+\alpha_{yy},\alpha_{zz}$
A_2	1	1	1	1	-1	-1	T_z,R_z	
B_1	1	-1	1	-1	1	-1		
B_2	1	-1	1	-1	-1	1		
E_1	2	1	-1	-2	0	0	$(T_x,T_y),(R_x,R_y)$	$(\alpha_{xz},\alpha_{yz})$
E_2	2	-1	-1	2	0	0		$(\alpha_{xx}-\alpha_{yy},\alpha_{xy})$

Table A.22

C_{2h}	I	C_2	i	σ_h		
A_g	1	1	1	1	R_z	$\alpha_{xx},\alpha_{yy},\alpha_{zz},\alpha_{xy}$
B_g	1	-1	1	-1	R_x,R_y	α_{xz},α_{yz}
A_u	1	1	-1	-1	T_z	
B_u	1	-1	-1	1	T_x,T_y	

Table A.23

C_{3h}	I	C_3	C_3^2	σ_h	S_3	S_3^5		
A'	1	1	1	1	1	1	R_z	$\alpha_{xx}+\alpha_{yy},\alpha_{zz}$
A''	1	1	1	-1	-1	-1	T_z	
E'	$\begin{cases}1\\1\end{cases}$	$\begin{matrix}\varepsilon\\\varepsilon^*\end{matrix}$	$\begin{matrix}\varepsilon^*\\\varepsilon\end{matrix}$	$\begin{matrix}1\\1\end{matrix}$	$\begin{matrix}\varepsilon\\\varepsilon^*\end{matrix}$	$\begin{matrix}\varepsilon^*\\\varepsilon\end{matrix}$	(T_x,T_y)	$(\alpha_{xx}-\alpha_{yy},\alpha_{xy})$
E''	$\begin{cases}1\\1\end{cases}$	$\begin{matrix}\varepsilon\\\varepsilon^*\end{matrix}$	$\begin{matrix}\varepsilon^*\\\varepsilon\end{matrix}$	$\begin{matrix}-1\\-1\end{matrix}$	$\begin{matrix}-\varepsilon\\-\varepsilon^*\end{matrix}$	$\begin{matrix}-\varepsilon^*\\-\varepsilon\end{matrix}$	(R^x,R_y)	$(\alpha_{xz},\alpha_{yz})$

$\varepsilon = \exp(2\pi i/3)$, $\varepsilon^* = \exp(-2\pi i/3)$

Table A.24

C_{4h}	I	C_4	C_2	C_4^3	i	S_4^3	σ_h	S_4		
A_g	1	1	1	1	1	1	1	1	R_z	$\alpha_{xx}+\alpha_{yy}, \alpha_{zz}$
B_g	1	-1	1	-1	1	-1	1	-1		$\alpha_{xx}-\alpha_{yy}, \alpha_{xy}$
E_g	$\Big\{{1 \atop 1}$	${i \atop -i}$	${-1 \atop -1}$	${-i \atop i}$	${1 \atop 1}$	${i \atop -i}$	${-1 \atop -1}$	${-i \atop i}\Big\}$	(R_x, R_y)	$(\alpha_{xz}, \alpha_{yz})$
A_u	1	1	1	1	-1	-1	-1	-1	T_z	
B_u	1	-1	1	-1	-1	1	-1	1		
E_u	$\Big\{{1 \atop 1}$	${i \atop -i}$	${-1 \atop -1}$	${-i \atop i}$	${-1 \atop -1}$	${-i \atop i}$	${1 \atop 1}$	${i \atop -i}\Big\}$	(T_x, T_y)	

Table A.25

C_{5h}	I	C_5	C_5^2	C_5^3	C_5^4	σ_h	S_5	S_5^7	S_5^3	S_5^9		
A'	1	1	1	1	1	1	1	1	1	1	R_z	$\alpha_{xx}+\alpha_{yy}, \alpha_{zz}$
E_1'	$\Big\{{1 \atop 1}$	${\varepsilon \atop \varepsilon^*}$	${\varepsilon^2 \atop \varepsilon^{2*}}$	${\varepsilon^{2*} \atop \varepsilon^2}$	${\varepsilon^* \atop \varepsilon}$	${1 \atop 1}$	${\varepsilon \atop \varepsilon^*}$	${\varepsilon^2 \atop \varepsilon^{2*}}$	${\varepsilon^{2*} \atop \varepsilon^2}$	${\varepsilon^* \atop \varepsilon}\Big\}$	(T_x, T_y)	
E_2'	$\Big\{{1 \atop 1}$	${\varepsilon^2 \atop \varepsilon^{2*}}$	${\varepsilon^{2*} \atop \varepsilon^2}$	${\varepsilon \atop \varepsilon^*}$	${\varepsilon^{2*} \atop \varepsilon^2}$	${1 \atop 1}$	${\varepsilon^2 \atop \varepsilon^{2*}}$	${\varepsilon^{2*} \atop \varepsilon^2}$	${\varepsilon \atop \varepsilon^*}$	${\varepsilon^{2*} \atop \varepsilon^2}\Big\}$		$(\alpha_{xx}-\alpha_{yy}, \alpha_{xy})$
A''	1	1	1	1	1	-1	-1	-1	-1	-1	T_z	
E_1''	$\Big\{{1 \atop 1}$	${\varepsilon \atop \varepsilon^*}$	${\varepsilon^2 \atop \varepsilon^{2*}}$	${\varepsilon^{2*} \atop \varepsilon^2}$	${\varepsilon^* \atop \varepsilon}$	${-1 \atop -1}$	${-\varepsilon \atop -\varepsilon^*}$	${-\varepsilon^2 \atop -\varepsilon^{2*}}$	${-\varepsilon^{2*} \atop -\varepsilon^2}$	${-\varepsilon^* \atop -\varepsilon}\Big\}$	(R_x, R_y)	$(\alpha_{xz}, \alpha_{yz})$
E_2''	$\Big\{{1 \atop 1}$	${\varepsilon^2 \atop \varepsilon^{2*}}$	${\varepsilon^{2*} \atop \varepsilon^2}$	${\varepsilon \atop \varepsilon^*}$	${\varepsilon^{2*} \atop \varepsilon^2}$	${-1 \atop -1}$	${-\varepsilon^2 \atop -\varepsilon^{2*}}$	${-\varepsilon^{2*} \atop -\varepsilon^2}$	${-\varepsilon \atop -\varepsilon^*}$	${-\varepsilon^{2*} \atop -\varepsilon^2}\Big\}$		

$\varepsilon = \exp(2\pi i/5), \; \varepsilon^* = \exp(-2\pi i/5)$

Table A.26

C_{6h}	I	C_6	C_3	C_2	C_3^2	C_6^5	i	S_3^5	S_6^5	σ_h	S_6	S_3		
A_g	1	1	1	1	1	1	1	1	1	1	1	1	R_z	$\alpha_{xx}+\alpha_{yy},\alpha_{zz}$
B_g	1	-1	1	-1	1	-1	1	-1	1	-1	1	-1		
E_{1g}	$\begin{Bmatrix}1\\1\end{Bmatrix}$	$\begin{matrix}\varepsilon\\\varepsilon^*\end{matrix}$	$\begin{matrix}-\varepsilon^*\\-\varepsilon\end{matrix}$	$\begin{matrix}-1\\-1\end{matrix}$	$\begin{matrix}-\varepsilon\\-\varepsilon^*\end{matrix}$	$\begin{matrix}\varepsilon^*\\\varepsilon\end{matrix}$	$\begin{matrix}1\\1\end{matrix}$	$\begin{matrix}\varepsilon\\\varepsilon^*\end{matrix}$	$\begin{matrix}-\varepsilon^*\\-\varepsilon\end{matrix}$	$\begin{matrix}-1\\-1\end{matrix}$	$\begin{matrix}-\varepsilon\\-\varepsilon^*\end{matrix}$	$\begin{Bmatrix}\varepsilon^*\\\varepsilon\end{Bmatrix}$	(R_x,R_y)	$(\alpha_{xz},\alpha_{yz})$
E_{2g}	$\begin{Bmatrix}1\\1\end{Bmatrix}$	$\begin{matrix}-\varepsilon^*\\-\varepsilon\end{matrix}$	$\begin{matrix}-\varepsilon\\-\varepsilon^*\end{matrix}$	$\begin{matrix}1\\1\end{matrix}$	$\begin{matrix}-\varepsilon^*\\-\varepsilon\end{matrix}$	$\begin{matrix}-\varepsilon\\-\varepsilon^*\end{matrix}$	$\begin{matrix}1\\1\end{matrix}$	$\begin{matrix}-\varepsilon^*\\-\varepsilon\end{matrix}$	$\begin{matrix}-\varepsilon\\-\varepsilon^*\end{matrix}$	$\begin{matrix}1\\1\end{matrix}$	$\begin{matrix}-\varepsilon^*\\-\varepsilon\end{matrix}$	$\begin{Bmatrix}-\varepsilon\\-\varepsilon^*\end{Bmatrix}$		$(\alpha_{xx}-\alpha_{yy},\alpha_{xy})$
A_u	1	1	1	1	1	1	-1	-1	-1	-1	-1	-1	T_z	
B_u	1	-1	1	-1	1	-1	-1	1	-1	1	-1	1		
E_{1u}	$\begin{Bmatrix}1\\1\end{Bmatrix}$	$\begin{matrix}\varepsilon\\\varepsilon^*\end{matrix}$	$\begin{matrix}-\varepsilon^*\\-\varepsilon\end{matrix}$	$\begin{matrix}-1\\-1\end{matrix}$	$\begin{matrix}-\varepsilon\\-\varepsilon^*\end{matrix}$	$\begin{matrix}\varepsilon^*\\\varepsilon\end{matrix}$	$\begin{matrix}-1\\-1\end{matrix}$	$\begin{matrix}-\varepsilon\\-\varepsilon^*\end{matrix}$	$\begin{matrix}\varepsilon^*\\\varepsilon\end{matrix}$	$\begin{matrix}1\\1\end{matrix}$	$\begin{matrix}\varepsilon\\\varepsilon^*\end{matrix}$	$\begin{Bmatrix}-\varepsilon^*\\-\varepsilon\end{Bmatrix}$	(T_x,T_y)	
E_{2u}	$\begin{Bmatrix}1\\1\end{Bmatrix}$	$\begin{matrix}-\varepsilon^*\\-\varepsilon\end{matrix}$	$\begin{matrix}-\varepsilon\\-\varepsilon^*\end{matrix}$	$\begin{matrix}1\\1\end{matrix}$	$\begin{matrix}-\varepsilon^*\\-\varepsilon\end{matrix}$	$\begin{matrix}-\varepsilon\\-\varepsilon^*\end{matrix}$	$\begin{matrix}-1\\-1\end{matrix}$	$\begin{matrix}\varepsilon^*\\\varepsilon\end{matrix}$	$\begin{matrix}\varepsilon\\\varepsilon^*\end{matrix}$	$\begin{matrix}-1\\-1\end{matrix}$	$\begin{matrix}\varepsilon^*\\\varepsilon\end{matrix}$	$\begin{Bmatrix}\varepsilon\\\varepsilon^*\end{Bmatrix}$		

$\varepsilon = \exp(2\pi i/6),\ \varepsilon^* = \exp(-2\pi i/6)$

Table A.27

D_{2d}	I	$2S_4$	C_2	$2C_2'$	$2\sigma_d$		
A_1	1	1	1	1	1		$\alpha_{xx} + \alpha_{yy}, \alpha_{zz}$
A_2	1	1	1	-1	-1	R_z	
B_1	1	-1	1	1	-1		$\alpha_{xx} - \alpha_{yy}$
B_2	1	-1	1	-1	1	T_z	α_{xy}
E	2	0	-2	0	0	$(T_x, T_y), (R_x, R_y)$	$(\alpha_{xz}, \alpha_{yz})$

Table A.28

D_{3d}	I	$2C_3$	$3C_2$	i	$2S_6$	$3\sigma_d$		
A_{1g}	1	1	1	1	1	1		$\alpha_{xx} + \alpha_{yy}, \alpha_{zz}$
A_{2g}	1	1	-1	1	1	-1	R_z	
E_g	2	-1	0	2	-1	0	(R_x, R_y)	$(\alpha_{xx} - \alpha_{yy}, \alpha_{xy}), (\alpha_{xz}, \alpha_{yz})$
A_{1u}	1	1	1	-1	-1	-1		
A_{2u}	1	1	-1	-1	-1	1	T_z	
E_u	2	-1	0	-2	1	0	(T_x, T_y)	

Table A.29

D_{4d}	I	$2S_8$	$2C_4$	$2S_8^3$	C_2	$4C_2'$	$4\sigma_d$		
A_1	1	1	1	1	1	1	1		$\alpha_{xx} + \alpha_{yy}, \alpha_{zz}$
A_2	1	1	1	1	1	-1	-1	R_z	
B_1	1	-1	1	-1	1	1	-1		
B_2	1	-1	1	-1	1	-1	1	T_z	
E_1	2	$\sqrt{2}$	0	$-\sqrt{2}$	-2	0	0	(T_x, T_y)	
E_2	2	0	-2	0	2	0	0		$(\alpha_{xx} - \alpha_{yy}, \alpha_{xy})$
E_3	2	$-\sqrt{2}$	0	$\sqrt{2}$	-2	0	0	(R_x, R_y)	$(\alpha_{xz}, \alpha_{yz})$

Table A.30

D_{5d}	I	$2C_5$	$2C_5^2$	$5C_2$	i	$2S_{10}^3$	$2S_{10}$	$5\sigma_d$		
A_{1g}	1	1	1	1	1	1	1	1		$\alpha_{xx}+\alpha_{yy},\alpha_{zz}$
A_{2g}	1	1	1	-1	1	1	1	-1	R_z	
E_{1g}	2	$2\cos72°$	$2\cos144°$	0	2	$2\cos72°$	$2\cos144°$	0	(R_x,R_y)	$(\alpha_{xz},\alpha_{yz})$
E_{2g}	2	$2\cos144°$	$2\cos72°$	0	2	$2\cos144°$	$2\cos72°$	0		$(\alpha_{xx}-\alpha_{yy},\alpha_{xy})$
A_{1u}	1	1	1	1	-1	-1	-1	-1		
A_{2u}	1	1	1	-1	-1	-1	-1	1	T_z	
E_{1u}	2	$2\cos72°$	$2\cos144°$	0	-2	$-2\cos72°$	$-2\cos144°$	0	(T_x,T_y)	
E_{2u}	2	$2\cos144°$	$2\cos72°$	0	-2	$-2\cos144°$	$-2\cos72°$	0		

Table A.31

D_{6d}	I	$2S_{12}$	$2C_6$	$2S_4$	$2C_3$	$2S_{12}^5$	C_2	$6C_2'$	$6\sigma_d$		
A_1	1	1	1	1	1	1	1	1	1		$\alpha_{xx}+\alpha_{yy},\alpha_{zz}$
A_2	1	1	1	1	1	1	1	-1	-1	R_z	
B_1	1	-1	1	-1	1	-1	1	1	-1		
B_2	1	-1	1	-1	1	-1	1	-1	1	T_z	
E_1	2	$\sqrt{3}$	1	0	-1	$-\sqrt{3}$	-2	0	0	(T_x,T_y)	
E_2	2	1	-1	-2	-1	1	2	0	0		$(\alpha_{xx}-\alpha_{yy},\alpha_{xy})$
E_3	2	0	-2	0	2	0	-2	0	0		
E_4	2	-1	-1	2	-1	-1	2	0	0		
E_5	2	$-\sqrt{3}$	1	0	-1	$\sqrt{3}$	-2	0	0	(R_x,R_y)	$(\alpha_{xz},\alpha_{yz})$

Table A.32

D_{2h}	I	$C_2(z)$	$C_2(y)$	$C_2(x)$	i	$\sigma(xy)$	$\sigma(xz)$	$\sigma(yz)$		
A_g	1	1	1	1	1	1	1	1		$\alpha_{xx},\alpha_{yy},\alpha_{zz}$
B_{1g}	1	1	-1	-1	1	1	-1	-1	R_z	α_{xy}
B_{2g}	1	-1	1	-1	1	-1	1	-1	R_y	α_{xz}
B_{3g}	1	-1	-1	1	1	-1	-1	1	R_x	α_{yz}
A_u	1	1	1	1	-1	-1	-1	-1		
B_{1u}	1	1	-1	-1	-1	-1	1	1	T_z	
B_{2u}	1	-1	1	-1	-1	1	-1	1	T_y	
B_{3u}	1	-1	-1	1	-1	1	1	-1	T_x	

Table A.33

D_{3h}	I	$2C_3$	$3C_2$	σ_h	$2S_3$	$3\sigma_v$		
A_1'	1	1	1	1	1	1		$\alpha_{xx}+\alpha_{yy},\alpha_{zz}$
A_2'	1	1	-1	1	1	-1	R_z	
E'	2	-1	0	2	-1	0	(T_x,T_y)	$(\alpha_{xx}-\alpha_{yy},\alpha_{xy})$
A_1''	1	1	1	-1	-1	-1		
A_2''	1	1	-1	-1	-1	1	T_z	
E''	2	-1	0	-2	1	0	(R_x,R_y)	$(\alpha_{xz},\alpha_{yz})$

Table A.34

D_{4h}	I	$2C_4$	C_2	$2C_2'$	$2C_2''$	i	$2S_4$	σ_h	$2\sigma_v$	$2\sigma_d$		
A_{1g}	1	1	1	1	1	1	1	1	1	1		$\alpha_{xx}+\alpha_{yy},\alpha_{zz}$
A_{2g}	1	1	1	-1	-1	1	1	1	-1	-1	R_z	
B_{1g}	1	-1	1	1	-1	1	-1	1	1	-1		$\alpha_{xx}-\alpha_{yy}$
B_{2g}	1	-1	1	-1	1	1	-1	1	-1	1		α_{xy}
E_g	2	0	-2	0	0	2	0	-2	0	0	(R_x,R_y)	$(\alpha_{xz},\alpha_{yz})$
A_{1u}	1	1	1	1	1	-1	-1	-1	-1	-1		
A_{2u}	1	1	1	-1	-1	-1	-1	-1	1	1	T_z	
B_{1u}	1	-1	1	1	-1	-1	1	-1	-1	1		
B_{2u}	1	-1	1	-1	1	-1	1	-1	1	-1		
E_u	2	0	-2	0	0	-2	0	2	0	0	(T_x,T_y)	

Table A.35

D_{5h}	I	$2C_5$	$2C_5^2$	$5C_2$	σ_h	$2S_5$	$2S_5^2$	$5\sigma_v$		
A_1'	1	1	1	1	1	1	1	1		$\alpha_{xx}+\alpha_{yy},\alpha_{zz}$
A_2'	1	1	1	-1	1	1	1	-1	R_z	
E_1'	2	$2\cos 72°$	$2\cos 144°$	0	2	$2\cos 72°$	$2\cos 144°$	0	(T_x,T_y)	
E_2'	2	$2\cos 144°$	$2\cos 72°$	0	2	$2\cos 144°$	$2\cos 72°$	0		$(\alpha_{xx}-\alpha_{yy},\alpha_{xy})$
A_1''	1	1	1	1	-1	-1	-1	-1		
A_2''	1	1	1	-1	-1	-1	-1	1	T_z	
E_1''	2	$2\cos 72°$	$2\cos 144°$	0	-2	$-2\cos 72°$	$-2\cos 144°$	0	(R_x,R_y)	$(\alpha_{xz},\alpha_{yz})$
E_2''	2	$2\cos 144°$	$2\cos 72°$	0	-2	$-2\cos 144°$	$-2\cos 72°$	0		

Table A.36

D_{6h}	I	$2C_6$	$2C_3$	C_2	$3C_2'$	$3C_2''$	i	$2S_3$	$2S_6$	σ_h	$3\sigma_d$	$3\sigma_v$		
A_{1g}	1	1	1	1	1	1	1	1	1	1	1	1		$\alpha_{xx}+\alpha_{yy},\alpha_{zz}$
A_{2g}	1	1	1	1	-1	-1	1	1	1	1	-1	-1	R_z	
B_{1g}	1	-1	1	-1	1	-1	1	-1	1	-1	1	-1		
B_{2g}	1	-1	1	-1	-1	1	1	-1	1	-1	-1	1		
E_{1g}	2	1	-1	-2	0	0	2	1	-1	-2	0	0	(R_x,R_y)	$(\alpha_{xz},\alpha_{yz})$
E_{2g}	2	-1	-1	2	0	0	2	-1	-1	2	0	0		$(\alpha_{xx}-\alpha_{yy},\alpha_{xy})$
A_{1u}	1	1	1	1	1	1	-1	-1	-1	-1	-1	-1		
A_{2u}	1	1	1	1	-1	-1	-1	-1	-1	-1	1	1	T_z	
B_{1u}	1	-1	1	-1	1	-1	-1	1	-1	1	-1	1		
B_{2u}	1	-1	1	-1	-1	1	-1	1	-1	1	1	-1		
E_{1u}	2	1	-1	-2	0	0	-2	-1	1	2	0	0	(T_x,T_y)	
E_{2u}	2	-1	-1	2	0	0	-2	1	1	-2	0	0		

Table A.37

$D_{\infty h}$	I	$2C_\infty^\phi$	$\cdots$	$\infty\sigma_v$	i	$2S_\infty^\phi$	$\cdots$	∞C_2		
$A_{1g}\equiv\Sigma_g^+$	1	1	$\cdots$	1	1	1	$\cdots$	1		$\alpha_{xx}+\alpha_{yy},\alpha_{zz}$
$A_{2g}\equiv\Sigma_g^-$	1	1	$\cdots$	-1	1	1	$\cdots$	-1	R_z	
$E_{1g}\equiv\Pi_g$	2	$2\cos\phi$	$\cdots$	0	2	$-2\cos\phi$	$\cdots$	0	(R_x,R_y)	$(\alpha_{xz},\alpha_{yz})$
$E_{2g}\equiv\Delta_g$	2	$2\cos 2\phi$	$\cdots$	0	2	$2\cos 2\phi$	$\cdots$	0		$(\alpha_{xx}-\alpha_{yy},\alpha_{xy})$
$E_{3g}\equiv\Phi_g$	2	$2\cos 3\phi$	$\cdots$	0	2	$-2\cos 3\phi$	$\cdots$	0		
$\cdots$	$\cdots$	$\cdots$		$\cdots$	$\cdots$	$\cdots$		$\cdots$		
$A_{2u}\equiv\Sigma_u^+$	1	1	$\cdots$	1	-1	-1	$\cdots$	-1	T_z	
$A_{1u}\equiv\Sigma_u^-$	1	1	$\cdots$	-1	-1	-1	$\cdots$	-1		
$E_{1u}\equiv\Pi_u$	2	$2\cos\phi$	$\cdots$	0	-2	$2\cos\phi$	$\cdots$	0	(T_x,T_y)	
$E_{2u}\equiv\Delta_u$	2	$2\cos 2\phi$	$\cdots$	0	-2	$-2\cos 2\phi$	$\cdots$	0		
$E_{3u}\equiv\Phi_u$	2	$2\cos 3\phi\cdots$		0	-2	$2\cos 3\phi$	$\cdots$	$0\cdots$		
$\cdots$										

Table A.38

S_4	I	S_4	C_2	S_4^3		
A	1	1	1	1	R_z T_z	$\alpha_{xx}+\alpha_{yy},\alpha_{zz}$
B	1	-1	1	-1		$\alpha_{xx}-\alpha_{yy},\alpha_{xy}$
E	$\left\{\begin{matrix}1\\1\end{matrix}\right.$	$\begin{matrix}i\\-i\end{matrix}$	$\begin{matrix}-1\\-1\end{matrix}$	$\left.\begin{matrix}-i\\i\end{matrix}\right\}$	$(T_x,T_y),(R_x,R_y)$	$(\alpha_{xz},\alpha_{yz})$

Table A.39

S_6	I	C_3	C_3^2	i	S_6^5	S_6		
A_g	1	1	1	1	1	1	R_z	$\alpha_{xx}+\alpha_{yy},\alpha_{zz}$
E_g	$\left\{\begin{matrix}1\\1\end{matrix}\right.$	$\begin{matrix}\varepsilon\\\varepsilon^*\end{matrix}$	$\begin{matrix}\varepsilon^*\\\varepsilon\end{matrix}$	$\begin{matrix}1\\1\end{matrix}$	$\begin{matrix}\varepsilon\\\varepsilon^*\end{matrix}$	$\left.\begin{matrix}\varepsilon^*\\\varepsilon\end{matrix}\right\}$	(R_x,R_y)	$(\alpha_{xx}-\alpha_{yy},\alpha_{xy}),(\alpha_{xz},\alpha_{yz})$
A_u	1	1	1	-1	-1	-1	T_z	
E_u	$\left\{\begin{matrix}1\\1\end{matrix}\right.$	$\begin{matrix}\varepsilon\\\varepsilon^*\end{matrix}$	$\begin{matrix}\varepsilon^*\\\varepsilon\end{matrix}$	$\begin{matrix}-1\\-1\end{matrix}$	$\begin{matrix}-\varepsilon\\-\varepsilon^*\end{matrix}$	$\left.\begin{matrix}-\varepsilon^*\\-\varepsilon\end{matrix}\right\}$	(T_x,T_y)	

$\varepsilon=\exp(2\pi i/3),\ \varepsilon^*=\exp(-2\pi i/3)$

Table A.40

S_8	I	S_8	C_4	S_8^3	C_2	S_8^5	C_4^3	S_8^7		
A	1	1	1	1	1	1	1	1	R_z T_z	$\alpha_{xx}+\alpha_{yy},\alpha_{zz}$
B	1	-1	1	-1	1	-1	1	-1		
E_1	$\left\{\begin{matrix}1\\1\end{matrix}\right.$	$\begin{matrix}\varepsilon\\\varepsilon^*\end{matrix}$	$\begin{matrix}i\\-i\end{matrix}$	$\begin{matrix}\varepsilon^*\\\varepsilon\end{matrix}$	$\begin{matrix}-1\\-1\end{matrix}$	$\begin{matrix}-\varepsilon\\-\varepsilon^*\end{matrix}$	$\begin{matrix}-i\\i\end{matrix}$	$\left.\begin{matrix}\varepsilon^*\\\varepsilon\end{matrix}\right\}$	(T_x,T_y)	
E_2	$\left\{\begin{matrix}1\\1\end{matrix}\right.$	$\begin{matrix}i\\-i\end{matrix}$	$\begin{matrix}-1\\-1\end{matrix}$	$\begin{matrix}-i\\i\end{matrix}$	$\begin{matrix}1\\1\end{matrix}$	$\begin{matrix}i\\-i\end{matrix}$	$\begin{matrix}-1\\-1\end{matrix}$	$\left.\begin{matrix}-i\\i\end{matrix}\right\}$		$(\alpha_{xx}-\alpha_{yy},\alpha_{xy})$
E_3	$\left\{\begin{matrix}1\\1\end{matrix}\right.$	$\begin{matrix}-\varepsilon^*\\-\varepsilon\end{matrix}$	$\begin{matrix}-i\\i\end{matrix}$	$\begin{matrix}\varepsilon\\\varepsilon^*\end{matrix}$	$\begin{matrix}-1\\-1\end{matrix}$	$\begin{matrix}\varepsilon^*\\\varepsilon\end{matrix}$	$\begin{matrix}i\\-i\end{matrix}$	$\left.\begin{matrix}-\varepsilon\\-\varepsilon^*\end{matrix}\right\}$	(R_x,R_y)	$(\alpha_{xz},\alpha_{yz})$

$\varepsilon=\exp(2\pi i/8),\ \varepsilon^*=\exp(-2\pi i/8)$

Table A.41

T_d	I	$8C_3$	$3C_2$	$6S_4$	$6\sigma_d$		
A_1	1	1	1	1	1		$\alpha_{xx} + \alpha_{yy} + \alpha_{zz}$
A_2	1	1	1	-1	-1		
E	2	-1	2	0	0		$(\alpha_{xx} + \alpha_{yy} - 2\alpha_{zz}, \alpha_{xx} - \alpha_{yy})$
$T_1 \equiv F_1$	3	0	-1	1	-1	(R_x, R_y, R_z)	
$T_2 \equiv F_2$	3	0	-1	-1	1	(T_x, T_y, T_z)	$(\alpha_{xy}, \alpha_{xz}, \alpha_{yz})$

Table A.42

T	I	$4C_3$	$4C_3^2$	$3C_2$		
A	1	1	1	1		$\alpha_{xx} + \alpha_{yy} + \alpha_{zz}$
E	$\begin{Bmatrix} 1 \\ 1 \end{Bmatrix}$	$\begin{matrix} \varepsilon \\ \varepsilon^* \end{matrix}$	$\begin{matrix} \varepsilon^* \\ \varepsilon \end{matrix}$	$\begin{Bmatrix} 1 \\ 1 \end{Bmatrix}$		$(\alpha_{xx} + \alpha_{yy} - 2\alpha_{zz}, \alpha_{xx} - \alpha_{yy})$
$T \equiv F$	3	0	0	-1	$(T_x, T_y, T_z), (R_x, T_y, R_z)$	$(\alpha_{xy}, \alpha_{xz}, \alpha_{yz})$

$\varepsilon = \exp(2\pi i/3)$, $\varepsilon^* = \exp(-2\pi i/3)$

Table A.43

O_h	I	$8C_3$	$6C_2$	$6C_4$	$3C_2(=3C_4^2)$	i	$6S_4$	$8S_6$	$3\sigma_h$	$6\sigma_d$		
A_{1g}	1	1	1	1	1	1	1	1	1	1		$\alpha_{xx}+\alpha_{yy}+\alpha_{zz}$
A_{2g}	1	1	-1	-1	1	1	-1	1	1	-1		
E_g	2	-1	0	0	2	2	0	-1	2	0		$(\alpha_{xx}+\alpha_{yy}-2\alpha_{zz},\alpha_{xx}-\alpha_{yy})$
$T_{1g}\equiv F_{1g}$	3	0	-1	1	-1	3	1	0	-1	-1	(R_x,R_y,R_z)	
$T_{2g}\equiv F_{2g}$	3	0	1	-1	-1	3	-1	0	-1	1		$(\alpha_{xz},\alpha_{yz},\alpha_{xy})$
A_{1u}	1	1	1	1	1	-1	-1	-1	-1	-1		
A_{2u}	1	1	-1	-1	1	-1	1	-1	-1	1		
E_u	2	-1	0	0	2	-2	0	1	-2	0		
$T_{1u}\equiv F_{1u}$	3	0	-1	1	-1	-3	-1	0	1	1	(T_x,T_y,T_z)	
$T_{2u}\equiv F_{2u}$	3	0	1	-1	-1	-3	1	0	1	-1		

Table A.44

O	I	$8C_3$	$6C_2$	$6C_4$	$3C_2'(=3C_4^2)$		
A_1	1	1	1	1	1		$\alpha_{xx}+\alpha_{yy}+\alpha_{zz}$
A_2	1	1	-1	-1	1		
E	2	-1	0	0	2		$(\alpha_{xx}+\alpha_{yy}-2\alpha_{zz},\alpha_{xx}-\alpha_{yy})$
$T_1\equiv F_1$	3	0	-1	1	-1	$(T_x,T_y,T_z)(R_x,R_y,R_z)$	
$T_2\equiv F_2$	3	0	1	-1	-1		$(\alpha_{xy},\alpha_{xz},\alpha_{yz})$

Table A.45

K_h	I	$\infty C_\infty^\phi \cdots$	$\infty S_\infty^\phi \cdots$	i		
S_g	1	1	1	1		$\alpha_{xx}+\alpha_{yy}+\alpha_{zz}$
S_u	1	1	-1	-1		
P_g	3	$1+2\cos\phi$	$1-2\cos\phi$	1	(R_x,R_y,R_z)	
P_u	3	$1+2\cos\phi$	$-1+2\cos\phi$	-1	(T_x,T_y,T_z)	
D_g	5	$1+2\cos\phi+2\cos 2\phi$	$1-2\cos\phi+2\cos 2\phi$	1		$(\alpha_{xx}+\alpha_{yy}-2\alpha_{zz},\alpha_{xx}-\alpha_{yy},\alpha_{xy},\alpha_{xz},\alpha_{yz})$
D_u	5	$1+2\cos\phi+2\cos 2\phi$	$-1+2\cos\phi-2\cos 2\phi$	-1		
F_g	7	$1+2\cos\phi+2\cos 2\phi+2\cos 3\phi$	$1-2\cos\phi+2\cos 2\phi-2\cos 3\phi$	1		
F_u	7	$1+2\cos\phi+2\cos 2\phi+2\cos 3\phi$	$-1+2\cos\phi-2\cos 2\phi+2\cos 3\phi$	-1		
$\cdots$	$\cdots$	$\cdots$	$\cdots$	$\cdots$		

Index of Atoms and Molecules

The system of indexing molecules is, first, according to the number of atoms in the molecule. Then, with the chemical formula written in what seems a natural way, they are ordered alphabetically in order of the atoms as they appear in the formula.

The system of labelling isotopically substituted molecules is the same as that used in the text. Except for a very few cases the only nuclei labelled are those which are *not* the most abundant species.

For most molecules with more than three atoms the name of the molecule is given in parentheses after the chemical formula. In most cases, only the common and not the systematic names are given so that, for example, C_2H_4 is called ethylene and not ethene.

Subject Index